Microwave Imaging
Methods and Applications

For a complete listing of titles in the
Artech House Microwave Library,
turn to the back of this book.

Microwave Imaging
Methods and Applications

Matteo Pastorino

Andrea Randazzo

ARTECH
HOUSE

BOSTON | LONDON
artechhouse.com

Library of Congress Cataloging-in-Publication Data
A catalog record for this book is available from the U.S. Library of Congress.

British Library Cataloguing in Publication Data
A catalogue record for this book is available from the British Library.

Cover design by John Gomes

ISBN 13: 978-1-63081-348-2

685 Canton Street
Norwood, MA 02062

10 9 8 7 6 5 4 3 2 1

Contents

Preface

Writing a book on microwave imaging is quite a complex task. This short-range, noninvasive, and nondestructive inspection modality has been considered a potentially effective diagnostic technique for a long time, and to some extent, it has been applied in real situations. Actually, it represents an ever-expanding research area, which is characterized by a strong multidisciplinary nature. The electromagnetic inverse problems associated with microwave imaging exhibit great mathematical complexities, further increased when nonlinear formulations have to be considered. Ill-posedness issues represent critical aspects that have to be properly handled. In addition, fast and effective numerical procedures should be applied to render the inverse scattering equations suitable for digital processing. However, before the inversion can take place, data should first be collected. In general, the data of interest is due to the interactions of a microwave radiation, generated by a proper source, and the target under test. Sophisticated acquisition systems have to be designed and realized, in which antennas are usually one of the critical elements. In fact, they frequently have to work in close proximity to the target, often with low levels of radiated and received power, and they are subject to many other practical limitations. We are sure that the reader already has in mind several critical issues related to instrumentation, measurements, and data processing. Nevertheless, besides all the mentioned theoretical and practical difficulties, one of the most powerful motivations for approaching and studying this topic comes from the potential applications.

Actually, the range of applicative fields in which microwave imaging techniques may be an effective choice—both as stand-alone methods and combined with other, more conventional diagnostic modalities—is very wide and even still growing, although without the same potentialities in all areas.

It is quite difficult to treat all the concepts, methods, apparatuses, and applications in a comprehensive and fascinating way. Therefore, we apologize in advance if some topics of interest for the reader are not treated with the level of depth and completeness they would have expected and desired. At the same time, we apologize if some approaches, methods, systems, or applicative areas (and the related key references) have not been mentioned or have been cited without the emphasis they deserve. However, although complex, our task is really challenging and exciting, because microwave imaging is so elegant and intriguing!

Although each aspect of this multidisciplinary diagnostic modality could be covered by one or more comprehensive monographs, and some excellent books written by prominent scientists are already available, we believe that an overview about methods and applications could stimulate the reader along different directions. First of all, one may wish to consider applying these techniques in the area in which he or she is currently involved, possibly not yet included in the microwave imaging world. Secondly, new ideas on how to improve the existing systems and techniques could emerge by comparing what has been done in other contexts or applicative fields. The detailed description of the book content and its organization is reported in the Introduction (Chapter 1).

We want to conclude by acknowledging some precious contributions. First of all, we would like to thank Dr. Alessandro Fedeli (University of Genoa, Italy) for helping us in organizing the book material and for performing part of the simulations used throughout the book. The contributions of the mathematician Professor Claudio Estatico (University of Genoa, Italy) have also been fundamental for the development of some methods used by the authors and described in various chapters. Finally, the authors are indebted to the team of the Department of Innovative Technologies of the University of Applied Sciences and Arts of Southern Switzerland (SUPSI) in Manno, Switzerland, for the joint activity on microwave imaging of wood.

1

Introduction

Microwave imaging denotes a series of noninvasive and nondestructive techniques aimed at sensing materials or targets in order to retrieve some physical properties, and/or to deduce information about the conditions of the structures under test. Usually, these techniques are based on short-range measurement systems able to collect the electromagnetic field resulting from the interaction between interrogating waves (at microwave frequencies) and the materials or targets to be inspected.

Engineers, scientists, and professionals from industries, research centers, universities, and many other institutions are continuously focused in developing new and innovative solutions in the multidisciplinary areas of imaging systems and techniques.

The use of electromagnetic fields at microwave frequencies for inspecting unknown targets has been proposed for years. However, theoretical, numerical, and practical aspects make this imaging modality quite difficult and incredibly challenging at the same time. The original enthusiasm that can be drawn from the pioneering papers and books on this subject has been followed by a certain disillusion. Nevertheless, in the last decade, the interest about these techniques has grown exponentially, in both the research community and in the real world. This is unequivocally attested by the ever-growing number of new approaches and systems that are continuously proposed in books, scientific papers, and conferences at international levels.

In this context, this book tries to provide a comprehensive overview about the most important and up-to-date techniques proposed for inspecting structures and bodies by using microwaves. Several different applicative scenarios are considered, covering civil and industrial engineering, subsurface prospection, security applications, and medical imaging. Throughout the book we preferred to avoid cumbersome mathematical details, which can be found in the specific literature on the subject. Conversely, we chose to have an engineering perspective when introducing the concepts and tools needed to understand the different approaches for the applications discussed in the various chapters.

This book begins with some basic introductory chapters. More specifically, Chapter 2 introduces the main concepts related to a microwave imaging system, including a preliminary classification concerning the various types of illuminating/receiving configurations. Then, the electromagnetic characterization of materials is treated in the framework of the Maxwellian theory, by introducing the constitutive equations for dielectrics and conductors, as well as the concepts of dispersion, losses, propagation, attenuation, and so on. Some dispersion models for dielectric materials, such as the Debye and Cole-Cole models, are also reported. The main aim of this chapter is to provide a common framework for discussing the methods that will be described in the subsequent sections of the book, and defining the specific notation used in the field of microwave imaging.

Chapter 3 provides an introductory discussion about the electromagnetic scattering theory, which represents the basic formulation for various imaging techniques. This chapter will focus on two- and three-dimensional scattering, integral equations, and Green's functions. Some basic concepts of inverse scattering are also included, classifying microwave imaging methods on the basis of both the considered configuration (e.g., tomographic imaging and buried object detection) and the objective of the inspection (i.e., qualitative and quantitative approaches). Some visual examples of the field scattered by simple targets are also reported for readers not yet acquainted with these imaging modalities. Moreover, as an example, an iterative reconstruction approach developed in Banach spaces is explained in details.

Chapters 4 through 7 concern the use of microwave imaging techniques in the previously outlined applicative scenarios. For the considered applications, potentialities and limitations of microwave imaging are discussed with reference to the particular objectives of the inspection. Moreover, several reconstruction procedures applied in these fields are outlined, reporting some significant results. Practical implementations of the illumination and measurement apparatuses are also reviewed and some examples of experimental prototypes are provided.

In particular, Chapter 4 (Microwave Imaging in Civil Engineering and Industrial Applications) reports the most recent techniques applied, or potentially applicable, in civil and industrial areas. This comprises imaging of cement structures, plastic materials, wood targets, fruits and vegetables, and materials flowing in pipes and vessels. Considerations about the dielectric properties of the materials encountered in these applicative areas are also included in separate sections. For these applications, some significant inversion methods—such as beamforming approaches, linear sampling method, and Gauss-Newton schemes—are described by also reporting some examples of recent results. Apparatuses for the illumination of the targets under test and the measurement of the resulting electromagnetic field are examined too, with reference to some examples of developed microwave imaging systems.

Chapter 5 (Microwave Imaging for Biomedical Applications) deals with methods and apparatuses specifically developed for medical imaging purposes. Two of the main uses of microwave imaging in this field—breast imaging and brain stroke detection—are discussed. A section devoted to the dielectric properties of biological materials at microwave frequencies is also included. Recently developed numerical and experimental phantoms, which can be used to test and validate imaging procedures, are examined as well. Concerning the reconstruction methods, both beamforming techniques (such as the delay-and-sum approach) and quantitative procedures (e.g., based on the distorted Born iterative method) are reported, introducing some convenient modifications for their use in the biomedical field. Finally, other recently proposed diagnostic usages are also outlined.

Chapter 6 (Microwave Imaging for Subsurface Prospection) is devoted to methods and systems for shallow subsurface detection, including the retrieval of pipes, tunnels, and other buried targets. In this framework, ground penetrating radar (GPR) apparatuses are widely employed for acquiring the scattered field data. Commonly used measurement configurations are introduced, illustrating standard processing methods for GPR (e.g., those based on migration approaches). Moreover, inverse scattering techniques (with reference to linearized approaches, nonlinear schemes, and qualitative methods) are treated by recalling the peculiar formulations of the scattering problem related to half-space configurations. Some examples of application of such approaches are reported. Finally, an overview of some practical implementations of GPR systems is also included.

Chapter 7 (Microwave Imaging for Security Purposes) considers, by following a similar approach, recently proposed microwave imaging applications for addressing security issues. In particular, techniques for the so-called through-the-wall imaging are examined (with reference to the migration/beamforming methods and inverse scattering techniques described in the other chapters). Furthermore, some ways for tackling the problems introduced by the presence of the wall are described. The advantages that can be obtained from the development of inverse scattering-based imaging approaches are also outlined. A section of the chapter is also devoted to the problem of concealed targets detection.

Finally, Chapter 8 presents some considerations concerning possible new trends and challenges that, in the authors' opinions, could stimulate new ideas and future developments in the field of microwave imaging.

2

Basic Concepts of Microwave Imaging Systems and Electromagnetic Characterization of Materials

This chapter introduces the main concepts related to a microwave imaging system, including some preliminary classification of the imaging approaches adopted in practical systems as well as in recently proposed solutions. The chapter also discusses the basic concepts concerning the electromagnetic characterization of the materials, which will be particularized in each of the following chapters concerning the materials involved in the various specific applications.

2.1 Active Imaging Systems

An active imaging system is usually composed of several elements. An electromagnetic source is used to generate the interrogating radiation, which is sent toward the body under test. Then there is a collecting system, which acquires the radiation modified by the presence of the target. One or more signals carrying the information about the target (which is usually located inside a test region that will be specified in the following sections) are sent to a processing unit, which elaborates them in order to extract information about the target, often in form of an image. Post-processing techniques can be successively adopted to improve the obtained results and to highlight particular features of interest. The above conceptual configuration is sketched in Figure 2.1.

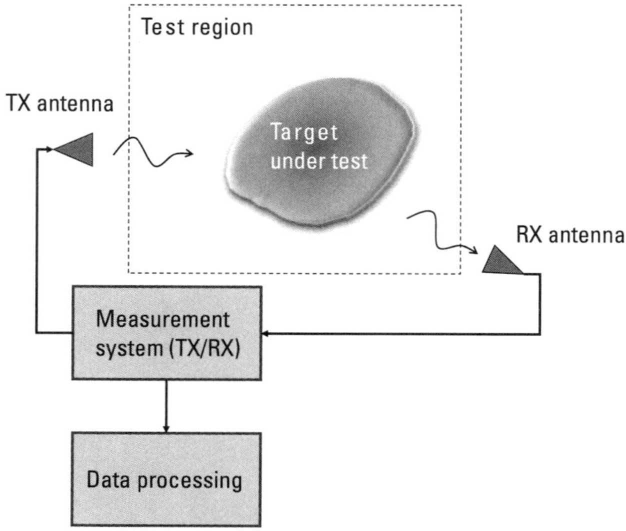

Figure 2.1 Schematic representation of a microwave imaging system.

2.1.1 Microwave Frequencies

The term microwave usually includes radiation with frequencies ranging from 300 MHz to 300 GHz. Therefore, since the wavelength in vacuum is related to the frequency by the following relation

$$\lambda_0 = \frac{v_0}{f} \tag{2.1}$$

where v_0 denotes the speed of electromagnetic waves in vacuum and f is the frequency, the result is that the wavelength of microwave signals range between 1 mm to 1m. Due to the fact that in practical applications these wavelengths are usually comparable with the geometrical dimensions of the target under test, the interactions between the incident radiation and the target produce a scattering effect, which represents the key mathematical challenge of microwave imaging systems. Chapter 3 will be devoted to describing the scattering problem associated with microwave imaging systems.

According to the band designation introduced in the field of radar techniques, microwave components and systems are often designated by letters that specify the frequency range of operation. Table 1.1 reports the band designations as indicated by the IEEE standard [1], as well as their relations with the frequency band designations of the International Telecommunication Union (ITU) [2].

Table 2.1
Microwave Frequency Bands

IEEE Radar nomenclature			ITU Nomenclature		
Band Designation	**Frequency Range**	**Wavelength Range (Vacuum)**	**Frequency Range**	**Wavelength Range (Vacuum)**	**Band Designation**
Ultrahigh Frequency (UHF)	300–1000 MHz	30 cm–1m	300–3000 MHz	0.1–1m	Ultrahigh Frequency (UHF)
L	1–2 GHz	15–30 cm			
S	2–4 GHz	7.5–15 cm	3–30 GHz	1–10 cm	Superhigh Frequency (SHF)
C	4–8 GHz	3.75–7.5 cm			
X	8–12 GHz	2.5–3.75 cm			
Ku	12–18 GHz	1.67–2.5 cm			
K	18–27 GHz	1.11–1.67 cm			
Ka	27–40 GHz	0.75–1.11 cm	30–300 GHz	0.1–1 cm	Extremely High Frequency (EHF)
V	40–75 GHz	4–7.5 mm			
W	75–110 GHz	2.73–4 mm			
mm	110–300 GHz	1–2.72 mm			

2.1.2 Illuminating Systems

As previously mentioned, an imaging system includes a source which, at microwave frequencies, is constituted by an antenna. In some cases, there is only one illuminating antenna, which produces the so-called incident field (i.e., the field that is present in space when no targets occupy the test region).

In time-harmonic fields, the incident electric and magnetic field vectors are denoted by $\mathbf{E}_{inc}(\mathbf{r})$ and $\mathbf{H}_{inc}(\mathbf{r})$, respectively, and $\mathbf{r}$ indicates the position vector. They can be known from measurements or simulations. Since the incident electromagnetic field plays a key role in the reconstruction process associated with imaging systems, as it will be seen in the following chapter, it is important to have a model for describing it. In some cases it is assumed that, far from the source, the incident field can be represented by a plane wave, i.e.

$$\mathbf{E}_{inc}^{PW}(\mathbf{r}) = E_0 e^{-jk_b \hat{\mathbf{d}} \cdot \mathbf{r}} \hat{\mathbf{p}} \tag{2.2}$$

$$\mathbf{H}_{inc}^{PW}(\mathbf{r}) = H_0 e^{-jk_b \hat{\mathbf{d}} \cdot \mathbf{r}} \hat{\mathbf{d}} \times \hat{\mathbf{p}} \tag{2.3}$$

where E_0 is the electric field amplitude (V/m), $\hat{\mathbf{p}}$ is the unit vector denoting the polarization of the wave, $\hat{\mathbf{d}}$ is the unit vector defining the direction of propagation, and k_b is the wavenumber given by

$$k_b = \omega\sqrt{\mu_b \varepsilon_b} \tag{2.4}$$

ε_b and μ_b being the dielectric permittivity (F/m) and the magnetic permeability (H/m) of the propagation medium, respectively, and $\omega = 2\pi f$ the angular frequency. In (2.3) the amplitude of the magnetic field H_0 (A/m) is given by $H_0 = \eta_b^{-1} E_0, \eta_b = \sqrt{\mu_b \varepsilon_b^{-1}}$ being the intrinsic impedance of the propagation medium (Ω). Please note that, where not otherwise specified, throughout this book we assume a time dependence for monochromatic fields and sources of the type $exp\,(j\omega t)$.

However, this assumption of plane wave incidence is not common, since we refer here to essentially short-range imaging systems and techniques for which far field radiation conditions are rarely fulfilled. In fact, they require that the following relations must be satisfied

$$r > 10\lambda_b \tag{2.5}$$

$$r > 10d \tag{2.6}$$

$$r > 2\frac{d^2}{\lambda_b} \tag{2.7}$$

where $r = |\mathbf{r}|$ denotes the distance between the source and the field point, λ_b is the wavelength in the propagation medium, and d is the maximum linear dimension of the source (i.e., the diameter of the minimum sphere that can contain it).

In tomographic applications, the incident field is often approximated with a cylindrical wave, as the one produced by an infinite line current, for which the incident electric field, in the transverse plane, is given by

$$\mathbf{E}_{inc}^{LS}(\mathbf{r}_t, \mathbf{r}_s) = -I\frac{\omega\mu_b}{4} H_0^{(2)}\left(k_b\sqrt{(x-x_s)^2 + (y-y_s)^2}\right)\hat{\mathbf{z}} \tag{2.8}$$

where $\mathbf{r}_t = x\hat{\mathbf{x}} + y\hat{\mathbf{y}}$ and $\mathbf{r}_s = x_s\hat{\mathbf{x}} + y_s\hat{\mathbf{y}}$ denote the position (in the transverse plane) of the line source, I is the amplitude of the current density, and $H_0^{(2)}(\bullet)$ is the Hankel function of second kind and zeroth order.

Equation (2.8) allows one to model omnidirectional antennas, that is, whose radiation pattern is isotropic in the transverse plane. Although this is a good approximation in several imaging applications (e.g., when the real illuminating antennas are dipoles), there are cases in which this model is not suitable (e.g., when using horn antennas). A basic way to model such behavior is to use a focused line-source model, in which the incident field is given by

$$\mathbf{E}_{inc}^{FLS}\left(\mathbf{r}_t,\mathbf{r}_s\right)=\mathbf{E}_{inc}^{LS}\left(\mathbf{r}_t,\mathbf{r}_s\right)\cos^{\alpha}\left(\phi_s-\phi_b\right) \tag{2.9}$$

where, ϕ_s is the angle between $\mathbf{r}_t$ and $\mathbf{r}_s$, ϕ_b is the direction of main radiation (in the transverse plane), and the exponent α controls the aperture of the radiation lobe. More complex models can also be obtained by modeling the incident field as a weighted combination of several elementary sources (e.g., line sources in different positions [3]). The incident field due to certain specific radiating elements can be also numerically computed by using proper commercial or custom codes.

The imaging configuration with only one transmitting antenna used to produce the incident radiation is indicated in the rest of the book as a single illumination configuration. If the target is illuminated from different positions (e.g., when the transmitting antenna move around it), we denote the imaging set up as a multi-illumination configuration. More complex systems may include a transmitting apparatus composed of an array of antennas, which can illuminate the region under test at the same time or sequentially. The use of transmitting arrays significantly complicates the illuminating system.

In several cases, different frequencies are used to inspect a given target. This modality is usually denoted as multifrequency imaging. Another possible approach is to involve the illumination of the field in time domain. In this case, the incident field is denoted by $\mathbf{e}_{inc}(\mathbf{r},t)$ and $\mathbf{h}_{inc}(\mathbf{r},t)$. Usually the incident field has a pulse nature and propagates toward the scatterer. The incident pulse is of course characterized by its frequency band. Often ultrawideband (UWB) fields (i.e., with bandwidth that exceeds 500 MHz or with fraction bandwidth greater than 20%) are used for imaging purposes.

Concerning the radiating elements, they depend on the imaging modalities. If monochromatic signals are used, they are usually $\lambda_b/2$ dipoles or small horns. In the cases in which a wideband illumination is required, they are often printed antennas, such as Vivaldi or other wideband elements. Several examples of antennas used in microwave imaging systems will be mentioned in the following chapters.

2.1.3 Receiving Systems

Due to the interaction between the incident electromagnetic field and the target, a scattered field is produced, whose electric and magnetic field vectors are denoted by $\mathbf{E}_{scatt}(\mathbf{r})$ and $\mathbf{H}_{scatt}(\mathbf{r})$, respectively. The sum of the incident and scattered fields is indicated in the following as the total fields, that is

$$\mathbf{E}_{tot}(\mathbf{r}) = \mathbf{E}_{inc}(\mathbf{r}) + \mathbf{E}_{scatt}(\mathbf{r}) \tag{2.10}$$

$$\mathbf{H}_{tot}(\mathbf{r}) = \mathbf{H}_{inc}(\mathbf{r}) + \mathbf{H}_{scatt}(\mathbf{r}) \tag{2.11}$$

It is important to note that the scattered field contains the information about the target under test, but only the total field can be measured in the presence of the target. Clearly, the total field coincides with the incident field if the scatterer is not present, and therefore, $\mathbf{E}_{scatt}(\mathbf{r}) = 0$ and $\mathbf{H}_{scatt}(\mathbf{r}) = 0$. To deduce the scattered field we usually need to know, or estimate, the incident field (when the target is not present) at the same points in which the total field is measured.

In general, the total field is measured by proper antennas in an observation domain. Similarly to the transmitting system, the receiving system can be composed by a single or multiple antennas. It is seldom that the receiving antenna coincides with the transmitting one (a monostatic configuration). In some cases, a single receiving antenna moves around the target and collects information from different angles (a bistatic configuration). In other cases, a set of different antennas is used to illuminate the inspected scenario and to collect the field due to the scattering phenomena (a multistatic configuration). Especially in tomographic applications, the illuminating and receiving system jointly rotate around the target. In this case, the set up is often denoted as a multi-illumination/multiview configuration.

For approaches developed in time-domain, the received signal is a modified version of the incident signal, in which the amplitude is changed after the interaction with the target, and the wavefront is delayed depending on the distance between the transmitting and receiving antenna elements and on the propagation velocity inside the target.

It should be noticed that in most cases a calibration phase is required, since in general the inversion procedures require the knowledge of field quantities at the measurement points, whereas the receiving apparatuses are able to provide the measurements of the scattering parameters (*S*-parameters). This aspect will be further mentioned through the book.

2.1.4 Interaction Between the Incident Field and the Target to be Inspected

At microwave frequencies, the interaction between the incident field produced by the transmitting apparatus and the unknown target is governed by the electromagnetic scattering laws, which will be discussed in Chapter 3. They take into account several phenomena, such as reflection, transmission, absorption, diffraction, and others.

It must be noted that in some applications the external shape of the target is known. Therefore, the imaging process is focused on retrieving information about the inner part of the target. The most ambitious objective is the retrieval of maps of the dielectric parameters of the whole internal structure (i.e., the distributions of dielectric permittivity, electric conductivity, and magnetic permeability). In other cases, the searched information may be limited to just some characteristics of the body under test (e.g., the position of the target inside a test region, the external shape, possible defects in the structure, and so forth).

Potentially, microwave imaging techniques are able to retrieve the three-dimensional (3-D) structure of the body, basing the reconstruction process on 3-D measurements performed around the target. However, often—mainly for computational reasons—the imaging process pursues the objective of reconstructing two-dimensional (2-D) maps of the same parameters (e.g., in 2-D tomographic imaging). These maps are usually slices of the target or images of its cross section in the case of cylindrical targets.

If the external shape of the target is not known, the imaging process usually assumes that the target is included in a fixed space region, which in the following is denoted as investigation domain (i.e., a volume in the 3-D case and a surface in the 2-D case). The investigation domain indicates the test region in which the imaging system acts. In such cases, the external shape of the target constitutes, in general, a problem unknown.

The interaction between the incident field and the target depends of course not only on the incident wave, but also on the physical and geometrical properties of the target. Therefore, it is very important to know the constitutive relationships governing the behaviors of the various materials when subjected to electromagnetic fields. This will be discussed in the following section of this chapter.

The interaction also depends on the considered propagation medium. In some cases, it can be assumed that the target is irradiated under free-space radiation condition. In this case, the fundamental mathematical tool is represented, for 3-D problems, by the free space dyadic Green's function, $\overline{\overline{G}}_b(\mathbf{r},\mathbf{r}')$, and, for 2-D problems, by the scalar Green's function, $g_b(\mathbf{r}_t, \mathbf{r}_t')$, which represent the electromagnetic responses to an elementary source and to an infinite line-current in the assumed infinite and homogeneous propagation medium, respectively. Such functions are expressed by the following equations

$$\overline{\overline{G}}_b(\mathbf{r},\mathbf{r}') = -(\overline{\overline{I}} + \frac{1}{k_b^2}\nabla\nabla)\frac{e^{-jk_b|\mathbf{r}-\mathbf{r}'|}}{4\pi|\mathbf{r}-\mathbf{r}'|} \tag{2.12}$$

$$g_b(\mathbf{r}_t,\mathbf{r}_t') = \frac{j}{4}H_0^{(2)}(k_b|\mathbf{r}_t - \mathbf{r}_t'|) \tag{2.13}$$

where $\bar{\bar{\mathbf{I}}}$ denotes the unit dyadic function. In the above equations, $\mathbf{r}$ and $\mathbf{r}'$ denote again the field and source points, respectively. Moreover, $\mathbf{r}_t$ and $\mathbf{r}_t'$ are the corresponding field and source points in the transverse plane in the 2-D case.

When the investigation domain is located in a half-space (e.g., in the case of an object buried in a two-layer structure), the proper Green's function and tensor must be adopted. The same holds true in the case of a presence of a stratified medium. When the investigation domain is located inside a more complex scenario, usually a numerically computed Green's function is necessary. The proper Green's functions used in the various imaging approaches will be recalled in the following chapters when necessary.

In several cases, the target to be inspected can be located inside a microwave chamber (which physically represents the investigation domain). This chamber may have plastic walls as well as metallic walls. Several examples of these chambers will be mentioned in the following chapters.

Depending of the considered reconstruction procedure, the incident field must be known not only inside the observation domain, but also inside the investigation domain. While the measurement of this field can be simply performed at the measurement points (without any target to be inspected), the same measurements inside the investigation domain are usually uneasy in practical applications. Therefore, this additional information is commonly obtained by using numerical simulators employing proper models of the sources, such as the ones mentioned in Section 2.1.2. It has also been proposed to obtain the incident field inside the investigation domain by solving a specific inverse source problem, starting from the incident field measured in the observation domain. This approach is interesting, but increases the complexity of the reconstruction process.

2.2 Dielectric Properties of the Medium

The propagation of electromagnetic waves in a given medium is, as it is well known, governed by Maxwell's equations, which, for time-harmonic fields, are given by

$$\nabla \bullet \mathbf{D}(\mathbf{r}) = \rho_e(\mathbf{r}) \tag{2.14}$$

$$\nabla \bullet \mathbf{B}(\mathbf{r}) = 0 \tag{2.15}$$

$$\nabla \times \mathbf{E}(\mathbf{r}) = -j\omega \mathbf{B}(\mathbf{r}) \tag{2.16}$$

$$\nabla \times \mathbf{H}(\mathbf{r}) = j\omega \mathbf{D}(\mathbf{r}) + \mathbf{J}_e(\mathbf{r}) \tag{2.17}$$

where $\mathbf{D}(\mathbf{r})$ and $\mathbf{B}(\mathbf{r})$ denote the electric displacement (C/m^2) and the magnetic induction (T) vectors, respectively. The sources are represented by the electric charge density $\rho_e(\mathbf{r})$ (C/m^3) and by the electric current density vector $\mathbf{J}_e(\mathbf{r})$ (A/m^2).

In vacuum, it results that

$$\mathbf{D}(\mathbf{r}) = \varepsilon_0 \mathbf{E}(\mathbf{r}) \tag{2.18}$$

$$\mathbf{B}(\mathbf{r}) = \mu_0 \mathbf{H}(\mathbf{r}) \tag{2.19}$$

where $\varepsilon_0 \cong 8.85 \times 10^{-12}$ F/m is the vacuum dielectric permittivity and $\mu_0 = 4\pi \times 10^{-7}$ H/m denotes the vacuum magnetic permeability.

For the aim of this book, the relationships between field vectors in materials different from vacuum are most relevant. Basically, two physical phenomena must be taken into account when a given material is subjected to an electromagnetic field (i.e., polarization and conduction [4]). At a macroscopic level, polarization can be represented in terms of the following relationships between the field vectors

$$\mathbf{D} = F_D(\mathbf{E},\mathbf{H}),\ \mathbf{B} = F_B(\mathbf{E},\mathbf{H}) \tag{2.20}$$

where F_D and F_B represent two functionals, which can have complex forms in certain situations (some example will be discussed in the following). However, for most linear, stationary, isotropic, and spatially nondispersive dielectric material, we can assume that

$$\mathbf{D}(\mathbf{r}) = \hat{\varepsilon}(\mathbf{r},\omega)\mathbf{E}(\mathbf{r}) \tag{2.21}$$

$$\mathbf{B}(\mathbf{r}) = \mu(\mathbf{r},\omega)\mathbf{H}(\mathbf{r}) \tag{2.22}$$

where $\hat{\varepsilon}$ (F/m) is the dielectric permittivity and μ (H/m) is the magnetic permeability of the material. At microwave frequencies, these parameters usually depend on frequencies due to their temporal dispersion.

Concerning relation (2.22), it can be found that for certain materials, called diamagnetic materials, the amplitude of the magnetic permittivity is slightly less than that of a vacuum, whereas for other materials it is slightly greater. In general, for these materials, we assume (in microwave imaging applications) that $\mu(\mathbf{r}, \omega) \cong \mu_0$. These materials are also denoted as nonmagnetic materials. However, for ferromagnetic materials, (2.22) is no longer valid, since the relation between the magnetic field vector and the magnetic induction vector is nonlinear and characterized by a hysteretic behavior.

Moreover, due to the time dispersive nature of most dielectric materials, it turns out that the dielectric permittivity in (2.21) not only depends on frequency, but is also a complex valued quantity, that is,

$$\hat{\varepsilon}(\mathbf{r},\omega)=\hat{\varepsilon}'(\mathbf{r},\omega)-j\hat{\varepsilon}''(\mathbf{r},\omega) \tag{2.23}$$

Conduction phenomena generate induced currents due to the movement of free electrons inside the materials. These currents can be expressed, for a large class of materials, as

$$\mathbf{J}_{ohm}(\mathbf{r})=\hat{\sigma}(\mathbf{r},\omega)\mathbf{E}(\mathbf{r}) \tag{2.24}$$

where $\hat{\sigma}(\mathbf{r}, \omega)$ is the electric conductivity (S/m) of the material, which again, if time dispersion is taken into account, can be modeled as a complex quantity, that is,

$$\hat{\sigma}(\mathbf{r},\omega)=\hat{\sigma}'(\mathbf{r},\omega)+j\hat{\sigma}''(\mathbf{r},\omega) \tag{2.25}$$

It is however worth remarking that, in most practical cases, for frequencies up to microwaves, the electric conductivity can be assumed to be a real valued and frequency-independent quantity.

Substituting (2.21) and (2.24) in (2.17), and taking into account that the total current density is the sum of the impressed and induced current densities, that is,

$$\mathbf{J}_e(\mathbf{r})=\mathbf{J}_0(\mathbf{r})+\mathbf{J}_{ohm}(\mathbf{r}) \tag{2.26}$$

we obtain

$$\nabla\times\mathbf{H}(\mathbf{r})=j\omega\left(\hat{\varepsilon}(\mathbf{r},\omega)-j\frac{\hat{\sigma}(\mathbf{r},\omega)}{\omega}\right)\mathbf{E}(\mathbf{r})+\mathbf{J}_0(\mathbf{r}) \tag{2.27}$$

It is thus possible to define an equivalent complex dielectric permittivity as

$$\begin{aligned}\varepsilon(\mathbf{r},\omega)&=\hat{\varepsilon}(\mathbf{r},\omega)-j\frac{\hat{\sigma}(\mathbf{r},\omega)}{\omega}\\&=\left[\hat{\varepsilon}'(\mathbf{r},\omega)+\frac{\hat{\sigma}''(\mathbf{r},\omega)}{\omega}\right]-j\left[\hat{\varepsilon}''(\mathbf{r},\omega)+\frac{\hat{\sigma}'(\mathbf{r},\omega)}{\omega}\right]\end{aligned} \tag{2.28}$$

In the scientific literature, (2.28) is often rewritten as

$$\varepsilon(\mathbf{r},\omega)=\varepsilon'(\mathbf{r},\omega)-j\varepsilon''(\mathbf{r},\omega)=\varepsilon'(\mathbf{r},\omega)-j\frac{\sigma(\mathbf{r},\omega)}{\omega} \tag{2.29}$$

where the equivalent (real) dielectric permittivity and electric conductivity are given by

$$\varepsilon'(\mathbf{r},\omega)=\hat{\varepsilon}'(\mathbf{r},\omega)+\frac{\hat{\sigma}''(\mathbf{r},\omega)}{\omega} \tag{2.30}$$

$$\varepsilon''(\mathbf{r},\omega)=\hat{\varepsilon}''(\mathbf{r},\omega)+\frac{\hat{\sigma}'(\mathbf{r},\omega)}{\omega} \tag{2.31}$$

$$\sigma(\mathbf{r},\omega)=\hat{\sigma}'(\mathbf{r},\omega)+\omega\varepsilon''(\mathbf{r},\omega) \tag{2.32}$$

As a consequence, a scattering problem involving a dielectric (dispersive) material with a certain electric conductivity can be treated as a scattering problem involving only the equivalent dielectric permittivity, according to the following relationship

$$\nabla\times\mathbf{H}(\mathbf{r})=j\omega\varepsilon(\mathbf{r},\omega)\mathbf{E}(\mathbf{r})+\mathbf{J}_0(\mathbf{r}) \tag{2.33}$$

Clearly, it is necessary to have correctly in mind the real meaning of this assumption in order to avoid confusion between quantities associated to different physical phenomena. However, in practice, it is often not known if the presence of the imaginary part of the equivalent dielectric permittivity is due to the dispersion effect or to the conductivity of the medium. The only known quantity is often the value of the imaginary part of $\varepsilon(\mathbf{r},\omega)$.

Moreover, we assume a given material to be considered as an ideal dielectric material if $\varepsilon''\approx 0$. Furthermore, a good conductor is such that $\sigma(\mathbf{r},\omega) \gg \omega|\varepsilon'(\mathbf{r},\omega)|$, that is, the conduction phenomenon is prominent with respect to the polarization phenomenon, which, in this case, is therefore negligible. Of course, under this condition we have

$$\varepsilon(\mathbf{r},\omega)\approx -j\frac{\sigma(\mathbf{r},\omega)}{\omega} \tag{2.34}$$

A perfect electric conducting (PEC) material is an ideal material in which the electric conductivity σ approaches an infinite value.

The presence of a non-null equivalent conductivity (either deriving from real conduction phenomena or from dispersion) produces an attenuation (depending upon the frequency) of the electromagnetic fields inside the materials. In particular, in these cases, the wavenumber assumes complex values and thus it can be written as $k_b = \beta - j\alpha$, where the attenuation α and the propagation constant β are given by

$$\alpha = \omega\sqrt{\mu_0 \varepsilon'}\left\{\frac{1}{2}\left[\sqrt{1+\left(\frac{\sigma}{\omega\varepsilon'}\right)^2}-1\right]\right\}^{1/2} \quad (2.35)$$

$$\beta = \omega\sqrt{\mu_0 \varepsilon'}\left\{\frac{1}{2}\left[\sqrt{1+\left(\frac{\sigma}{\omega\varepsilon'}\right)^2}+1\right]\right\}^{1/2} \quad (2.36)$$

In order to quantify the penetration of the fields inside the material, usually the skin depth is employed. Such quantity, which is defined as the distance after which the amplitude of the propagating electric field is reduced by a factor of e^{-1}, is given by

$$\delta = \frac{1}{\alpha} = \frac{1}{\omega\sqrt{\mu_0 \varepsilon'}}\left\{\frac{1}{2}\left[\sqrt{1+\left(\frac{\sigma}{\omega\varepsilon'}\right)^2}-1\right]\right\}^{-1/2} \quad (2.37)$$

The mean power density associated with an electromagnetic wave is provided by the Poynting vector, which is given by

$$\mathbf{S}(\mathbf{r}) = \frac{1}{2}\mathbf{E}(\mathbf{r}) \times \mathbf{H}^*(\mathbf{r}) \quad (2.38)$$

where * denotes the complex conjugate value. For a plane wave, as the one reported in (2.2), it results that the power density inside the medium is

$$p(\mathbf{r}) = \mathrm{Re}\left\{\mathbf{S}(\mathbf{r}) \bullet \hat{\mathbf{d}}\right\} = \frac{1}{2}\mathrm{Re}\{\eta_b\}\left|\mathbf{E}(\mathbf{r})\right|^2 = \frac{1}{2}\mathrm{Re}\{\eta_b\}\left|E(0)\right|^2 e^{-2\alpha\hat{\mathbf{d}}\cdot r} \quad (2.39)$$

In general, the dissipated power density (i.e., transformed in heat) inside the material is given by

$$p_{ab}(\mathbf{r}) = \frac{1}{2}\sigma(\mathbf{r})|\mathbf{E}(\mathbf{r})|^2 = \frac{1}{2}\left[\hat{\sigma}'(\mathbf{r}) + \omega\varepsilon''(\mathbf{r})\right]|\mathbf{E}(\mathbf{r})|^2 \tag{2.40}$$

which, especially for biomedical applications, provides the value of the specific absorption rate (*SAR*) (W/kg), given by

$$SAR = \frac{1}{V}\int_V \frac{p_{ab}(\mathbf{r})}{\rho(\mathbf{r})} d\mathbf{r} = \frac{1}{V}\int_V \frac{1}{2}\frac{\sigma(\mathbf{r})|\mathbf{E}(\mathbf{r})|^2}{\rho(\mathbf{r})} d\mathbf{r} \tag{2.41}$$

where ρ denotes the mass density (kg/m^3) and V is the volume over which the *SAR* is evaluated (usually defined in the specific protection regulations).

The previous relationships will be used in the following chapters to describe the materials involved in the various considered applications. It should be mentioned that these concepts are relevant not only for studying the targets under test, but also with reference to the materials which are often adopted to construct matching structures for reducing the reflections of the external surfaces of the targets and improving the delivered electromagnetic energy.

2.2.1 Parametric Models of the Dielectric Properties

As discussed in the previous section, the vast majority of materials exhibit a temporal dispersion, which produces a frequency-dependence in the dielectric properties. Although the real behavior can only be obtained by measurements, in most cases there is the need to model such behavior in a numerical way (e.g., for performing forward simulations aimed at studying the response of the material or for implementing data inversion schemes).

To this end, several models have been developed in the past for describing the dielectric properties of the materials. The Debye model and the Cole-Cole model are reported in the following sections. It is worth noting that other models have also been developed for specific materials (some of them are mentioned in the next chapters). Moreover, mixing formulas [5] are also often used to describe the dielectric properties of mixtures of materials.

2.2.1.1 Debye Model

The Debye model has been introduced in [6] for describing the behavior of the dielectric properties of polar molecules. It is often used to model the dispersive and conduction behaviors of dielectric materials. By using the Debye model, the complex dielectric permittivity of the material is approximated with the following parametric formula

$$\varepsilon(\omega)=\varepsilon_\infty+\frac{\Delta\varepsilon}{1+j\omega\tau}-j\frac{\sigma_s}{\omega} \tag{2.42}$$

where ε_∞, σ_s, and τ are real-valued parameters that depend upon the specific material. From (2.42) it is possible to explicitly write the equivalent real dielectric permittivity and electric conductivity as

$$\varepsilon'(\omega)=\varepsilon_\infty+\mathrm{Re}\left\{\frac{\Delta\varepsilon}{1+j\omega\tau}\right\}=\varepsilon_\infty+\frac{\Delta\varepsilon}{1+(\omega\tau)^2} \tag{2.43}$$

$$\sigma(\omega)=\sigma_s-\omega\mathrm{Im}\left\{\frac{\Delta\varepsilon}{1+j\omega\tau}\right\}=\sigma_s+\frac{\omega^2\tau\Delta\varepsilon}{1+(\omega\tau)^2} \tag{2.44}$$

It is worth noting that when $\omega\to+\infty$, it shows that $\varepsilon\to\varepsilon_\infty$, whereas, for $\omega\to 0$, we have $\sigma\to\sigma_s$. Consequently, the two parameters, ε_∞ and σ_s, represent the asymptotic values of the dielectric permittivity and the electric conductivity, respectively. Moreover, since $\varepsilon'\to\varepsilon_\infty+\Delta\varepsilon$, for $\omega\to 0$, it also shows that $\Delta\varepsilon=\varepsilon_s-\varepsilon_\infty$, ε_s being the static dielectric permittivity.

In several cases, the model in (2.42) is not sufficient to accurately describe the behavior of the dielectric properties over the whole range of frequencies of interest. In order to overcome such limitation, it has been empirically extended by considering the following multipole Debye model

$$\varepsilon(\omega)=\varepsilon_\infty+\sum_{n=1}^{N}\frac{\Delta\varepsilon_n}{1+j\omega\tau_n}-j\frac{\sigma_s}{\omega} \tag{2.45}$$

where ε_∞, σ_s, N, $\Delta\varepsilon_n$, and τ_n are again parameters depending upon the specific material.

As a significant example, we consider here the dielectric properties of water, whose content determines the behaviour of several materials, for example, biological tissues and soils. The complex dielectric permittivity of water (which has been investigated for a long time [6–8]), for frequencies lower than about 100 GHz, can be described by using a single-pole Debye model. However, it has been found that the parameters of the model depend upon the temperature [7]. To overcome such limitation, empirical models relating the Debye parameters to the temperature have been developed (for example, see [8, 9]). More recently, models employing a two-pole structure have also been developed for increasing the accuracy (for example, the ones proposed by Stogryn et al. [10] and Meissner and Wentz [11]).

An example of the behavior of the real and imaginary parts of the dielectric permittivity of pure water in the microwave frequency band for a temperature of 25°C is shown in Figure 2.2. Both single-pole [9] and double-pole [11] models are reported. Moreover, the experimental data provided in [12] are superimposed for reference purposes. As can be seen, both models are able to describe the frequency behavior of the dielectric permittivity, although, as expected, the double-pole model approximates the measured data more accurately. It is worth noting that in both cases the absorption peak due to the Debye-type relaxation caused by the water molecules (which, at 25°C is approximately located at 18 GHz) is correctly modeled.

2.2.1.2 Cole-Cole Model

The Debye model has been empirically extended by Cole and Cole in [13]. In particular, the complex dielectric permittivity is described by the following relationship

$$\varepsilon(\omega) = \varepsilon_\infty + \frac{\Delta\varepsilon}{1+(j\omega\tau)^{1-\alpha}} - j\frac{\sigma_s}{\omega} \tag{2.46}$$

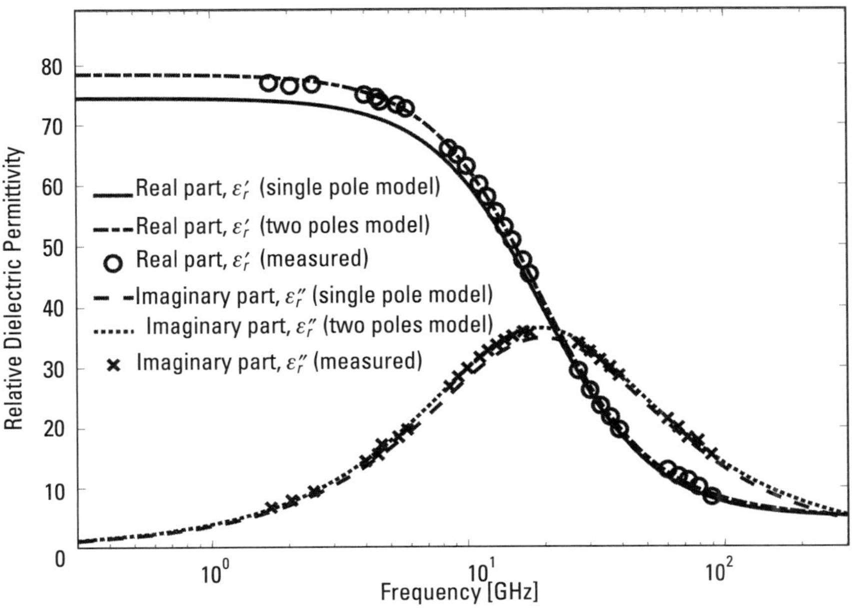

Figure 2.2 Relative dielectric permittivity (real and imaginary parts) of pure water (at 25°C) in the microwave frequency range. Single- and two-pole Debye models and comparisons with measured data [12].

where ε_∞ is the real dielectric permittivity for $\omega \to +\infty$, σ_s is the static electric conductivity, and $\Delta\varepsilon = \varepsilon_s - \varepsilon_\infty$, ε_s being the static dielectric permittivity. Moreover, τ is a generalized relaxation constant. The newly introduced constant value α varies between 0 and 1. It is worth noting that if $\alpha = 0$, the Cole-Cole model reduces to the Debye one.

The model has been also generalized by considering multiple poles, that is, the complex dielectric permittivity can be expressed as

$$\varepsilon(\omega) = \varepsilon_\infty + \sum_{n=1}^{N} \frac{\Delta\varepsilon_n}{1 + (j\omega\tau_n)^{1-\alpha_n}} - j\frac{\sigma_s}{\omega} \tag{2.47}$$

where ε_∞ and σ_s are the asymptotic values of the dielectric permittivity and the electric conductivity, and $\Delta\varepsilon_n$, τ_n, and α_n, $n = 1, \ldots, N$, are the parameters of the N poles.

The previously reported Cole-Cole models have been employed for describing the dispersion effects in different types of materials, including liquid mixtures and biological tissues. The latter case, as discussed in detail in Chapter 5, is of particular importance for microwave imaging. An example of the relative dielectric permittivity (real and imaginary parts) of a biological tissue modeled

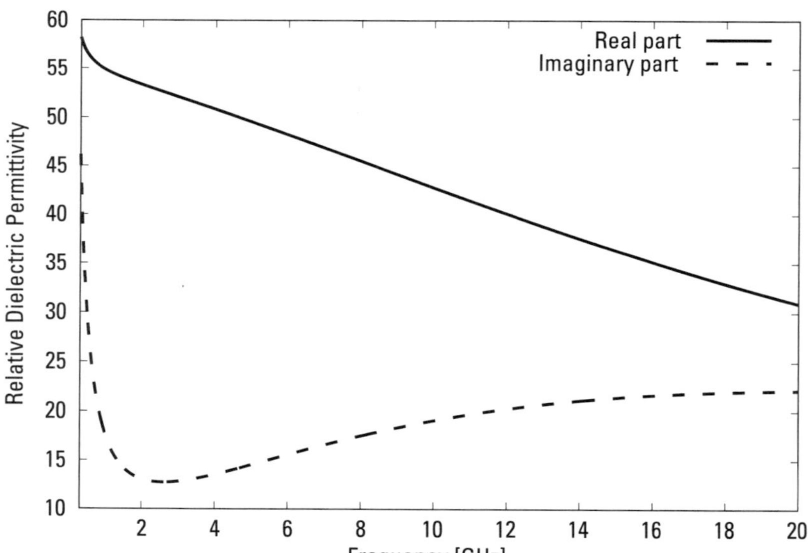

Figure 2.3 Relative dielectric permittivity (real and imaginary parts) of muscle tissue in the frequency range of 300 MHz–20 GHz, obtained by using a multipole Cole-Cole model, with N = 4, derived from the data reported in [14].

by using the multipole Cole-Cole relationship in (2.47) is shown in Figure 2.3. In particular, the curves refer to muscle tissues and have been obtained by using the four-pole model developed in [14].

References

[1] "IEEE Standard Letter Designations for Radar-Frequency Bands," *IEEE Std 521-2002*, 2003.

[2] "Nomenclature of the Frequency and Wavelength Bands Used in Telecommunications," *ITU Recomm. V431*, 2015.

[3] Nounouh, S., C. Eyraud, H. Tortel, and A. Litman, "Modeling of the Antenna Effects and Calibration for Subsurface Probing," *Micro. Opt. Technol. Lett.*, Vol. 56, No. 11, Nov. 2014, pp. 2516–2522.

[4] Von Hippel, A. R., *Dielectrics and Waves*, Boston: Artech House, 1995.

[5] Sihvola, A. H., *Electromagnetic Mixing Formulas and Applications*, London: Institution of Engineering and Technology, 1999.

[6] Debye, P. J. W., *Polar Molecules*, New York: The Chemical Catalog Company, Inc., 1929.

[7] Kaatze, U., "Complex Permittivity of Water as a Function of Frequency and Temperature," *J. Chem. Eng. Data*, Vol. 34, No. 4, Oct. 1989, pp. 371–374.

[8] Ulaby, F. T., R. K. Moore, and A. K. Fung, *Microwave Remote Sensing: Active and Passive*, Norwood, MA: Artech House, 1986.

[9] Klein, L., and C. Swift, "An Improved Model for the Dielectric Constant of Sea Water at Microwave Frequencies," *IEEE Trans. Antennas Propag.*, Vol. 25, No. 1, January 1977, pp. 104–111.

[10] Stogryn, A. P., H. T. Bull, K. Ruayi, and S. Iravanchy, "The Microwave Permittivity of Sea and Fresh Water," *Aeroj. Intern. Rep.*, 1996.

[11] Meissner, T., and F. J. Wentz, "The Complex Dielectric Constant of Pure and Sea Water from Microwave Satellite Observations," *IEEE Trans. Geosci. Remote Sens.*, Vol. 42, No. 9, September 2004, pp. 1836–1849.

[12] Barthel, J., et al.,"A Computer-controlled System of Transmission Lines for the Determination of the Complex Permittivity of Lossy Liquids between 8.5 and 90 GHz," *Berichte Bunsenges. Für Phys. Chem.*, Vol. 95, No. 8, August 1991, pp. 853–859.

[13] Cole, K. S., and R. H. Cole, "Dispersion and Absorption in Dielectrics I. Alternating Current Characteristics," *J. Chem. Phys.*, Vol. 9, No. 4, p. 341, 1941.

[14] Gabriel, S., R. W. Lau, and C. Gabriel, "The Dielectric Properties of Biological Tissues: III. Parametric Models for the Dielectric Spectrum of Tissues," *Phys. Med. Biol.*, Vol. 41, No. 11, November 1996, pp. 2271–2293.

3

Fundamentals of Electromagnetic Imaging and Inverse Scattering

There are several comprehensive books describing the mathematical formulation of the inverse electromagnetic scattering problem [1–10]. In this chapter, we wish to recall the basic equations adopted by engineers in order to develop systems and techniques in a wide range of microwave imaging applications.

3.1 Electromagnetic Scattering

Let us consider the configuration shown in Figure 3.1. We refer here to a three-dimensional (3-D) geometry with a dielectric target (possibly lossy) embedded in a homogeneous nonmagnetic background medium characterized by a complex dielectric permittivity ε_b.

As already introduced in Chapter 2, the incident electromagnetic field is defined as the field produced by the transmitting antennas when no targets are present in the scenario. The incident field is usually a known quantity, since it can be measured or simulated. Moreover, it can be chosen in the system design phase.

Let us now consider the case in which an inhomogeneous (possibly lossy) dielectric target is located inside the investigation domain V. The target is characterized by a space-dependent complex dielectric permittivity denoted as $\varepsilon(\mathbf{r})$, $\mathbf{r} \in V_{obj}$, where $V_{obj} \subset V$ is the volume occupied by the target. Nonmagnetic objects are considered here.

The scattered electric field vector, defined in (2.10), can be expressed as [10]

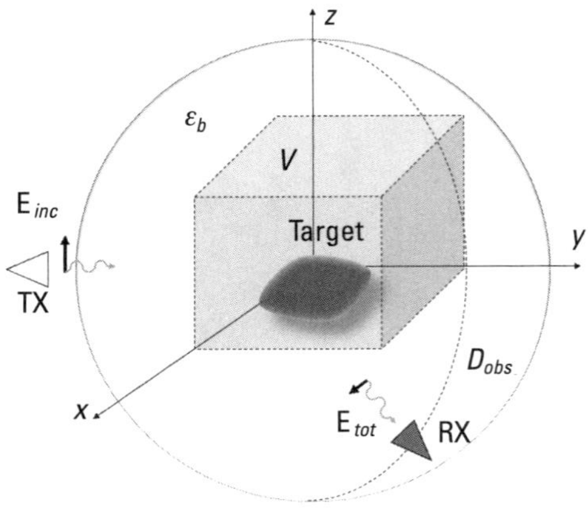

Figure 3.1 Schematic representation of the imaging configuration.

$$\mathbf{E}_{scatt}(\mathbf{r}) = j\omega\mu_0 \int_V \mathbf{J}_{eq}(\mathbf{r}') \bullet \bar{\bar{\mathbf{G}}}_b(\mathbf{r},\mathbf{r}')d\mathbf{r}' \tag{3.1}$$

where $\bar{\bar{\mathbf{G}}}_b$ is the dyadic Green's function for the considered configuration (see [2.12]) and $\mathbf{J}_{eq}$ is an equivalent current density defined as [11]

$$\mathbf{J}_{eq}(\mathbf{r}) = j\omega(\varepsilon(\mathbf{r}) - \varepsilon_b)\mathbf{E}_{tot}(\mathbf{r}) \tag{3.2}$$

It is worth noting that $\mathbf{J}_{eq}(\mathbf{r}) = 0$ for $\mathbf{r} \notin V_{obj}$, since outside the target it results in $\varepsilon(\mathbf{r}) = \varepsilon_b$. By substituting (3.2) in (3.1), the scattered electric field can be explicitly written as

$$\mathbf{E}_{scatt}(\mathbf{r}) = -k_b^2 \int_V c(\mathbf{r}')\mathbf{E}_{tot}(\mathbf{r}') \bullet \bar{\bar{\mathbf{G}}}_b(\mathbf{r},\mathbf{r}')d\mathbf{r}' \tag{3.3}$$

where the contrast function is defined as

$$c(\mathbf{r}) = \frac{\varepsilon(\mathbf{r})}{\varepsilon_b} - 1 \tag{3.4}$$

Finally, according to (2.10), the total electric field vector is obtained as

$$\mathbf{E}_{tot}(\mathbf{r}) = \mathbf{E}_{inc}(\mathbf{r}) - k_b^2 \int_V c(\mathbf{r}')\mathbf{E}_{tot}(\mathbf{r}') \bullet \bar{\bar{\mathbf{G}}}_b(\mathbf{r},\mathbf{r}')d\mathbf{r}' \tag{3.5}$$

Equation (3.5) is the fundamental relation for the imaging process under the assumptions made in this section. Actually, since the incident field is a known quantity, if one measures the total field in the observation domain, D_{obs}, (3.5)—which turns out to be a Fredholm integral equation of the first kind—could be theoretically solved in order to retrieve the unknown contrast function.

It should be noted that in the case that a target exhibits a magnetic permeability different from the one of vacuum, a new contribution to the scattering field can be added to the solving equation, which is expressed in terms of an equivalent magnetic current density [5].

There are two well-known problems associated with this equation. The first one concerns nonlinearity, since both the contrast and the internal total field are unknown quantities. The second one concerns the ill-posedness issues associated to a Fredholm equation of first kind. The first problem will be discussed with reference to the specific solving procedures considered in the various applications. The second one will be briefly mentioned here, but the reader is referred to books and papers addressing inverse scattering from a stronger mathematical viewpoint (some of them have been mentioned at the beginning of this chapter).

When dealing with weakly scattering targets (i.e., objects which represent, due to their dielectric parameters and dimensions, weak discontinuities with respect to the propagation medium), (3.5) can be linearized by using the so-called first order Born approximation, for which the scattered electric field is expressed in terms of the incident field only [12], that is,

$$\mathbf{E}_{tot}^{B_1}(\mathbf{r}) = \mathbf{E}_{inc}(\mathbf{r}) - k_b^2 \int_V c(\mathbf{r}')\mathbf{E}_{inc}(\mathbf{r}') \bullet \bar{\bar{\mathbf{G}}}_b(\mathbf{r},\mathbf{r}')d\mathbf{r}' \tag{3.6}$$

It is worth remarking that, since the total field is expressed only in terms of the incident field, multiple scattering is neglected. Such approximation is valid when the following condition is satisfied [13]:

$$k_b a c_{max} < 2\pi\gamma \tag{3.7}$$

where a is the radius of the minimum sphere enclosing the scatterer, $c_{max} = \max_{|\mathbf{r}|<a} |c(\mathbf{r})|$, and γ is a constant. In the scientific literature, different values of γ in the range [0,1] has been derived, for example, $\gamma = 0.16$ in [14], and $\gamma =$ 0.25 in [15].

Higher-order Born approximations are sometimes adopted. In particular, the nth order approximation is given by

$$\mathbf{E}_{tot}^{B_n}(\mathbf{r}) = \mathbf{E}_{inc}(\mathbf{r}) - k_b^2 \int_V c(\mathbf{r}')\mathbf{E}_{tot}^{B_{n-1}}(\mathbf{r}') \bullet \bar{\bar{\mathbf{G}}}_b(\mathbf{r},\mathbf{r}')d\mathbf{r}' \tag{3.8}$$

where $\mathbf{E}_{tot}^{B_{n-1}}$ is the total internal field obtained by using a Born approximation of order $(n-1)$. The second order approximation, $(n=2)$, has been used for an example in [16, 17]. The field vectors in (3.8) are the terms of the Neumann series for the total electric field expansion, which converges for weak scatterers only. Other linearization schemes can be used, some of them are mentioned with reference to two-dimensional (2-D) configurations.

Let us consider again (3.5). When $\mathbf{r} \in D_{obs}$, this equation constitutes the so-called data equation. To address the nonlinearity of the inverse problem, usually another equation is associated to this one, that is, the same (3.5) but evaluated for $\mathbf{r} \in V$. By introducing the linear operator $G_{state}(\cdot)$, defined by

$$\mathbf{G}_{state}(\bullet)(\mathbf{r}) = -k_b^2 \int_V (\bullet)(\mathbf{r}') \bullet \overline{\overline{\mathbf{G}}}_b(\mathbf{r},\mathbf{r}')d\mathbf{r}', \ \mathbf{r} \in V \tag{3.9}$$

from (3.5) we obtain

$$\mathbf{E}_{tot}(\mathbf{r}) = \mathbf{E}_{inc}(\mathbf{r}) + \mathbf{G}_{state}(c\mathbf{E}_{tot})(\mathbf{r}), \quad \mathbf{r} \in V \tag{3.10}$$

It is worth noting that (3.10) is usually called the state equation. Such an equation represents an additional relationship between the two unknowns (i.e., the internal total electric field and the contrast function).

Analogously, a similar operator relationship can be written for the points outside the target, that is, for the data equation we obtain

$$\mathbf{E}_{tot}(\mathbf{r}) = \mathbf{E}_{inc}(\mathbf{r}) + \mathbf{G}_{data}(c\mathbf{E}_{tot})(\mathbf{r}), \quad \mathbf{r} \in D_{obs} \tag{3.11}$$

where

$$\mathbf{G}_{data}(\bullet)(\mathbf{r}) = -k_b^2 \int_V (\bullet)(\mathbf{r}') \bullet \overline{\overline{\mathbf{G}}}_b(\mathbf{r},\mathbf{r}')d\mathbf{r}', \ \mathbf{r} \in D_{obs} \tag{3.12}$$

Consequently, the scattering problem can be expressed by the following system of integral equations

$$\begin{cases} \mathbf{E}_{tot}(\mathbf{r}) = \mathbf{E}_{inc}(\mathbf{r}) + \mathbf{G}_{state}(c\mathbf{E}_{tot})(\mathbf{r}) & \mathbf{r} \in V \\ \mathbf{E}_{tot}(\mathbf{r}) = \mathbf{E}_{inc}(\mathbf{r}) + \mathbf{G}_{data}(c\mathbf{E}_{tot})(\mathbf{r}) & \mathbf{r} \in D_{obs} \end{cases} \tag{3.13}$$

By defining the linear operator $\mathbf{C}(\mathbf{E}_{tot})(\mathbf{r}) = c(\mathbf{r})\mathbf{E}_{tot}(\mathbf{r})$, the solution to (3.10) can be formally written as

$$\mathbf{E}_{tot}(\mathbf{r}) = (\mathbf{I} - \mathbf{G}_{state}\mathbf{C})^{-1}\mathbf{E}_{inc}(\mathbf{r}), \ \mathbf{r} \in V \tag{3.14}$$

where **I** is the identity vector operator. Such relationship can be substituted into (3.11), leading to

$$\mathbf{E}_{tot}(\mathbf{r}) = \mathbf{E}_{inc}(\mathbf{r}) + \mathbf{G}_{data}\mathbf{C}(\mathbf{I} - \mathbf{G}_{state}\mathbf{C})^{-1}\mathbf{E}_{inc}(\mathbf{r}),\ \mathbf{r} \in D_{obs} \tag{3.15}$$

This equation can be solved numerically after discretization of the continuous model. Several solving procedures, currently applied, will be discussed in the following chapters with reference to specific applications.

It is worth noting that other equivalent representations can be used. For example, by defining the contrast source current density $\mathbf{J}_{cs}$ as [18]

$$\mathbf{J}_{cs}(\mathbf{r}) = c(\mathbf{r})\mathbf{E}_{tot}(\mathbf{r}),\ \mathbf{r} \in V \tag{3.16}$$

(3.11) can be rewritten as

$$\mathbf{E}_{tot}(\mathbf{r}) = \mathbf{E}_{inc}(\mathbf{r}) + \mathbf{G}_{data}(\mathbf{J}_{cs})(\mathbf{r}),\ \mathbf{r} \in D_{obs} \tag{3.17}$$

Moreover, by multiplying both terms of (3.10) by $c(\mathbf{r})$, the following equation is obtained

$$\mathbf{J}_{cs}(\mathbf{r}) = c(\mathbf{r})\mathbf{E}_{inc}(\mathbf{r}) + c(\mathbf{r})\mathbf{G}_{state}(\mathbf{J}_{cs})(\mathbf{r}),\ \mathbf{r} \in V \tag{3.18}$$

Consequently, the scattering problem is now described by the following system of equations

$$\begin{cases} \mathbf{E}_{tot}(\mathbf{r}) = \mathbf{E}_{inc}(\mathbf{r}) + \mathbf{G}_{data}(\mathbf{J}_{cs})(\mathbf{r}) & \mathbf{r} \in D_{obs} \\ \mathbf{J}_{cs}(\mathbf{r}) = c(\mathbf{r})\mathbf{E}_{inc}(\mathbf{r}) + c(\mathbf{r})\mathbf{G}_{state}(\mathbf{J}_{cs})(\mathbf{r}) & \mathbf{r} \in V \end{cases} \tag{3.19}$$

in which the first equation turns out to be a linear equation.

As already mentioned, alternative formulations can be adopted. As an example, in [19] a new equation has been derived by manipulating the contrast source state equation. In particular, let us define the scalar function $p(\mathbf{r})$ as

$$p(\mathbf{r}) = \frac{c(\mathbf{r})}{1 - c(\mathbf{r}) f_m(\mathbf{r})} \tag{3.20}$$

where $f_m(\mathbf{r})$ is a scalar function equal to the average value of the diagonal terms of the dyadic function $\overline{\overline{\mathbf{F}}}(\mathbf{r})$ given by

$$\overline{\overline{\mathbf{F}}}(\mathbf{r}) = -k_b^2 \int_V \overline{\overline{\mathbf{G}}}_b(\mathbf{r}, \mathbf{r}')\, d\mathbf{r}' \tag{3.21}$$

It has been proven in [19] that the state equation can be then rewritten as

$$\mathbf{J}_{cs}(\mathbf{r}) = p(\mathbf{r})\mathbf{E}_{inc}(\mathbf{r}) + p(\mathbf{r})\mathbf{G}_{mod}(\mathbf{J}_{cs})(\mathbf{r}) \tag{3.22}$$

where the vector operator $\mathbf{G}_{mod}$ is defined as

$$\mathbf{G}_{mod}(\mathbf{J})(\mathbf{r}) = -k_b^2 \int_V [\mathbf{J}(\mathbf{r}') - \mathbf{J}(\mathbf{r})] \bullet \overline{\overline{\mathbf{G}}}_b(\mathbf{r},\mathbf{r}')d\mathbf{r}' + \mathbf{J}(\mathbf{r}) \bullet \Delta\overline{\overline{\mathbf{F}}} \tag{3.23}$$

being

$$\Delta\overline{\overline{\mathbf{F}}}(\mathbf{r}) = \overline{\overline{\mathbf{F}}}(\mathbf{r}) - f_m(\mathbf{r})\overline{\overline{\mathbf{I}}} \tag{3.24}$$

It has been claimed in [19] that this formulation "leads, in many cases of practical interest, to a reduction of the degree of nonlinearity of the relationship among parameters embedding dielectric characteristics and scattered fields as compared to the traditional model, thus allowing increased robustness against false solutions in the inversion process."

The scattering equations can also be directly expressed in terms of the *S*-parameters related to the transmitting and receiving antennas. Actually, in practical applications, the available measurements are often scalar quantities (in time or frequency domains) representing the output of the receiving antennas. For example, when using a vector network analyzer, usually such quantities are just the *S*-parameters [2]. In this case, according to [20], a possible way to relate the *S*-parameters to the distribution of the dielectric properties in the investigation domain is to rewrite the integral scattering equations as

$$S_{scatt}^{ik} = \frac{j\omega\varepsilon_b}{2a_i a_k} \int_V c(\mathbf{r}')\mathbf{E}_{inc}^i(\mathbf{r}') \bullet \mathbf{E}_{tot}^k(\mathbf{r}')d\mathbf{r}' \tag{3.25}$$

where S_{scatt}^{ik} is the scattering contribution of the *S*-parameter measured in the *i*th port when the target is illuminated by using the *k*th port, a_i and a_k are the root-power-waves incoming in the *i*th and *k*th ports, $\mathbf{E}_{tot}^k$ is the total electric field due to the illumination from the *k*th antenna, and $\mathbf{E}_{inc}^i$ is the incident electric field due to the *i*th antenna working in transmission mode. It is worth noting that (3.25) refers only to the scattered field. A similar incident term S_{inc}^{ik} representing the incident field can be derived [20, 21]. Other possible formulations to directly link the measured *S*-parameters to the dielectric properties of the inspected region have also been proposed in the microwave imaging literature. For example in [22], a new kernel for the integral scattering equation (referred as vector Green's function) is obtained by including a model of the antennas based on the source-scattering matrix formulation.

In order to provide a visual idea of the scattering phenomenon in near field, a very simple example is provided next. A dielectric object composed by a homogeneous beam with square cross section is considered. The external medium is vacuum, therefore $\eta_b = \eta_0$ and $\lambda_b = \lambda_0$. The beam has a square cross section with sides $L_x = L_y = 0.5\lambda_0$ and finite height L_z. The dielectric properties (lossless material) has been set equal to $\varepsilon_r = 4$. The center of the beam is supposed to be in the origin of the reference system. The incident electric field is generated by a *z*-directed short dipole of length $h = 0.025\lambda_0$. When the dipole is located in the origin of the coordinate system, the incident field vector is given by

$$\mathbf{E}_{inc}^{SD}(\mathbf{r}) = \frac{\eta_b I_{SD} h e^{-jk_b r}}{4\pi r}[2\left(\frac{1}{r} - j\frac{1}{k_b r^2}\right)\cos\theta\hat{\mathbf{r}} + \left(jk_b + \frac{1}{r} - j\frac{1}{k_b r^2}\right)\sin\theta\hat{\boldsymbol{\theta}}] \tag{3.26}$$

A standard spherical coordinate system is used, where the azimuth angle Φ is the angle in the *x*-*y* plane computed with reference to the *x* axis, and the elevation angle θ is computed with respect to the *z* axis. In the present case the source is located at point in $(\lambda_0, 0, 0)$. Therefore, a coordinate transformation $\mathbf{R} = \mathbf{r} - \lambda_0\hat{\mathbf{x}}$ has been applied to (3.26). Moreover, we assumed $I_{SD} = 0.04$A. The scattered electric field is computed on a circumference with radius $R_m = \lambda_0$ located in the *x*-*y* plane and centered at the origin of the coordinate system. The field values have been numerically computed by solving the scattering equations by means of the method of moments, with pulse basis functions and Dirac's delta weighting functions [23]. In particular, the object has been discretized into cubic cells of side $L = 0.025\lambda_0$.

The behaviors of the computed scattered fields (amplitudes and phases) for different values of the height L_z in the range $[0.5\lambda_0, 10\lambda_0]$ are shown in Figure 3.2. As can be seen in this case, the scattered field tends to concentrate in the forward direction (corresponding to $\Phi = 180°$). For comparison purposes, the corresponding field due to an infinite cylinder with the same cross section and illuminated by a line-current source (see [2.8]) located in the transverse plane at position $(\lambda_0, 0)$ is also reported (the source current has been properly scaled for obtaining comparable incident fields values [i.e., $I = 0.74$ mA]). As can be seen, as L_z increases, the solution for the 3-D case approaches the one obtained under the 2-D approximation, although the illumination conditions are not really the same in the two cases. Moreover, it can be seen, at least in this simple configuration, that a height of few wavelengths is sufficient for obtaining similar values of the scattered electric field.

So far we consider the case in which the target to be inspected is made of dielectric material, possibly with a finite electric conductivity. If the object

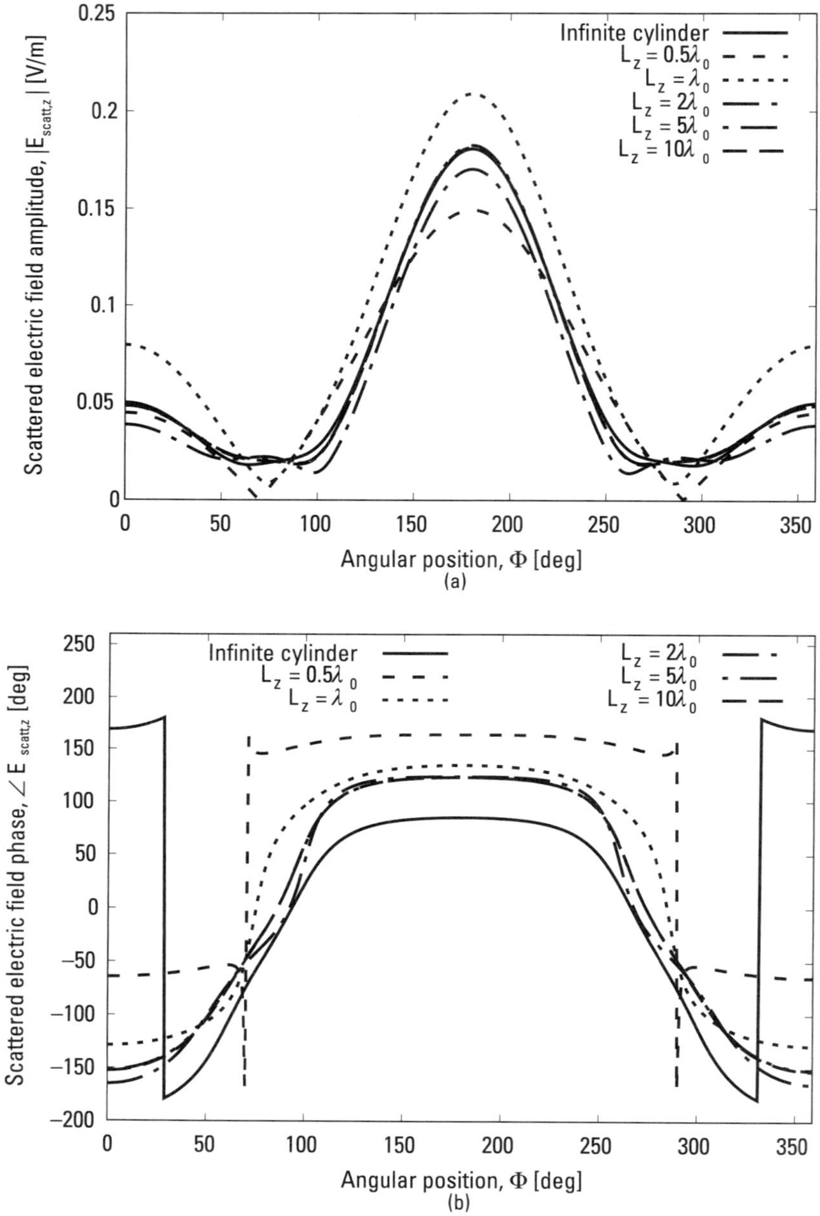

Figure 3.2 Behavior of the scattered electric field in presence of a 3-D target with a parallelepiped shape. Comparison with the field computed by using the 2-D cylindrical approximation in the case of an infinite cylinder with the same cross section. (a) Amplitude and (b) phase of the scattered electric field (*z*-component) computed on a circumference surrounding the object.

can be assumed a perfect electric conducting (PEC) target, the waves cannot penetrate inside it. Therefore, the induced current density may be only limited to the external surface. Accordingly, an integral equation similar to (3.5) can be written, where the volume integral is replaced by a surface integral. If the external shape of the target is not known, the domain of the integral is another problem unknown. Often the induced surface current density can be approximated by using the Kirchhoff approximation [5, 24], which state that the current density can be written in terms of the incident field only, that is,

$$\mathbf{J}_s^K(\mathbf{r}) = 2\hat{\mathbf{n}} \times \mathbf{H}_{inc}(\mathbf{r}), \ \mathbf{r} \in S_{inc} \tag{3.27}$$

where S_{inc} denotes the illuminated part of the external surface of the scatterer, whereas in the remaining part of this surface, it is assumed that $\mathbf{J}_s^K(\mathbf{r}) = 0$.

3.2 Two-Dimensional Inverse Scattering

In several applications, the targets are characterized by elongated shapes. In these cases, it is possible to use a simplified electromagnetic model of the scattering phenomena. In particular, let us suppose that the following conditions are satisfied

1. The target is assumed to be infinite along one direction (e.g., the z axis).
2. The dielectric properties of the inspected region are independent from the z-coordinate, that is,

$$\varepsilon(\mathbf{r}) = \varepsilon(\mathbf{r}_t), \quad \mathbf{r} \in V \tag{3.28}$$

   $\mathbf{r}_t = x\hat{\mathbf{x}} + y\hat{\mathbf{y}}$ being, as usual, the position vector in the transverse plane.
3. The investigation domain is infinite along the z axis and its cross section, denoted by D, is uniform along the z coordinate.
4. The transmitting antenna generates a z-polarized transverse-magnetic (TM) field, that is,

$$\mathbf{E}_{inc}(\mathbf{r}) = E_{inc,z}(\mathbf{r}_t)\hat{\mathbf{z}} \tag{3.29}$$

The considered configuration is sketched in Figure 3.3.

Under such conditions, the total field, and consequently also the scattered field, are independent from the z-coordinate and z-polarized [11], that is,

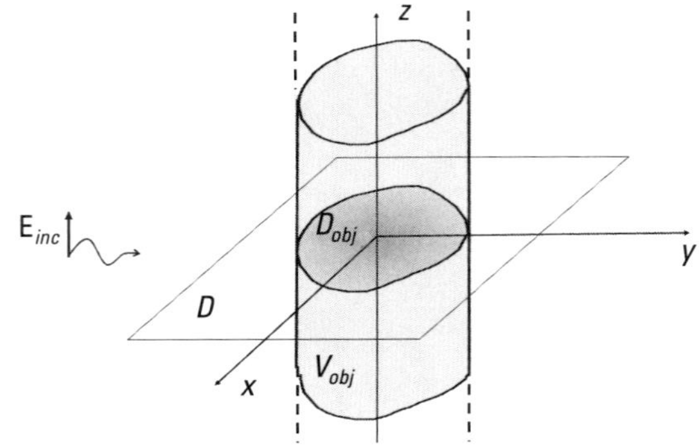

Figure 3.3 Schematic representation of a 2-D imaging configuration involving an infinite cylinder under transverse-magnetic illumination conditions.

$$\mathbf{E}_{tot}(\mathbf{r}) = E_{tot,z}(\mathbf{r}_t)\hat{\mathbf{z}},\ \mathbf{E}_{scatt}(\mathbf{r}) = E_{scatt,z}(\mathbf{r}_t)\hat{\mathbf{z}} \tag{3.30}$$

Consequently, the scattering equation (3.5) reduces to the scalar equation [10]

$$E_{tot,z}(\mathbf{r}_t) = E_{inc,z}(\mathbf{r}_t) - k_b^2 \int_D c(\mathbf{r}_t')E_{tot,z}(\mathbf{r}_t')g_b(\mathbf{r}_t,\mathbf{r}_t')d\mathbf{r}_t' \tag{3.31}$$

where g_b is the 2-D Green's function for the considered background medium. In the case of free space, it is given by the relationship reported in (2.13). Other forms for different configurations are considered in the following chapters.

It is worth remarking that an alternative 2-D formulation can be obtained by considering transverse electric (TE) illuminations. In this case, the incident electric field can be expressed as

$$\mathbf{E}_{inc,t}(\mathbf{r}_t) = E_{inc,x}(\mathbf{r}_t)\hat{\mathbf{x}} + E_{inc,y}(\mathbf{r}_t)\hat{\mathbf{y}} \tag{3.32}$$

and the corresponding scattering equation for the electric field becomes [25]

$$\mathbf{E}_{tot,t}(\mathbf{r}_t) = \mathbf{E}_{inc,t}(\mathbf{r}_t) - k_b^2 \int_D c(\mathbf{r}_t')\mathbf{E}_{tot,t}(\mathbf{r}_t') \bullet \left(\mathbf{I} + \frac{\nabla_t \nabla_t}{k_b^2}\right) g_b(\mathbf{r}_t,\mathbf{r}_t')d\mathbf{r}_t' \tag{3.33}$$

where ∇_t is the nabla operator referred to the transverse coordinates (x, y) and $\mathbf{E}_{tot,t}(\mathbf{r}_t) = E_{tot,x}(\mathbf{r}_t)\hat{\mathbf{x}} + E_{tot,y}(\mathbf{r}_t)\hat{\mathbf{y}}$. It is worth noting that in this case the involved field quantities are still vectors. A scalar equation similar to the one of the TM case can be obtained by considering the z-component of the magnetic field. However, it seems that the vector electric-field equation can provide better performances in imaging problems [26]. The 2-D TE model has been successfully used in several applications (for example in [25, 27–32]) and, in some cases, it has been found to be able to provide better results than its TM counterpart.

Linearization of (3.31) can also be applied, as for the 3-D case, for example, by using the first order Born approximation, for which

$$E_{tot,z}^{B_1}(\mathbf{r}_t) = E_{inc,z}(\mathbf{r}_t) - k_b^2 \int_D c(\mathbf{r}_t') E_{inc,z}(\mathbf{r}_t') g_b(\mathbf{r}_t, \mathbf{r}_t') d\mathbf{r}_t' \quad (3.34)$$

A very simple example of application of the Born approximation is discussed in the following. A homogeneous infinite lossless circular cylinder with radius $a = 0.5\lambda_0$ and a cross section centered in the origin of the reference system is assumed to be located in vacuum. The value of the relative dielectric permittivity has been changed in the range [1,3]. The incident electric field is due to a line-current source located in $(0.75\lambda_0, 0)$ with an excitation current given by $I = 1\text{mA}$. The total field is evaluated in a single measurement point of coordinates $(-0.75\lambda_0, 0)$. In order to evaluate the limitations of the Born approximation in this simple case, the contrast function is estimated by using the following relation (which is directly derived from [3.34])

$$c \cong \frac{E_{tot,z}(\mathbf{r}_t) - E_{inc,z}(\mathbf{r}_t)}{-k_b^2 \int_D E_{inc,z}(\mathbf{r}_t') g_b(\mathbf{r}_t, \mathbf{r}_t') d\mathbf{r}_t'} \quad (3.35)$$

where $E_{tot,z}$ is the exact total electric field (i.e., the one that would be effectively measured). For numerically solving the integral, the cross section of the target has been discretized into $N = 3125$ square subdomains of side $L = 0.16\lambda_0$. The estimated values versus the actual values are shown in Figure 3.4. As can be seen, for values of the contrast function below 0.5 (i.e., $\varepsilon_r < 1.5$), there is a very good agreement between the estimated and the actual values. However, when the contrast increases, the estimated values tend to diverge with respect to the actual one.

Another approximation that has been used in microwave imaging is the Rytov approximation [15, 33, 34]. In this case, the total electric field is approximated as

$$E_{tot,z}^{R}(\mathbf{r}_t) = E_{inc,z}(\mathbf{r}_t) e^{\Phi^{\mathrm{R}}(\mathbf{r}_t)} \quad (3.36)$$

where the phase term $\Phi^{\mathrm{R}}(\mathbf{r}_t)$ depends linearly from the contrast function, that is,

$$\Phi^{\mathrm{R}}(\mathbf{r}_t) = \frac{E_{scatt,z}^{B_1}(\mathbf{r}_t)}{E_{inc,z}(\mathbf{r}_t)} = -\frac{k_b^2}{E_{inc,z}(\mathbf{r}_t)} \int_D c(\mathbf{r}_t') E_{inc,z}(\mathbf{r}_t') g_b(\mathbf{r}_t, \hat{\mathbf{r}}_t) d\mathbf{r}_t' \quad (3.37)$$

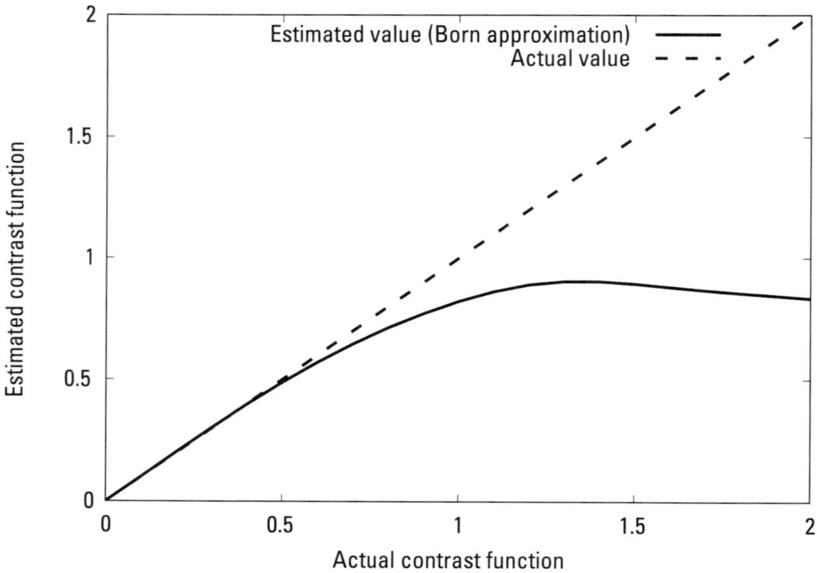

Figure 3.4 Example of behavior of the contrast function estimated by using the first-order Born approximation versus the actual value. A homogeneous target has been considered.

In (3.37), $E_{scatt,z}^{B_1}(\mathbf{r}_t) = E_{tot,z}^{B_1}(\mathbf{r}_t) - E_{inc,z}(\mathbf{r}_t)$ is the scattered field computed by using the first-order Born approximation. When $\left|E_{scatt,z}^{B_1}(\mathbf{r}_t) / E_{inc,z}(\mathbf{r}_t)\right|$ << 1, (e.g., for small targets) the Rytov approximation reduces to the Born one. However, it has been found that the Rytov approximation can be applied for larger scatterers than those for which the Born approximation is valid, but only in the case that their dielectric profiles change slowly [15]. A discussion about the validity of the Born and Rytov approximations can be found in [34]. It should be noted that the Rytov approximation has been recently reconsidered in imaging techniques [21], due to above mentioned property that it exhibits as compared to the Born linearization.

Other linearizations can also be used. For example, an extended Born approximation has been proposed in [35], whereas in [36] a generalized approximation for linear scattering has been introduced. The Rayleigh approximation may also slightly outperform the Born approximation in some cases [37, 38].

Similarly to the 3-D case discussed in Section 3.1, it is again possible to formulate the scattering problem as a system of two equations. In particular, by defining the 2-D scalar counterparts of the state and data operators in (3.9) and (3.12) as

$$G_{data}(\bullet)(\mathbf{r}_t) = -k_b^2 \int_D (\bullet)(\mathbf{r}_t') g_b(\mathbf{r}_t, \mathbf{r}_t') d\mathbf{r}_t', \ \mathbf{r}_t \in D_{obs} \tag{3.38}$$

$$G_{state}(\bullet)(\mathbf{r}_t) = -k_b^2 \int_D (\bullet)(\mathbf{r}_t')\, g_b(\mathbf{r}_t, \mathbf{r}_t')\, d\mathbf{r}_t', \;\; \mathbf{r}_t \in D \tag{3.39}$$

the scattering problem can be formulated as the solution of

$$\begin{cases} E_{tot,z}(\mathbf{r}_t) = E_{inc,z}(\mathbf{r}_t) + G_{state}(cE_{tot,z})(\mathbf{r}_t) & \mathbf{r}_t \in D \\ E_{tot,z}(\mathbf{r}_t) = E_{inc,z}(\mathbf{r}_t) + G_{data}(cE_{tot,z})(\mathbf{r}_t) & \mathbf{r}_t \in D_{obs} \end{cases} \tag{3.40}$$

Moreover, the following relationship can be obtained

$$E_{tot,z}(\mathbf{r}_t) = E_{inc,z}(\mathbf{r}_t) + G_{data} C (I - G_{state} C)^{-1} E_{inc,z}(\mathbf{r}_t), \;\; \mathbf{r}_t \in D_{obs} \tag{3.41}$$

where I is the scalar identity operator and $C(E_{tot,z})(\mathbf{r}_t) = c(\mathbf{r}_t) E_{tot,z}(\mathbf{r}_t)$, $\mathbf{r}_t \in D$.

It is worth noting that in microwave imaging techniques a discretization of the continuous model is usually required, since analytical solutions are only available for very specific problems, usually involving canonical scatterers, which are of limited interest in practical applications. The simplest discretization of the integral equations is based on a pixelated representation of the problem. In particular, the measured scattered field is assumed to be available only at certain positions inside the observation domain (where the receiving antennas are assumed to be positioned), and the investigation domain is partitioned into a finite number of subdomains. The internal field quantities and the dielectric properties of the medium are assumed to be constant inside each subdomain. Details of this technique can be found in [5]. However, more sophisticated approaches can be followed, for example, by using full domain basis functions covering the entire investigation domain [39, 40] or wavelet representations [41, 42]. In any case, the original integral equations are usually reduced to algebraic equations allowing matrix representations.

As an example, the discretized forms of the 2-D data and state equations (3.40) are reported below, by using the simplest discretization approach previously mentioned. The investigation area D is discretized into N nonoverlapping square subdomains D_n of sides l and centers $\mathbf{r}_{inv,n} = (x_{inv,n}, y_{inv,n})$, $n = 1, \ldots N$. The measurement points that compose the observation domain are denoted by $\mathbf{r}_{meas,m} = (x_{meas,n}, y_{meas,n})$, $m = 1, \ldots, M$. Under such conditions, the discretized equations result to be

$$\mathbf{e}_{scatt}^{ext} = [\mathbf{G}_{data}]\,\mathrm{diag}(\mathbf{c})\,\mathbf{e}_{tot}^{int} \tag{3.42}$$

$$\mathbf{e}_{tot}^{int} = \mathbf{e}_{inc}^{int} + [\mathbf{G}_{state}]\,\mathrm{diag}(\mathbf{c})\,\mathbf{e}_{tot}^{int} \tag{3.43}$$

where the involved numerical arrays $\mathbf{e}_{scatt}^{ext}$ (of size M), $\mathbf{e}_{inc}^{int}$ (of size N), $\mathbf{e}_{tot}^{int}$ (of size N), and c (of size N) are defined as

$$\mathbf{e}_{scatt}^{ext} = \begin{bmatrix} E_{tot,z}\left(\mathbf{r}_{meas,1}\right) - E_{inc,z}\left(\mathbf{r}_{meas,1}\right) \\ \vdots \\ E_{tot,z}\left(\mathbf{r}_{meas,M}\right) - E_{inc,z}\left(\mathbf{r}_{meas,M}\right) \end{bmatrix} \tag{3.44}$$

$$\mathbf{e}_{tot}^{int} = \begin{bmatrix} E_{tot,z}\left(\mathbf{r}_{inv,1}\right) \\ \vdots \\ E_{tot,z}\left(\mathbf{r}_{inv,N}\right) \end{bmatrix} \tag{3.45}$$

$$\mathbf{e}_{inc}^{int} = \begin{bmatrix} E_{inc,z}\left(\mathbf{r}_{inv,1}\right) \\ \vdots \\ E_{inc,z}\left(\mathbf{r}_{inv,N}\right) \end{bmatrix} \tag{3.46}$$

$$\mathbf{c} = \begin{bmatrix} c\left(\mathbf{r}_{inv,1}\right) \\ \vdots \\ c\left(\mathbf{r}_{inv,N}\right) \end{bmatrix} \tag{3.47}$$

Moreover, in (3.42) and (3.43), diag($\mathbf{c}$) denotes a $N \times N$ diagonal matrix whose nonzero elements are the values contained in the array c, and the matrices [$\mathbf{G}_{data}$] (of dimensions $M \times N$) and [$\mathbf{G}_{state}$] (of dimensions $N \times N$), which are the discrete counterparts of the operators G_{data} and G_{state} in (3.38) and (3.39), are given by

$$\left[\mathbf{G}_{data}\right] = \begin{bmatrix} g_{11}^{ext} & \cdots & g_{1N}^{ext} \\ \vdots & \ddots & \vdots \\ g_{M1}^{ext} & \cdots & g_{MN}^{ext} \end{bmatrix}, \left[\mathbf{G}_{state}\right] = \begin{bmatrix} g_{11}^{int} & \cdots & g_{1N}^{int} \\ \vdots & \ddots & \vdots \\ g_{N1}^{int} & \cdots & g_{NN}^{int} \end{bmatrix} \tag{3.48}$$

where

$$g_{mn}^{ext} = -k_b^2 \int_{D_n} g_b\left(\mathbf{r}_{meas,m}, \mathbf{r}_t'\right) d\mathbf{r}_t', \ m = 1,\ldots,M, n = 1,\ldots,N \tag{3.49}$$

$$g_{mn}^{int} = -k_b^2 \int_{D_n} g_b\left(\mathbf{r}_{inv,m}, \mathbf{r}_t'\right) d\mathbf{r}_t', \ m, n = 1,\ldots,N \tag{3.50}$$

It is worth noting that special care must be devoted to the computation of the integrals, in order to deal with the singularity of the Green's function that arises in the diagonal terms of the state matrix. However, in standard configurations, closed-form formulas are available for the computation of (3.49) and (3.50) (see for example [43] for the 2-D and 3-D free-space case).

Equation (3.43) is well posed and can be solved with respect to $\mathbf{e}_{tot}^{int}$. By substituting the solution in the discretized data equation, the following nonlinear matrix form of (3.41) can be obtained

$$\mathbf{e}_{scatt}^{ext} = \left[\mathbf{G}_{data}\right]\mathrm{diag}(\mathbf{c})\left(\left[\mathbf{I}\right] - \left[\mathbf{G}_{state}\right]\mathrm{diag}(\mathbf{c})\right)^{-1} \mathbf{e}_{inc}^{int} \tag{3.51}$$

where $\mathbf{I}$ is the identity matrix. Similar discretized expression can be obtained when dealing with the 3-D case, where the dimensionalities of the involved vectors and arrays is increased due to the vector nature of the fields and by exploiting different basis fuctions for representing the unknowns.

Whatever the adopted discretization schema is, the discretized versions of the scattering equations maintain their nonlinear nature with respect to the contrast function (i.e., the unknown to be retrieved). Unfortunately, the ill-posed nature of the continuous model results in the poor conditioning of the discrete model. Therefore, the solution of the equations of the discrete model must also be found in a regularized sense.

To have just an idea about the significance of the scattering contribution due to a simple target, an infinite cylinder with a square cross section of sides equal to 30 cm and centered at the origin of a cylindrical coordinate system is considered. The sides of the cross section are parallel to the x and y axes, respectively. A void circular inclusion (simulating, for example, a defect in a homogeneous structure) is present. It is centered at point (5,5) cm and has a diameter d_{inc}, which has been changed in the range 0.5–10 cm (Figure 3.5[a]). The dielectric properties of the cylinder have been set equal to properties of the dry concrete predicted by the Debye model reported in Chapter 4 (i.e., $\varepsilon_r = 4.5$ and $\sigma = 3.9$ mS/m at 3 GHz). The working frequency has been varied in the range 0.5–5 GHz. The forward scattering problem has been solved by using the method of moments with pulse basis function and Dirac's delta weighting functions [23]. In particular, the target has been discretized into $N = 3{,}969$ square subdomains of side 4.8 mm. The structure is illuminated by an ideal line-current source (see [2.8]) located at point (25,0) cm with an excitation current $I = 1$mA.

The scattered electric field has been evaluated in $M = 360$ points equally spaced on a circumference of radius $R_m = 25$ cm, centered at the origin of the coordinate system, and located in the transverse plane. The following quantity

$$\Psi_{inc} = \frac{\left\| E_{scatt,z}^{void} - E_{scatt,z}^{no-void} \right\|_2}{\left\| E_{inc,z} \right\|_2} \tag{3.52}$$

has been used to quantify the contribution of the defect to the scattered field, where $E_{scatt,z}^{void}$ and $E_{scatt,z}^{no-void}$ are the scattered electric fields with and without the void defect, and $\|\bullet\|_2$ denotes the 2-norm computed on 360 points, which are equally spaced on the circumference on which the field is evaluated. The values of Ψ_{inc} with respect to the considered values of the frequency and of the defect size are reported in Figure 3.5(b). As can be seen, very small defects produce negligible perturbations of the scattered field. This fact can also be seen from Figure 3.6, which shows some examples of the scattered electric field for different values of the diameter of the inclusion and for a frequency equal to 2 GHz. For the smallest size of the defect reported in the figure (i.e., with diameter equal to 1 cm) the field with and without the inclusion are almost superimposed. As expected, when the size increases, more significant differences are instead present.

3.3 Inverse Scattering Solving Procedures: the Example of a Newton Method in L^p Banach Spaces

According to the formulation developed in the previous sections—which is widely used in currently-used microwave imaging methods—linear and nonlinear inverse scattering approaches turn out to be governed by integral equations that need, after discretization, to be solved by suitable algorithms. Several approaches for dealing with both the nonlinearity and the ill-posed nature of the problem have been proposed in the physical-mathematical literature. Methods based on Newton schemes are examples of widely applied approaches [44–47]. For illustrative purposes, an efficient approach based on an inexact Newton scheme is described in details in this section, with reference to the more general continuous model (therefore, different discretization rules can be applied.) In the considered approach, the nonlinear problem of finding c is solved by means of two nested iterative loops. In the outer one, which is composed by Newton iterations, the scattering equations are linearized through a first-order expansion around the current estimation of the contrast function c. In the inner loop, the obtained linearized scattering problem is solved in a regularized sense by performing a regularization in the framework of the L^p Banach spaces.

(a)

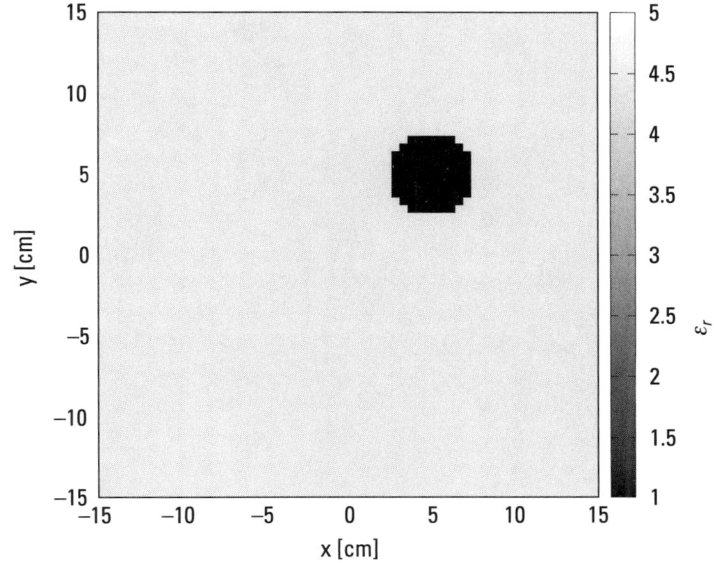

(b)

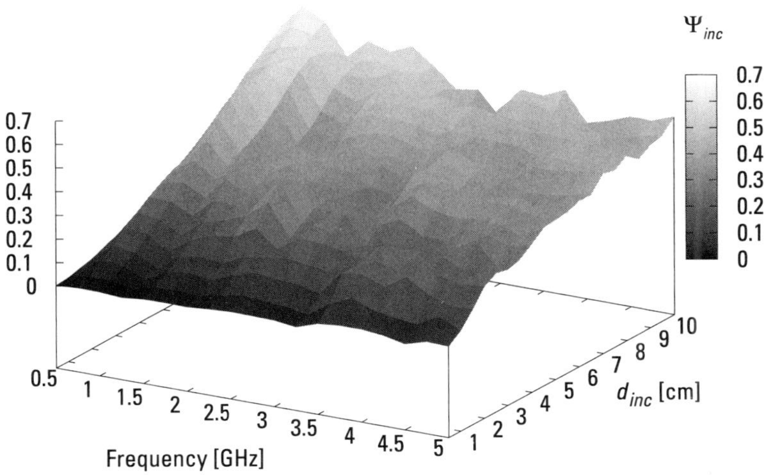

Figure 3.5 (a) Model of a square cylinder with a void inclusion (in this case, $d_{inc} = 2.5$ cm), and (b) behavior of the parameter Ψ_{inc} versus the frequency (0.5–5 GHz) and the d_{inc} (0.5–10 cm).

Let us consider again (3.41), which describe the exact inverse scattering problem formulated in terms of the combination of the data and state equations for a 2-D problem under TM illumination. This equation can be formally rewritten in terms of a nonlinear operator F such that

(a)

(b)

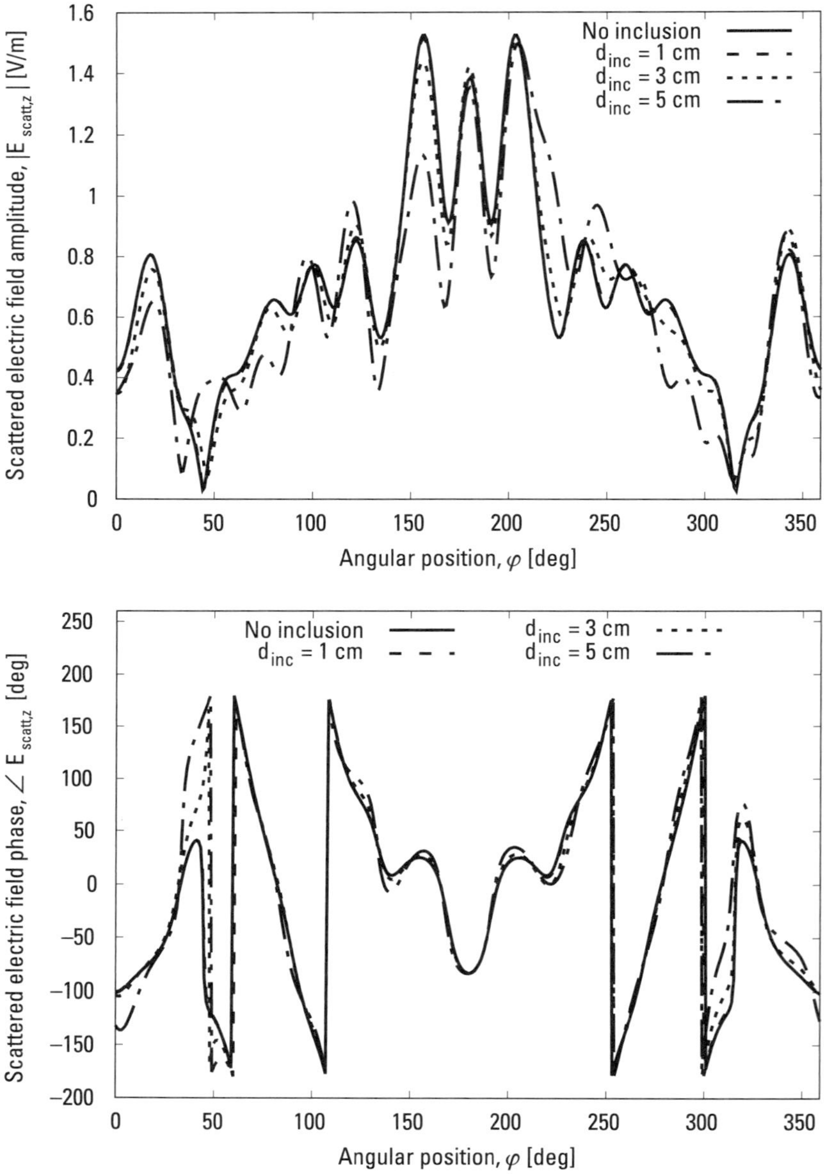

Figure 3.6 Behavior of the scattered electric field due to a square cylinder with a void inclusion for different values of the diameter of the inclusion. (a) Amplitude and (b) phase of the z-component on a circumference of radius 25 cm surrounding the target (360 equally spaced points).

$$E_{scatt,z}\left(\mathbf{r}_t\right) = F\left(c\right)\left(\mathbf{r}_t\right) \qquad \mathbf{r}_t \in D_{obs} \tag{3.53}$$

This relationship links the scattered field in the observation domain $E_{scatt,z}(\mathbf{r}_t)$, $\mathbf{r}_t \in D_{obs}$ (i.e., our data) with the contrast function inside the investigation domain (i.e., the unknown). In particular, in the framework of Banach spaces, we consider a scattered field $E_{scatt,z} \in L^p(D_{obs})$, a contrast function $c \in L^p(D)$, and a nonlinear scattering operator F: $L^p(D) \rightarrow L^p(D_{obs})$. The basic steps of the inexact Newton method in L^p Banach spaces are described in the following list.

1. First, the L^p Banach space in which the problem is solved is defined by fixing a parameter $p > 1$.
2. The Newton iterations are initialized at $n = 0$ (n being the outer iteration index) with a starting guess $c_0 \in L^p(D)$. In the frequent case in which no a priori information about the contrast function of the structure under test is available, the initial guess is taken as an empty investigation domain, which corresponds to $c_0 = 0$.
3. The first-order expansion of the operator F in (3.53) is computed by applying a Fréchet differentiation [48], linearizing F around the currently estimated contrast function $c_n \in L^p(D)$. The resulting linear equation is

$$F_n'h(\mathbf{r}_t) = E_n(\mathbf{r}_t), \quad \mathbf{r}_t \in D_{obs} \tag{3.54}$$

where $h \in L^p(D)$ is the unknown of the linearized problem, and $E_n = E_{scatt,z} - F(c_n) \in L^p(D_{obs})$ and $F_n' : L^p(D) \rightarrow L^p(D_{obs})$ represents the Fréchet derivative of the nonlinear operator F at c_n, which is defined as in [48]

$$F_n'h(\mathbf{r}_t) = G_{data}\left(I - C_n G_{state}\right)^{-1} H\left(I - G_{state} C_n\right)^{-1} E_{inc,z}(\mathbf{r}_t), \quad \mathbf{r}_t \in D_{obs} \tag{3.55}$$

with $C_n(f)(\mathbf{r}_t) = c_n(\mathbf{r}_t)f(\mathbf{r}_t)$, and $H(f)(\mathbf{r}_t) = h(\mathbf{r}_t)f(\mathbf{r}_t)$, $\mathrm{r}_t \in D$.

4. For each Newton iteration, the linearized equation (3.54) is solved by using a regularizing algorithm in L^p Banach spaces, which minimizes the square of the p-norm of the residual functional $\mathcal{R}_n : L^p(D_{obs}) \rightarrow \mathbb{R}$, given by

$$\mathcal{R}_n(h) = \frac{1}{2}\left\|F_n'h - E_n\right\|_p^2 \tag{3.56}$$

where the involved p-norm is defined as $\|a\|_p^p = \int_{D_{obs}} |a(\mathbf{r})|^p \, d\mathbf{r}$. For performing this minimization, the basic steps of two different approaches, namely the Landweber method and a conjugate-gradient like procedure are outlined here:

a. In both methods, the iterations start with $k = 0$ (k being the inner iteration index), defining an initial guess $h_0^* = 0 \in L^q(D)$, q being the Hölder conjugate of p. Additionally, in the conjugate-gradient procedure only, the quantity

$$u_0^* = -F_n'^* J_p\left(F_n' h_k - E_n\right) \in L^q(D) \tag{3.57}$$

is computed as the zeroth descent functional of (3.56). The symbol $F_n'^*$ represents the adjoint operator of F_n', whereas the duality map J_p: $L^p(D) \rightarrow L^q(D)$ is given by

$$J_p(f) = \|f\|_p^{2-p} |f|^{p-1} \operatorname{sign}(f) \tag{3.58}$$

where $\operatorname{sign}(f) = 0$ for $f = 0$, and $\operatorname{sign}(f) = e^{j\arg(f)}$ otherwise.

b. An inner iterative loop is then performed, computing

$$\begin{cases} h_{k+1}^* & = & h_k^* + \alpha_k u_k^* \\ h_{k+1} & = & J_q\left(h_{k+1}^*\right) \\ u_{k+1}^* & = & -F_n'^* J_p\left(F_n' h_{k+1} - E_n\right) + \beta_k u_k^* \end{cases} \quad k = 0, 1, 2, \ldots \tag{3.59}$$

until suitable stopping criteria is satisfied (e.g., the discrepancy principle). In the Landweber method, $\beta_k = 0$ and the constant step size α_k should lie in the range $0 < \alpha_k < 2 / \left\|F_n'^* F_n'\right\|^2$ for ensuring convergence, $\|\bullet\|$ being the spectral norm [48]. In the conjugate-gradient method, the step size α_k is computed for each iteration as $\alpha_k = \arg\min_{\alpha>0} \left\|F_n' J_q\left(h_k^* + \alpha u_k^*\right) - E_n\right\|_p^2$, and the real weights β_k are given by

$$\beta_k = \gamma \frac{\left\|F_n' h_{k+1} - E_n\right\|_p^2}{\left\|F_n' h_k - E_n\right\|_p^2} \tag{3.60}$$

Here, $h^*_{k+1}, u^*_{k+1} \in L^q(D)$, $h_{k+1} \in L^p(D)$, and $0 < \gamma < 0.5$ is a fixed relaxation parameter [49].

5. Once the inner (Landweber or conjugate-gradient) iterations terminate, a regularized solution $\overline{h}$ of the linearized equation (3.54) is found. This solution is then used to update the contrast function as

$$c_{n+1} = c_n + \overline{h} \tag{3.61}$$

6. The Newton procedure is repeated from step 3, updating the iteration index $n := n + 1$, until proper convergence conditions are satisfied.

An example of the application of the inexact Newton method is now described. It refers to the very helpful experimental dataset provided to the microwave imaging community by the Institute Frésnel, Marseille, France [50]. In particular, we consider the reconstruction of the so-called *TwinDielTM* target. This target is composed of two dielectric cylinders with circular cross sections of radius $R = 0.015$m, separated by a distance $D = 0.09$m between their centers. Both cylinders are characterized by a relative dielectric permittivity $\varepsilon_r = 3 \pm 0.3$. $M = 36$ positions of the source and $s = 49$ measurement points for each source location have been considered. More details about the measurement setup can be found in [50]. For retrieving the dielectric properties of the target, the inexact Newton/Landweber method has been employed. Both the inner and the outer loops have been terminated when the relative variation of the normalized residual in two subsequent iterations is under 5%. Figure 3.7 reports the reconstructed distributions of the relative dielectric permittivity in this case, computed by using the field data acquired at $f = 2$ GHz. As can be seen, the presented Banach space approach with $p = 1.2$ seems to outperform the more conventional Hilbert space method ($p = 2$), leading to a more accurate characterization of the targets. The behavior of the normalized data residual in the external Newton iterations is shown in Figure 3.8(a), whereas Figure 3.8(b) reports the normalized data residual in the inner Landweber iterations for the first Newton step. The average relative reconstruction errors computed on the whole investigation domain, on the background region, and on the region occupied by the objects are listed in Table 3.1 for some values of the norm parameter p.

The discussed Newton scheme in L^p Banach spaces is just one example of inversion procedures aimed at solving the full nonlinear scattering problem. A plethora of other methods have been proposed in the past years to solve the above inverse scattering problem. The main classification concerns qualitative and quantitative methods. The reader is referred to the specific books indicated

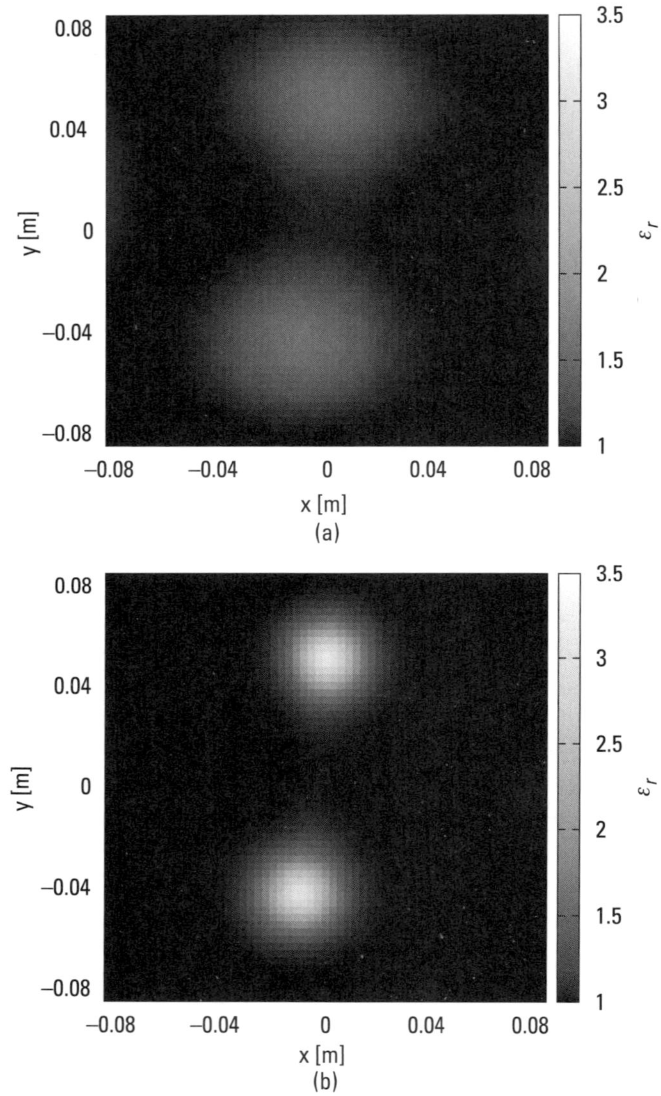

Figure 3.7 Reconstructed distributions of the relative dielectric permittivity of the Frésnel *TwoDielTM* target at $f = 2$ GHz by means of the inexact Newton/Landweber algorithm. (a) Hilbert space approach ($p = 2$); (b) Banach space approach with $p = 1.2$.

at the beginning of this chapter for a detailed description of the various proposed methods, which is beyond the scope of this book. Here we just mention some examples, which will be eventually reconsidered in the following chapters in view of the specific applications. The category of qualitative methods essentially includes two types of methods, that is, the methods aimed at retrieving only some specific information about the target under test (e.g., location,

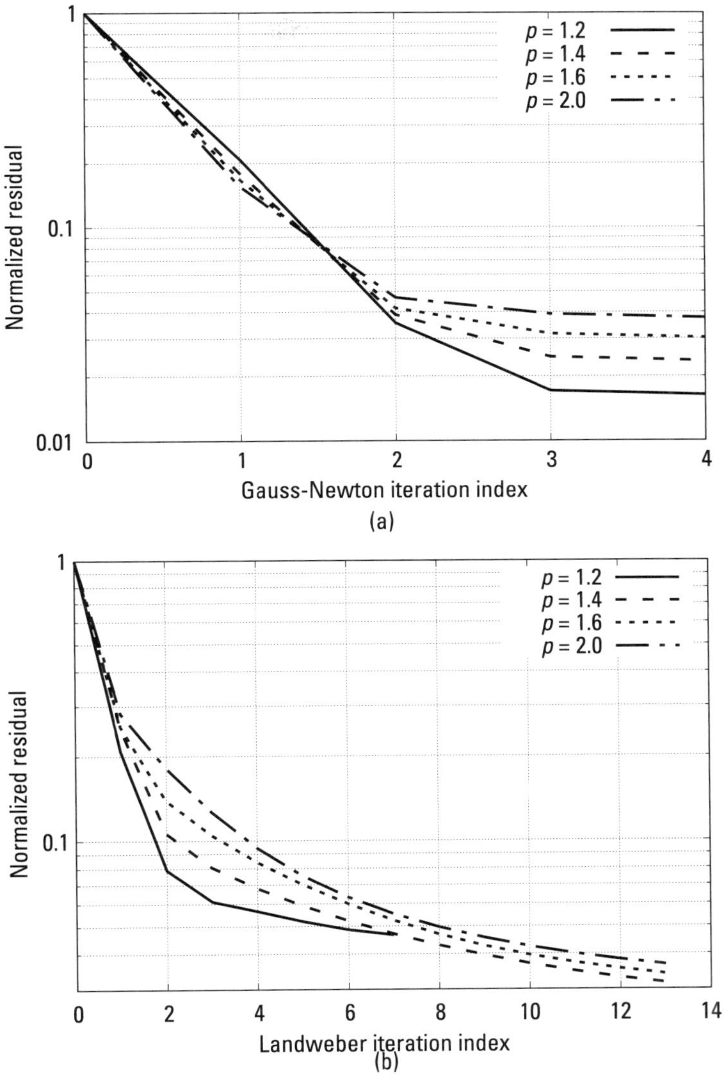

Figure 3.8 Normalized data residual for the inexact Newton/Landweber reconstruction of the Frésnel *TwoDielTM* target at $f = 2$ GHz, for different values of the parameter p. (a) Residual variation versus the Newton iteration index, (b) behavior of the residual inside the Landweber scheme for the first Newton iteration.

support, shape, presence of interfaces, and so forth) and methods based on specific approximations on the electromagnetic model. On the contrary, quantitative methods are generally aimed at retrieving the distributions of the physical parameters of the target, by solving the inverse scattering problem in its nonlinear exact formulation. In other words, quantitative methods are based on

Table 3.1
Inexact Newton/Landweber Reconstruction of the Frésnel *TwinDielTM* Target at $f = 2$ GHz

	$p = 2.0$ (Hilbert)	$p = 1.6$	$p = 1.4$	$p = 1.2$
Global Error	0.21	0.17	0.14	0.10
Background Error	0.20	0.16	0.13	0.09
Object Error	0.49	0.45	0.39	0.31

solving the problem without approximations different from those associated to the numerical techniques adopted for resolving the related equations.

Examples of qualitative methods are the linear sampling method (LSM) and similar approaches [51–57], as well as the level set method [58–64], which are all techniques essentially used to retrieve the support and the shape of the target. Diffraction tomography [15, 33, 34, 65–69] based the Born approximation and on the Fourier diffraction theorem (which states that the Fourier transform of the measured scattered electric field along a given probing line is related to the Fourier transform of the equivalent current density distribution along certain paths in the transformed plane) has also been widely studied. The Born approximation has also been employed in conjunction with other inversion schemes. A technique widely used in microwave imaging applications for inverting the resulting linear equation is based on the use of the truncated singular value decomposition (TSVD) of the linearized scattering operator [70–74]. In fact, by filtering out the singular values below a certain threshold (related to the noise on the data) it is possible to straightforwardly obtain a regularized solution of the problem [8]. It is worth noting that the TSVD method has also been applied to different linearization schemes, such as those based on Rytov and Kirchhoff approximations [24, 73]. Other qualitative methods are based on higher order Born approximations, iterative Born methods, Rytov approximations, microwave holography [75–79], and others. Several versions of the previous approaches have been discussed in the literature and applied in several different contexts (some of them will be recalled again thoughout this book).

As previously introduced, quantitative methods aim at solving the inverse problem without any approximation different from the ones adopted in the numerical generation and treatment of the corresponding discrete problem. These techniques are usually iterative techniques. They can be grouped into deterministic and stochastic ones. The deterministic techniques are usually solved by iterative procedures based on Newton methods (already mentioned) or on various versions of gradient and descent-type methods [80–87]. The stochastic techniques are used to find global optima for the considered problem by means of methods involving stochastic concepts [88]. In this framework, population-based iterative methods, such as genetic algorithms, differential evolution methods, particle swarm optimization approaches, ant colony optimization, and artificial bee colony methods have been widely employed and adapted to the specific applicative scenario to be inspected.

In several cases (both related to the adoption of deterministic and stochastic techniques), the problem's solution is recast as an optimization problem, in which a proper cost function is defined and optimized. The cost function measures, in a given metric space, the distance between the input data (usually the values of the measured scattered field—or the scattering parameters—in the observation domain) and the same quantities computed at each iteration, on the basis of the current retrieved physical or dielectric profiles of the target under test. Several different choices can be made in selecting the cost function. For example, if the scattering problem is formulated in terms of the system of equations (3.40) (which are related to 2-D configurations; a similar cost function can be used for the 3-D case, too), a possible choice is the following

$$\begin{aligned}&\mathcal{R}(c)=\mathcal{R}_{data}(c)+\mathcal{R}_{state}(c)=\\&\alpha_{data}\left\|E_{scatt,z}-G_{data}\left(cE_{tot,z}\right)\right\|^2_{L^2(D_{obs})}+\\&\alpha_{state}\left\|E_{inc,z}-E_{tot,z}+G_{state}\left(cE_{tot,z}\right)\right\|^2_{L^2(D_{inv})}\end{aligned} \tag{3.62}$$

where $\|\bullet\|^2_{L^2(D_{obs})}$ and $\|\bullet\|^2_{L^2(D_{inv})}$ are the L^2 norms computed on the observation and investigation domains, and $\alpha_{data}=\left\|E_{scatt,z}\right\|^{-2}_{L^2(D_{obs})}$ and $\alpha_{state}=\left\|E_{inc,z}\right\|^{-2}_{L^2(D_{inv})}$ are normalization constants. A similar cost function can be obtained by using the contrast source formulation [18, 89]. It is worth noting that, by using (3.62), the total internal field $E_{tot,z}$ is unknown, which can cause the dimension of the optimization problem to be very high. Alternatively, it is possible to use the combined form of the scattering equations (with reference to [3.53]), that is,

$$\mathcal{R}(c)=\frac{1}{2}\left\|E_{scatt,z}-F(c)\right\|^2_{L^2(D_{obs})} \tag{3.63}$$

Several additional terms, which could be additive or multiplicative, can be inserted in the cost function, including Tikonov regularization factors, edge preserving relations [90], and, more generally, any term able to take into account the a priori information available for the specific application. These additional and penalty terms are needed to face the ill-posedness of the problem and to reduce the search space of the iterative procedure. A priori information is easier to be included in stochastic methods, but their computational cost make them applicable only in certain practical cases, for example, for inspecting scenarios and scatterers which can be modeled by using a limited number of unknown parameters.

The cost functions previously discussed are usually defined in terms of the L^2 norm of the data misfit. Often the regularization terms are also expressed by using the same norm (e.g., in the Tikhonov regularization strategy). However, as it is well known, such choices usually produce oversmoothing and ringing effects in the reconstructed dielectric profiles, due to the low pass filtering of the regularization methods. Specific regularization strategies have been developed to address this drawback, for example, the total variation approaches [91, 92]. The previously described L^p Banach space procedure has also been found to be able to reduce the oversmoothing effects in the reconstructions.

Recently, compressive sensing (CS) [93] has been proposed as an effective technique for microwave imaging of sparse targets [94–104], for which oversmoothing effects could be very problematic. CS has been initially proposed in the field of signal processing for addressing the problems of using the minimum possible amount of data for representing a sparse signal and of recovering a signal from the available data. However, it is worth noting that CS cannot be directly applied to the nonlinear scattering equation, but it is necessary to exploit some linearization techniques (e.g., those discussed in the previous section). By denoting a linearized version of the scattering operator with F_{lin}, compressive sensing imaging can be expressed as the following minimization problem

$$\min \|c\|^2_{L^1(D_{inv})} \text{ subject to } \left\|E_{scatt,z} - F_{lin}(c)\right\|^2_{L^2(D_{obs})} < \varepsilon \tag{3.64}$$

where ε is a fixed threshold. It is worth mentioning that CS would require the use of the L^0 norm in the cost function, however this poses several theoretical and numerical problems. Consequently, it is common to approximate the compressive sensing problem by using the L^1 norm as indicated in (3.64). Moreover, CS assumes that the unknown function (the contrast function in the formulation reported above) is sparse, that is, it can be represented with few nonzero coefficients with respect to a proper set of basis functions. This conditions is often not satisfied with standard basis functions usually employed in microwave imaging (e.g., pulse basis functions), apart when dealing with very small targets. In those cases where the unknown is not naturally sparse with respect to such simple discretization schemes, it is necessary to define a proper set of basis functions able to sparsify the unknown. To this end, several approaches have been proposed in the literature, that is, by exploiting wavelet basis [42, 105, 106] or by working on the gradient of the contrast function [96]. For solving (3.64) several algorithms have been developed, that is, the iterative hard thresholding [107] and matching pursuit [108] techniques. However, it is worth noting that, from a theoretical point of view, CS methods require that the operator satisfy the restricted isometric property (RIP), which is usually very difficult to verify and enforce in microwave imaging problems [97]. Consequently, alternative formulations based on Bayesian CS [109] have also been proposed [95]. Approaches based on additional L^2-norm terms has been proposed as well [110].

The solution of the inverse scattering problem can also be treated in terms of a Bayesian approach and Markov Random Fields, involving probabilistic concepts [111–116]. Neural networks, support vector machines and other machine-learning concepts have also been applied for imaging purposes at microwave frequencies [117–120].

Hybrid approaches can be used as well. In such approaches, different classes of inversion methods are combined together in order to mitigate the limitations of the standalone techniques and to exploit their advantages. In particular, different hybridization schemes can be considered. A first type of hybridization is related to the use of qualitative techniques for providing suitable starting guesses for a quantitative deterministic inversion procedure or for constraining the updates during the iterations. In fact, it is well known that the quality of the results provided by deterministic approaches, such as those based on gradient or Newton schemes, often rely on the choice of an initial solution sufficiently near to the actual configuration. For example, in [121] such kind of hybridization is achieved by using the multistatic adaptive microwave imaging (MAMI) [122] technique in combination with an inverse scattering approach based on the distorted Born iterative methods [123] (discussed in Chapter 5). In [124], the LSM is used to estimate the support of the targets, which is then used for reconstructing the dielectric properties by using a contrast-source extended-Born inversion scheme. Other schemes belonging to this type have been proposed in the literature and some of them will be recalled in the following chapters with reference to specific applications. Hybridization schemes can also be applied to stochastic inversion algorithms for speeding up the reconstruction. For example, in [125] the LSM is used to find the support of the scatterers, and subsequently only the dielectric properties inside the identified regions are reconstructed by using the ant colony optimization method [126], with significant computational savings. Other possible hybrid approaches involving stochastic techniques consist in inserting quantitative deterministic procedures inside the population update mechanism in order to speed up the converge. An example of this kind of techniques is the memetic algorithm, which has been used in [127, 128] for reconstructing the geometrical and dielectric properties of elliptic cylinders.

The previously-discussed imaging techniques are based on the frequency-domain formulation of the electromagnetic scattering. However, it is worth remarking that in microwave imaging, radar-type time-domain illuminating fields are often used.

In these cases, a pulse (or a train of pulses) is usually sent to inspect the target, with bandwidth covering a certain range of frequencies [129–133]. Different choices of the interrogating time-domain signals are possible. Standard microwave imaging systems usually adopt Gaussian or Gaussian derivative pulsed waveforms, but pseudorandom noise (PRN) sequences can also be employed [134, 135].

When dealing with time domain systems, it is clearly possible to apply the Fourier transform to extract the electric field data at a certain number of fixed frequencies and consequently to apply the previously recalled inversion methods. However, methods directly working on time-domain data can be applied as well. Often in this framework techniques based on backpropagation of the received time signals have been proposed. Different versions of these techniques have been developed for the specific applicative fields, for example, the delay-and-sum method for breast imaging [136] or the migration algorithms for subsurface sensing [137]. The main concepts regarding such approaches will be discussed in the next chapters.

Another class of approaches based on backpropagation includes time reversal methods. In these methods, the signals received in the measurement points are backpropagated in the investigation domain by using time-reversed wave equations. These procedures have been found to provide qualitative images showing high values in correspondence to the positions where the scatterer (and, more in general, the radiating electromagnetic sources) are located. Time reversal methods can be developed directly in time domain, by employing the finite-difference time domain (FDTD) method for computing the time-reversed fields [138, 139]. Alternatively, they can be formulated in a frequency domain by employing the Fourier transform. In this case, assuming multistatic imaging setup, for each frequency in the bandwidth of the considered pulse, a time-reversal matrix is constructed as $[T] = [K]^*[K]$, where $[K]$ is the multistatic data matrix, whose (m, n) th element is the field measured by the mth antenna in receiving mode and due to the nth antenna operating in transmission mode, and * denotes conjugate transposition. The time-reversal matrix is usually processed by performing eigenvalue decomposition in order to obtain a qualitative map with peaks in correspondence to the scatterer positions. Popular approaches of these types are those based on the decomposition of the time reversal operator (DORT) method [140–142], the multiple signal classification (MUSIC) method [143–145], and their modified versions (e.g., beamspace [146, 147] and space frequency [148] DORT and MUSIC).

Other time-domain approaches that have been found to provide good results are those based on the forward-backward time-stepping (FBTS) procedure [149, 150]. In this case, the dielectric properties of the target under test are found by solving the following minimization problem

$$\min_{\mathbf{p}} \int_0^T \left\| \mathbf{u}(\mathbf{r},t) - \mathbf{F}_{TD}(\mathbf{p})(\mathbf{r},t) \right\|_{L^2(D_{obs})}^2 dt \tag{3.65}$$

where $\mathbf{u}$ is an array containing the components of the field vectors (e.g., in the TM case $\mathbf{u} = [E_z, H_x, H_y]^t$), $\mathbf{p}$ is an array containing the dielectric properties (e.g., the relative dielectric permittivity and the electric conductivity or the parameters of a Debye model), and $\mathbf{F}_{TD}$ is the forward time-domain operator, which can directly expressed in terms of the Maxwell equations and can be computed by using the FDTD method. The optimization problem in (3.65) is usually solved by using gradient-type methods, which require to compute the Fréchet derivative and its adjoint operator. The latter can be efficiently computed by using again a FDTD method with backward time stepping.

References

[1] Devaney, A. J., *Mathematical Foundations of Imaging, Tomography and Wavefield Inversion*, Cambridge, NY: Cambridge University Press, 2012.

[2] Nikolova, N. D., *Introduction to Microwave Imaging*, Cambridge: Cambridge University Press, 2017.

[3] Noghanian, S., A. Sabouni, T. Desell, and A. Ashtari, *Microwave Tomography: Global Optimization, Parallelization and Performance Evaluation*, New York City, NY: Springer New York, 2014.

[4] Conceição, R. C., J. J. Mohr, and M. O'Halloran (eds.), *An Introduction to Microwave Imaging for Breast Cancer Detection*, Cham, Switzerland: Springer International Publishing, 2016.

[5] Pastorino, M., *Microwave Imaging*, Hoboken, NJ: Wiley, 2010.

[6] Cakoni, F., and D. Colton, *Qualitative Methods in Inverse Scattering Theory*, New York: Springer-Verlag, 2006.

[7] Pike, E. R., and P. C. Sabatier (eds.), *Scattering: Scattering and Inverse Scattering in Pure and Applied Science*, San Diego, California: Academic Press, 2002.

[8] Bertero, M., and P. Boccacci, *Introduction to Inverse Problems in Imaging*, Bristol, UK: Institute of Physics Publishing, 1998.

[9] Vogel, C. R., *Computational Methods for Inverse Problems*, Philadelphia: Society for Industrial and Applied Mathematics, 2002.

[10] Chew, W. C., *Waves and Fields in Inhomogeneous Media*, Piscataway, NY: IEEE Press, 1995.

[11] Balanis, C. A., *Advanced Engineering Electromagnetics*, Hoboken, NJ: Wiley, 2012.

[12] Born, M., and E. Wolf, *Principles of Optics: Electromagnetic Theory of Propagation, Interference and Diffraction of Light*, 7th ed. Cambridge, NY: Cambridge University Press, 1999.

[13] Natterer, F., "An Error Bound for the Born Approximation," *Inverse Probl.*, Vol. 20, No. 2, April 2004, pp. 447–452.

[14] Chen, B., and J. J. Stamnes, "Validity of Diffraction Tomography Based on the First Born and the First Rytov Approximations," *Appl. Opt.*, Vol. 37, No. 14, May 1998, pp. 2996–3006.

[15] Slaney, M., A. C. Kak, and L. E. Larsen, "Limitations of Imaging with First-Order Diffraction Tomography," *IEEE Trans. Microw. Theory Tech.*, Vol. 32, No. 8, August 1984, pp. 860–874.

[16] Estatico, C., M. Pastorino, and A. Randazzo, "An Inexact-Newton Method for Short-Range Microwave Imaging within the Second-Order Born Approximation," *IEEE Trans. Geosci. Remote Sens.*, Vol. 43, No. 11, November 2005, pp. 2593–2605.

[17] Pierri, R., and G. Leone, "Inverse Scattering of Dielectric Cylinders by a Second-Order Born Approximation," *IEEE Trans. Geosci. Remote Sens.*, Vol. 37, No. 1, January 1999, pp. 374–382.

[18] van den Berg, P. M., and R. E. Kleinman, "A Contrast Source Inversion Method," *Inverse Probl.*, Vol. 13, No. 6, December 1997, pp. 1607–1620.

[19] Catapano, I., L. Crocco, M. D'Urso, and T. Isernia, "A Novel Effective Model for Solving 3-D Nonlinear Inverse Scattering Problems in Lossy Scenarios," *IEEE Geosci. Remote Sens. Lett.*, Vol. 3, No. 3, July 2006, pp. 302–306.

[20] Beaverstone, A. S., D. S. Shumakov, and N. K. Nikolova, "Frequency-Domain Integral Equations of Scattering for Complex Scalar Responses," *IEEE Trans. Microw. Theory Tech.*, Vol. 65, No. 4, April 2017, pp. 1120–1132.

[21] Shumakov, D. S., and N. K. Nikolova, "Fast Quantitative Microwave Imaging with Scattered-Power Maps," *IEEE Trans. Microw. Theory Tech.*, Vol. 66, No. 1, January 2018, pp. 439–449.

[22] Haynes, M., and M. Moghaddam, "Vector Green's Function for *S*-parameter Measurements of the Electromagnetic Volume Integral Equation," *IEEE Trans. Antennas Propag.*, Vol. 60, No. 3, March 2012, pp. 1400–1413.

[23] Harrington, R. F., *Field Computation by Moment Methods*, Piscataway, NJ: IEEE, 1993.

[24] Pierri, R., A. Liseno, and F. Soldovieri, "Shape Reconstruction from PO Multifrequency Scattered Fields via the Singular Value Decomposition Approach," *IEEE Trans. Antennas Propag.*, Vol. 49, No. 9, September 2001, pp. 1333–1343.

[25] Ma, J., W. C. Chew, C.-C. Lu, and J. Song, "Image Reconstruction from TE Scattering Data Using Equation of Strong Permittivity Fluctuation," *IEEE Trans. Antennas Propag.*, Vol. 48, No. 6, June 2000, pp. 860–867.

[26] Kooij, B. J., and P. M. van den Berg, "Nonlinear Inversion in TE Scattering," *IEEE Trans. Microw. Theory Tech.*, Vol. 46, No. 11, November 1998, pp. 1704–1712.

[27] Mojabi, P., and J. LoVetri, "Comparison of TE and TM Inversions in the Framework of the Gauss-Newton Method," *IEEE Trans. Antennas Propag.*, Vol. 58, No. 4, April 2010, pp. 1336–1348.

[28] Zakaria, A., and J. LoVetri, "The Finite-Element Method Contrast Source Inversion Algorithm for 2D Transverse Electric Vectorial Problems," *IEEE Trans. Antennas Propag.*, Vol. 60, No. 10, October 2012, pp. 4757–4765.

[29] Rekanos, I. T., T. V. Yioultsis, and C. S. Hilas, "An Inverse Scattering Approach Based on the Differential E-Formulation," *IEEE Trans. Geosci. Remote Sens.*, Vol. 42, No. 7, July 2004, pp. 1456–1461.

[30] Qing, A., "Electromagnetic Imaging of Two-Dimensional Perfectly Conducting Cylinders with Transverse Electric Scattered Field," *IEEE Trans. Antennas Propag.*, Vol. 50, No. 12, December 2002, pp. 1786–1794.

[31] Franceschini, D., M. Donell, G. Franceschini, and A. Massa, "Iterative Image Reconstruction of Two-Dimensional Scatterers Illuminated by TE Waves," *IEEE Trans. Microw. Theory Tech.*, Vol. 54, No. 4, June 2006, pp. 1484–1494.

[32] Hajihashemi, M. R., and M. El-Shenawee, "TE versus TM for the Shape Reconstruction of 2-D PEC Targets Using the Level-Set Algorithm," *IEEE Trans. Geosci. Remote Sens.*, Vol. 48, No. 3, March 2010, pp. 1159–1168.

[33] Kak, A. C., and M. Slaney, *Principles of Computerized Tomographic Imaging*, Philadelphia: Society for Industrial and Applied Mathematics, 2001.

[34] Keller, J. B., "Accuracy and Validity of the Born and Rytov Approximations*," *J. Opt. Soc. Am.*, Vol. 59, No. 8, August 1969, p. 1003.

[35] Isernia, T., L. Crocco, and M. D'Urso, "New Tools and Series for Forward and Inverse Scattering Problems in Lossy Media," *IEEE Geosci. Remote Sens. Lett.*, Vol. 1, No. 4, October 2004, pp. 327–331.

[36] Marengo, E. A., E. S. Galagarza, and R. Solimene, "Data-driven Linearizing Approach in Inverse Scattering," *J. Opt. Soc. Am.*, Vol. 34, No. 9, September 2017, pp. 1561–1576.

[37] Uno, T., Y. Miki, and S. Adachi, "One-Dimensional Imaging of Dielectric Bodies of Revolution Based on Rayleigh Approximation," in *Proc. 1991 Antennas and Propagation Society Symposium*, Ontario, Canada, pp. 1296–1299.

[38] Caorsi, S., G. L. Gragnani, and M. Pastorino, "Three-Dimensional Inverse-Scattering Numerical Solution within the Rayleigh Approximation: Preliminary Results," *Microw. Opt. Technol. Lett.*, Vol. 15, No. 3, June 1997, pp. 173–176.

[39] Isernia, T., V. Pascazio, and R. Pierri, "On the Local Minima in a Tomographic Imaging Technique," *IEEE Trans. Geosci. Remote Sens.*, Vol. 39, No. 7, July 2001, pp. 1596–1607.

[40] Winters, D. W., et al., "Three-Dimensional Microwave Breast Imaging: Dispersive Dielectric Properties Estimation Using Patient-Specific Basis Functions," *IEEE Trans. Med. Imaging*, Vol. 28, No. 7, July 2009, pp. 969–981.

[41] Li, M., O. Semerci, and A. Abubakar, "A Contrast Source Inversion Method in the Wavelet Domain," *Inverse Probl.*, Vol. 29, No. 2, February 2013, p. 025015.

[42] Anselmi, N., M. Salucci, G. Oliveri, and A. Massa, "Wavelet-Based Compressive Imaging of Sparse Targets," *IEEE Trans. Antennas Propag.*, Vol. 63, No. 11, November 2015, pp. 4889–4900.

[43] Gao, G., C. Torres-Verdin, and T. M. Habashy, "Analytical Techniques to Evaluate the Integrals of 3-D and 2-D Spatial Dyadic Green's Functions," *Prog. Electromagn. Res.*, Vol. 52, 2005, pp. 47–80.

[44] De Zaeytijd, Jü., A. Franchois, C. Eyraud, and J.-M. Geffrin, "Full-Wave Three-Dimensional Microwave Imaging with a Regularized Gauss–Newton Method—Theory and Experiment," *IEEE Trans. Antennas Propag.*, Vol. 55, No. 11, November 2007, pp. 3279–3292.

[45] Mojabi, P., and J. LoVetri, "Microwave Biomedical Imaging Using the Multiplicative Regularized Gauss–Newton Inversion," *IEEE Antennas Wirel. Propag. Lett.*, Vol. 8, 2009, pp. 645–648.

[46] Desmal, A., and H. Bacı, "A Preconditioned Inexact Newton Method for Nonlinear Sparse Electromagnetic Imaging," *IEEE Geosci. Remote Sens. Lett.*, Vol. 12, No. 3, March 2015, pp. 532–536.

[47] Rubek, T., P. M. Meaney, P. Meincke, and K. D. Paulsen, "Nonlinear Microwave Imaging for Breast-Cancer Screening Using Gauss-Newton's Method and the CGLS Inversion Algorithm," *IEEE Trans. Antennas Propag.*, Vol. 55, No. 8, August 2007, pp. 2320–2331.

[48] Estatico, C., M. Pastorino, and A. Randazzo, "A Novel Microwave Imaging Approach Based on Regularization in L^p Banach Spaces," *IEEE Trans. Antennas Propag.*, Vol. 60, No. 7, July 2012, pp. 3373–3381.

[49] Estatico, C., S. Gratton, F. Lenti, and D. Titley-Peloquin, "A Conjugate Gradient Like Method for p-Norm Minimization in Functional Spaces," *Numer. Math.*, 2017.

[50] Belkebir, K., and M. Saillard, "Guest Editors' Introduction," *Inverse Probl.*, Vol. 17, No. 6, 2001, p. 1565.

[51] Colton, D., H. Haddar, and M. Piana, "The Linear Sampling Method in Inverse Electromagnetic Scattering Theory," *Inverse Probl.*, Vol. 19, No. 6, December 2003, pp. S105–S137.

[52] Catapano, I., L. Crocco, and T. Isernia, "On Simple Methods for Shape Reconstruction of Unknown Scatterers," *IEEE Trans. Antennas Propag.*, Vol. 55, No. 5, May 2007, pp. 1431–1436.

[53] Brignone, M., G. Bozza, R. Aramini, M. Pastorino, and M. Piana, "A Fully No-Sampling Formulation of the Linear Sampling Method for Three-Dimensional Inverse Electromagnetic Scattering Problems," *Inverse Probl.*, Vol. 25, No. 1, January 2009, pp. 1–20 (015014).

[54] Potthast, R., "A Study on Orthogonality Sampling," *Inverse Probl.*, Vol. 26, No. 7, 2010, p. 074015.

[55] Griesmaier, R., "Multi-Frequency Orthogonality Sampling for Inverse Obstacle Scattering Problems," *Inverse Probl.*, Vol. 27, No. 8, 2011, p. 085005.

[56] Ito, K., B. Jin, and J. Zou, "A Direct Sampling Method for Inverse Electromagnetic Medium Scattering," *Inverse Probl.*, Vol. 29, No. 9, 2013, p. 095018.

[57] Akıncı, M. N., M. Çayören, and I. Akduman, "Near-Field Orthogonality Sampling Method for Microwave Imaging: Theory and Experimental Verification," *IEEE Trans. Microw. Theory Tech.*, Vol. 64, No. 8, August 2016, pp. 2489–2501.

[58] Dorn, O., E. L. Miller, and C. M. Rappaport, "A Shape Reconstruction Method for Electromagnetic Tomography Using Adjoint Fields and Level Sets," *Inverse Probl.*, Vol. 16, No. 5, October 2000, pp. 1119–1156.

[59] Ferraye, R., J.-Y. Dauvignac, and C. Pichot, "An Inverse Scattering Method Based on Contour Deformations by Means of a Level Set Method Using Frequency Hopping Technique," *IEEE Trans. Antennas Propag.*, Vol. 51, No. 5, May 2003, pp. 1100–1113.

[60] Dorn, O., and D. Lesselier, "Level Set Methods for Inverse Scattering," *Inverse Probl.*, Vol. 22, No. 4, August 2006, p. R67.

[61] Eskandari, M. R., M. Dehmollaian, and R. Safian, "Simultaneous Microwave Imaging and Parameter Estimation Using Modified Level-Set Method," *IEEE Trans. Antennas Propag.*, Vol. 64, No. 8, August 2016, pp. 3554–3564.

[62] Santosa, F., "A Level-Set Approach for Inverse Problems Involving Obstacles," *ESAIM Control Optim. Calc. Var.*, Vol. 1, 1996, pp. 17–33.

[63] Litman, A., D. Lesselier, and F. Santosa, "Reconstruction of a Two-Dimensional Binary Obstacle by Controlled Evolution of a Level-Set," *Inverse Probl.*, Vol. 14, No. 3, June 1998, p. 685.

[64] Hajihashemi, M. R., and M. El-Shenawee, "Level Set Algorithm for Shape Reconstruction of Non-Overlapping Three-Dimensional Penetrable Targets," *IEEE Trans. Geosci. Remote Sens.*, Vol. 50, No. 1, January 2012, pp. 75–86.

[65] Pan, S., and A. Kak, "A Computational Study of Reconstruction Algorithms for Diffraction Tomography: Interpolation versus Filtered-Backpropagation," *IEEE Trans. Acoust. Speech Signal Process.*, Vol. 31, No. 5, October 1983, pp. 1262–1275.

[66] Wolf, E., "Three-dimensional Structure Determination of Semi-Transparent Objects from Holographic Data," *Opt. Commun.*, Vol. 1, No. 4, September 1969, pp. 153–156.

[67] Devaney, A. J., "A Filtered Backpropagation Algorithm for Diffraction Tomography," *Ultrason. Imaging*, Vol. 4, No. 4, October 1982, pp. 336–350.

[68] Devaney, A. J., "A Computer Simulation Study of Diffraction Tomography," *IEEE Trans. Biomed. Eng.*, Vol. BME-30, No. 7, July 1983, pp. 377–386.

[69] Bolomey, J. C., et al., "Microwave Diffraction Tomography for Biomedical Applications," *IEEE Trans. Microw. Theory Tech.*, Vol. 30, No. 11, November 1982, pp. 1998–2000.

[70] Soldovieri, F., and R. Solimene, "Through-Wall Imaging via a Linear Inverse Scattering Algorithm," *IEEE Geosci. Remote Sens. Lett.*, Vol. 4, No. 4, October 2007, pp. 513–517.

[71] Bayat, N., and P. Mojabi, "A Mathematical Framework to Analyze the Achievable Resolution from Microwave Tomography," *IEEE Trans. Antennas Propag.*, Vol. 64, No. 4, April 2016, pp. 1484–1489.

[72] Gennarelli, G., and F. Soldovieri, "A Linear Inverse Scattering Algorithm for Radar Imaging in Multipath Environments," *IEEE Geosci. Remote Sens. Lett.*, Vol. 10, No. 5, September 2013, pp. 1085–1089.

[73] Pierri, R., A. Liseno, R. Solimene, and F. Soldovieri, "Beyond Physical Optics SVD Shape Reconstruction of Metallic Cylinders," *IEEE Trans. Antennas Propag.*, Vol. 54, No. 2, February 2006, pp. 655–665.

[74] Shea, J. D., B. D. Van Veen, and S. C. Hagness, "A TSVD Analysis of Microwave Inverse Scattering for Breast Imaging," *IEEE Trans. Biomed. Eng.*, Vol. 59, No. 4, April 2012, pp. 936–945.

[75] Tricoles, G., and N. H. Farhat, "Microwave Holography: Applications and Techniques," *Proc. IEEE*, Vol. 65, No. 1, 1977, pp. 108–121.

[76] Amineh, R. K., M. Ravan, A. Khalatpour, and N. K. Nikolova, "Three-Dimensional Near-Field Microwave Holography Using Reflected and Transmitted Signals," *IEEE Trans. Antennas Propag.*, Vol. 59, No. 12, December 2011, pp. 4777–4789.

[77] Sheen, D. M., D. L. McMakin, and T. E. Hall, "Three-Dimensional Millimeter-Wave Imaging for Concealed Weapon Detection," *IEEE Trans. Microw. Theory Tech.*, Vol. 49, No. 9, September 2001, pp. 1581–1592.

[78] Elsdon, M., D. Smith, M. Leach, and S. J. Foti, "Experimental Investigation of Breast Tumor Imaging Using Indirect Microwave Holography," *Microw. Opt. Technol. Lett.*, Vol. 48, No. 3, March 2006, pp. 480–482.

[79] Tsai, C.-H., J. Chang, L.-Y. O. Yang, and S.-Y. Chen, "Three-Dimensional Microwave Holographic Imaging with Probe and Phase Compensations," *IEEE Trans. Antennas Propag.*, 2017, p. 1.

[80] Kleinman, R. E., and P. M. van den Berg, "A Modified Gradient Method for Two-Dimensional Problems in Tomography," *J. Comput. Appl. Math.*, Vol. 42, No. 1, September 1992, pp. 17–35.

[81] Harada, H., D. J. N. Wall, T. Takenaka, and M. Tanaka, "Conjugate Gradient Method Applied to Inverse Scattering Problem," *IEEE Trans. Antennas Propag.*, Vol. 43, No. 8, August 1995, pp. 784–792.

[82] Liu, Q. H., et al., "Active Microwave Imaging. I. 2-D Forward and Inverse Scattering Methods," *IEEE Trans. Microw. Theory Tech.*, Vol. 50, No. 1, January 2002, pp. 123–133.

[83] Fhager, A., and M. Persson, "Comparison of Two Image Reconstruction Algorithms for Microwave Tomography," *Radio Sci.*, Vol. 40, No. 3, June 2005, pp. 1–15.

[84] Semenov, S. Y., et al., "Microwave-Tomographic Imaging of the High Dielectric-Contrast Objects Using Different Image-Reconstruction Approaches," *IEEE Trans. Microw. Theory Tech.*, Vol. 53, No. 7, July 2005, pp. 2284–2294.

[85] Caorsi, S., M. Donelli, D. Franceschini, and A. Massa, "A New Methodology Based on an Iterative Multiscaling for Microwave Imaging," *IEEE Trans. Microw. Theory Tech.*, Vol. 51, No. 4, April 2003, pp. 1162–1173.

[86] Rekanos, I. T., S. M. Panas, and T. D. Tsiboukis, "Microwave Imaging Using the Finite-Element Method and a Sensitivity Analysis Approach," *IEEE Trans. Med. Imaging*, Vol. 18, No. 11, November 1999, pp. 1108–1114.

[87] Belkebir, K., R. E. Kleinman, and C. Pichot, "Microwave Imaging-Location and Shape Reconstruction from Multifrequency Scattering Data," *IEEE Trans. Microw. Theory Tech.*, Vol. 45, No. 4, April 1997, pp. 469–476.

[88] Pastorino, M., "Stochastic Optimization Methods Applied to Microwave Imaging: A Review," *IEEE Trans. Antennas Propag.*, Vol. 55, No. 3, March 2007, pp. 538–548.

[89] van den Berg, P. M., and A. Abubakar, "Contrast Source Inversion Method: State of Art," *Prog. Electromagn. Res.*, Vol. 34, 2001, pp. 189–218.

[90] Charbonnier, P., L. Blanc-Feraud, G. Aubert, and M. Barlaud, "Deterministic Edge-Preserving Regularization in Computed Imaging," *IEEE Trans. Image Process.*, Vol. 6, No. 2, February 1997, pp. 298–311.

[91] van den Berg, P. M., and R. E. Kleinman, "A total variation enhanced modified gradient algorithm for profile reconstruction," *Inverse Probl.*, Vol. 11, No. 3, 1995, p. L5.

[92] Abubaker, A., and P. M. van den Berg, "Total Variation as a Multiplicative Constraint for Solving Inverse Problems," *IEEE Trans. Image Process.*, Vol. 10, No. 9, September 2001, pp. 1384–1392.

[93] Donoho, D. L., "Compressed Sensing," *IEEE Trans. Inf. Theory*, Vol. 52, No. 4, April 2006, pp. 1289–1306.

[94] Massa, A., "Compressive Sensing-Basics, State-of-the-Art, and Advances in Electromagnetic Engineering," in *Proc. 2015 9Th European Conference on Antennas and Propagation (EuCAP)*, pp. 1–4.

[95] Oliveri, G., P. Rocca, and A. Massa, "A Bayesian-Compressive-Sampling-Based Inversion for Imaging Sparse Scatterers," *IEEE Trans. Geosci. Remote Sens.*, Vol. 49, No. 10, October 2011, pp. 3993–4006.

[96] Oliveri, G., N. Anselmi, and A. Massa, "Compressive Sensing Imaging of Non-Sparse 2D Scatterers by a Total-Variation Approach within the Born Approximation," *IEEE Trans. Antennas Propag.*, Vol. 62, No. 10, October 2014, pp. 5157–5170.

[97] Oliveri, G., M. Salucci, N. Anselmi, and A. Massa, "Compressive Sensing as Applied to Inverse Problems for Imaging: Theory, Applications, Current Trends, and Open Challenges.," *IEEE Antennas Propag. Mag.*, Vol. 59, No. 5, October 2017, pp. 34–46.

[98] Ambrosanio, M., and V. Pascazio, "A Compressive-Sensing-Based Approach for the Detection and Characterization of Buried Objects," *IEEE J. Sel. Top. Appl. Earth Obs. Remote Sens.*, Vol. 8, No. 7, July 2015, pp. 3386–3395.

[99] Bevacqua, M. T., and R. Scapaticci, "A Compressive Sensing Approach for 3D Breast Cancer Microwave Imaging with Magnetic Nanoparticles as Contrast Agent," *IEEE Trans. Med. Imaging*, Vol. 35, No. 2, February 2016, pp. 665–673.

[100] Huang, Q., L. Qu, B. Wu, and G. Fang, "UWB Through-Wall Imaging Based on Compressive Sensing," *IEEE Trans. Geosci. Remote Sens.*, Vol. 48, No. 3, March 2010, pp. 1408–1415.

[101] Qiu, W., J. Zhou, H. Zhao, and Q. Fu, "Three-Dimensional Sparse Turntable Microwave Imaging Based on Compressive Sensing," *IEEE Geosci. Remote Sens. Lett.*, Vol. 12, No. 4, April 2015, pp. 826–830.

[102] Wu, Q., Y. D. Zhang, F. Ahmad, and M. G. Amin, "Compressive-Sensing-Based High-Resolution Polarimetric Through-The-Wall Radar Imaging Exploiting Target Characteristics," *IEEE Antennas Wirel. Propag. Lett.*, Vol. 14, 2015, pp. 1043–1047.

[103] Azghani, M., P. Kosmas, and F. Marvasti, "Microwave Medical Imaging Based on Sparsity and an Iterative Method with Adaptive Thresholding," *IEEE Trans. Med. Imaging*, Vol. 34, No. 2, February 2015, pp. 357–365.

[104] Takın, U., and Ö. Özdemir, "Sparsity Regularized Nonlinear Inversion for Microwave Imaging," *IEEE Geosci. Remote Sens. Lett.*, Vol. 14, No. 12, December 2017, pp. 2220–2224.

[105] Guo, L., and A. M. Abbosh, "Wavelet-Based Compressive Sensing for Head Imaging," in *Proc. 2015 International Symposium on Antennas and Propagation (ISAP)*, pp. 1–3.

[106] Ambrosanio, M., P. Kosmas, and V. Pascazio, "Exploiting Wavelet Decomposition to Enhance Sparse Recovery in Microwave Imaging," in *Proc. 11th European Conference on Antennas and Propagation (EuCAP)*, Paris, France, 2017, pp. 1607–1610.

[107] Blumensath, T., and M. E. Davies, "Iterative Hard Thresholding for Compressed Sensing," *Appl. Comput. Harmon. Anal.*, Vol. 27, No. 3, November 2009, pp. 265–274.

[108] Wang, J., S. Kwon, and B. Shim, "Generalized Orthogonal Matching Pursuit," *IEEE Trans. Signal Process.*, Vol. 60, No. 12, December 2012, pp. 6202–6216.

[109] Ji, S., Y. Xue, and L. Carin, "Bayesian Compressive Sensing," *IEEE Trans. Signal Process.*, Vol. 56, No. 6, June 2008, pp. 2346–2356.

[110] Shah, P., U. K. Khankhoje, and M. Moghaddam, "Inverse Scattering Using a Joint L1-L2 Norm-Based Regularization," *IEEE Trans. Antennas Propag.*, Vol. 64, No. 4, April 2016, pp. 1373–1384.

[111] Gharsalli, L., H. Ayasso, B. Duchêne, and A. Mohammad-Djafari, "Inverse Scattering in a Bayesian Framework: Application to Microwave Imaging for Breast Cancer Detection," *Inverse Probl.*, Vol. 30, No. 11, 2014, p. 114011.

[112] Caorsi, S., et al., "A Gibbs Random Field-Based Active Electromagnetic Method for Noninvasive Diagnostics in Biomedical Applications," *Radio Sci.*, Vol. 30, No. 1, January 1995, pp. 291–301.

[113] Nikolova, M., J. Idier, and A. Mohammad-Djafari, "Inversion of Large-Support Ill-Posed Linear Operators Using a Piecewise Gaussian MRF," *IEEE Trans. Image Process.*, Vol. 7, No. 4, April 1998, pp. 571–585.

[114] Carfantan, H., and A. Mohammad-Djafari, "An Overview of Nonlinear Diffraction Tomography within the Bayesian Estimation Framework," in *Inverse Problems of Wave Propagation and Diffraction*, Springer, Berlin, Heidelberg, 1997, pp. 107–124.

[115] Lobel, P., L. Blanc-Féraud, C. Pichot, and M. Barlaud, "A New Regularization Scheme for Inverse Scattering," *Inverse Probl.*, Vol. 13, No. 2, 1997, p. 403.

[116] Pascazio, V., and G. Ferraiuolo, "Statistical Regularization in Linearized Microwave Imaging through MRF-Based MAP Estimation: Hyperparameter Estimation and Image Computation," *IEEE Trans. Image Process.*, Vol. 12, No. 5, May 2003, pp. 572–582.

[117] Gong, X., and Y. Wang, "A Neural Network Approach to the Microwave Inverse Scattering Problem with Edge-Preserving Regularization," *Radio Sci.*, Vol. 36, No. 5, September 2001, pp. 825–832.

[118] Brovko, A. V., E. K. Murphy, and V. V. Yakovlev, "Waveguide Microwave Imaging: Neural Network Reconstruction of Functional 2-D Permittivity Profiles," *IEEE Trans. Microw. Theory Tech.*, Vol. 57, No. 2, February 2009, pp. 406–414.

[119] Rekanos, I. T., "Neural-Network-Based Inverse-Scattering Technique for Online Microwave Medical Imaging," *IEEE Trans. Magn.*, Vol. 38, No. 2, March 2002, pp. 1061–1064.

[120] Caorsi, S., and P. Gamba, "Electromagnetic Detection of Dielectric Cylinders by a Neural Network Approach," *IEEE Trans. Geosci. Remote Sens.*, Vol. 37, No. 2, March 1999, pp. 820–827.

[121] Ono, Y., and Y. Kuwahara, "An Analysis of Microwave Imaging Using a Combination of Multi-Static Radar Imaging and Inverse Scattering Tomography Methods," in *Proc. 2017 IEEE International Symposium on Antennas and Propagation & USNC/URSI National Radio Science Meeting*, San Diego, CA, USA, 2017, pp. 2385–2386.

[122] Xie, Y., et al., "Multistatic Adaptive Microwave Imaging for Early Breast Cancer Detection," *IEEE Trans. Biomed. Eng.*, Vol. 53, No. 8, August 2006, pp. 1647–1657.

[123] Chew, W. C., and Y. M. Wang, "Reconstruction of Two-Dimensional Permittivity Distribution Using the Distorted Born Iterative Method," *IEEE Trans. Med. Imaging*, Vol. 9, No. 2, June 1990, pp. 218–225.

[124] Catapano, I., L. Crocco, M. D. Urso, and T. Isernia, "3D Microwave Imaging via Preliminary Support Reconstruction: Testing on the Frésnel 2008 Database," *Inverse Probl.*, Vol. 25, No. 2, February 2009, p. 024002.

[125] Brignone, M., et al., "A Hybrid Approach to 3D Microwave Imaging by Using Linear Sampling and ACO," *IEEE Trans. Antennas Propag.*, Vol. 56, No. 10, October 2008, pp. 3224–3232.

[126] Socha, K., and M. Dorigo, "Ant Colony Optimization for Continuous Domains," *Eur. J. Oper. Res.*, Vol. 185, No. 3, March 2008, pp. 1155–1173.

[127] Caorsi, S., A. Massa, M. Pastorino, M. Raffetto, and A. Randazzo, "Detection of Buried Inhomogeneous Elliptic Cylinders by a Memetic Algorithm," *IEEE Trans. Antennas Propag.*, Vol. 51, No. 10, October 2003, pp. 2878–2884.

[128] Caorsi, S., A. Massa, M. Pastorino, and A. Randazzo, "Electromagnetic Detection of Dielectric Scatterers Using Phaseless Synthetic and Real Data and the Memetic Algorithm," *IEEE Trans. Geosci. Remote Sens.*, Vol. 41, No. 12, December 2003, pp. 2745–2753.

[129] Zeng, X., et al., "Development of a Time Domain Microwave System for Medical Diagnostics," *IEEE Trans. Instrum. Meas.*, Vol. 63, No. 12, Decemeber 2014, pp. 2931–2939.

[130] Santorelli, A., et al., "A Time-Domain Microwave System for Breast Cancer Detection Using a Flexible Circuit Board," *IEEE Trans. Instrum. Meas.*, Vol. 64, No. 11, November 2015, pp. 2986–2994.

[131] Zeng, X., et al., "Experimental Investigation of the Accuracy of an Ultrawideband Time-Domain Microwave-Tomographic System," *IEEE Trans. Instrum. Meas.*, Vol. 60, No. 12, December 2011, pp. 3939–3949.

[132] Helbig, M., et al., "Design and Test of an Imaging System for UWB Breast Cancer Detection," *Frequenz*, Vol. 66, No. 11–12, Jan. 2012.

[133] Papadopoulos, T. G., and I. T. Rekanos, "Time-Domain Microwave Imaging of Inhomogeneous Debye Dispersive Scatterers," *IEEE Trans. Antennas Propag.*, Vol. 60, No. 2, Feb. 2012, pp. 1197–1202.

[134] Zeng, X., et al., "Investigation of Stimulus Signals for a Time Domain Microwave Imaging System," *IET Microw. Antennas Propag.*, Vol. 11, No. 11, 2017, pp. 1636–1643.

[135] Zhang, X., X. Xi, M. Li, and D. Wu, "Comparison of Impulse Radar and Spread-Spectrum Radar in Through-Wall Imaging," *IEEE Trans. Microw. Theory Tech.*, Vol. 64, No. 3, March 2016, pp. 699–706.

[136] Fear, E. C., X. Li, S. C. Hagness, and M. A. Stuchly, "Confocal Microwave Imaging for Breast Cancer Detection: Localization of Tumors in Three Dimensions," *IEEE Trans. Biomed. Eng.*, Vol. 49, No. 8, 2002, pp. 812–822.

[137] Sakamoto, T., T. Sato, P. J. Aubry, and A. G. Yarovoy, "Ultra-Wideband Radar Imaging Using a Hybrid of Kirchhoff Migration and Stolt F-K Migration with an Inverse Boundary Scattering Transform," *IEEE Trans. Antennas Propag.*, Vol. 63, No. 8, August 2015, pp. 3502–3512.

[138] P. Kosmas and C. M. Rappaport, "FDTD-Based Time Reversal for Microwave Breast Cancer Detection-Localization in Three Dimensions," *IEEE Trans. Microw. Theory Tech.*, Vol. 54, No. 4, June 2006, pp. 1921–1927.

[139] P. Kosmas and C. M. Rappaport, "Time Reversal with the FDTD Method for Microwave Breast Cancer Detection," *IEEE Trans. Microw. Theory Tech.*, Vol. 53, No. 7, pp. 2317–2323, Jul. 2005.

[140] Prada, C., and M. Fink, "Eigenmodes of the Time Reversal Operator: A Solution to Selective Focusing in Multiple-Target Media," *Wave Motion*, Vol. 20, No. 2, September 1994, pp. 151–163.

[141] Minonzio, J.-G., et al., "Theory of the Time-Reversal Operator for a Dielectric Cylinder Using Separate Transmit and Receive Arrays," *IEEE Trans. Antennas Propag.*, Vol. 57, No. 8, August 2009, pp. 2331–2340.

[142] Yavuz, M. E., and F. L. Teixeira, "Full Time-Domain DORT for Ultrawideband Electromagnetic Fields in Dispersive, Random Inhomogeneous Media," *IEEE Trans. Antennas Propag.*, Vol. 54, No. 8, August 2006, pp. 2305–2315.

[143] Marengo, E. A., F. K. Gruber, and F. Simonetti, "Time-Reversal MUSIC Imaging of Extended Targets," *IEEE Trans. Image Process.*, Vol. 16, No. 8, August 2007, pp. 1967–1984.

[144] Kirsch, A., "The MUSIC-Algorithm and the Factorization Method in Inverse Scattering Theory for Inhomogeneous Media," *Inverse Probl.*, Vol. 18, No. 4, August 2002, pp. 1025–1040.

[145] Solimene, R., and A. Dell'Aversano, "Some Remarks on Time-Reversal MUSIC for Two-Dimensional Thin PEC Scatterers," *IEEE Geosci. Remote Sens. Lett.*, Vol. 11, No. 6, June 2014, pp. 1163–1167.

[146] Hossain, M. D., A. S. Mohan, and M. J. Abedin, "Beamspace Time-Reversal Microwave Imaging for Breast Cancer Detection," *IEEE Antennas Wirel. Propag. Lett.*, Vol. 12, 2013, pp. 241–244.

[147] Hossain, M. D., and A. S. Mohan, "Cancer Detection in Highly Dense Breasts Using Coherently Focused Time-Reversal Microwave Imaging," *IEEE Trans. Comput. Imaging*, Vol. 3, No. 4, December 2017, pp. 928–939.

[148] Zhong, X., C. Liao, and W. Lin, "Space-Frequency Decomposition and Time-Reversal Imaging," *IEEE Trans. Antennas Propag.*, Vol. 63, No. 12, Decemeber 2015, pp. 5619–5628.

[149] Tanaka, T., T. Takenaka, and S. He, "An FDTD Approach to the Time-Domain Inverse Scattering Problem for an Inhomogeneous Cylindrical Object," *Microw. Opt. Technol. Lett.*, Vol. 20, No. 1, January 1999, pp. 72–77.

[150] Johnson, J. E., et al., "Advances in the 3-D Forward–Backward Time-Stepping (FBTS) Inverse Scattering Technique for Breast Cancer Detection," *IEEE Trans. Biomed. Eng.*, Vol. 56, No. 9, September 2009, pp. 2232–2243.

4

Microwave Imaging in Civil Engineering and Industrial Applications

Some of the most relevant applications of microwave imaging techniques in the field of civil and industrial engineering are discussed in this chapter. They concern nondestructive testing and evaluations, and are based both on qualitative and quantitative methods. Several examples are reported.

4.1 Potentialities and Limitations of Microwave Imaging in Civil Engineering and Industrial Applications

Microwaves are extensively applied in civil and industrial engineering, and their potentialities in this field manifest from different points of view. Essentially, they are widely used for material characterization as well as for nondestructive testing and industrial process monitoring. This is of course due to their capability to penetrate dielectric materials, such as coatings, in order to provide information about the internal structures of samples and devices. Moreover, microwaves are also able to heat structures from the power supplied and internally dissipated due to the lossy nature of materials, and this concept can be applied in order to dry or disinfect objects and structures, and cure polymers or composite materials. Monitoring the evolution and effectiveness of these processes is another important task that can be tackled by exploiting microwave imaging techniques.

The book by R. Zoughi [1] provides a comprehensive reference about classical microwave methods for material characterization and nondestructive testing techniques. Most of the standard approaches in this field are based on reflection and transmission concepts. These approaches use propagating or radiating waves that impinge on samples or structures, and exploit the information about materials or possible defects inside the structure that is included in reflected or transmitted waves. This information is often contained in the associated scattering parameters, that is, the reflection coefficient, S_{11}, and the transmission coefficient, S_{21}. Microwave measurements represent a challenging but well-established task, which plays a fundamental role in all the applications related to microwave diagnostics. Reviews of microwave measurements can be found in the related scientific literature (see, for example, [2–7] and the reference therein). Direct measurements of the dielectric parameters characterizing materials are also of interest for our purposes. Such parameters, which are usually frequency dependent, have been discussed in Chapter 2. By using proper samples of the materials of interest, their dielectric properties can be retrieved by exploiting several techniques (typically classified as resonant and nonresonant methods) that often rely on the use of dielectric resonators, cavities, waveguides, and coaxial and microstrip transmission lines [4].

More recently, imaging techniques have been proposed for material characterization and nondestructive testing, and they are the main subject of this chapter. Essentially, two different inspection approaches can be adopted. The first approach concerns the use of scanning elements that can provide high resolution images based on some approximations (e.g., the Born approximation discussed in Chapter 3). The reconstruction algorithms are usually based on synthetic data processing, similar to the one used in synthetic aperture radar (SAR) or in the synthetic aperture focusing technique (SAFT) (which will be better discussed in Chapter 6). These techniques are generally denoted as near field microwave imaging and can be efficiently applied for inspecting composite and layered materials, for the evaluation and monitoring of paint thickness, for the identification of corrosion under paint, and for surface crack detection in metallic surfaces [8]. It is worth noting that in these cases usually it is necessary to acquire the scattered field data in a set of measurement points located on a line or on a surface. Clearly, this requires both to be able to physically put the required number of antennas (usually with a $\lambda_0/2$ spacing) on the acquisition plane and to use a switch for scanning the various elements. A possible alternative is the use of the modulated scattering technique [9–12], which is based on the fact that a passive probe (e.g., a dipole or a slot), when irradiated by an electromagnetic wave, scatters a field that is related to the impinging electromagnetic radiation. In this way, by using an array of properly loaded probes, it is possible to acquire a large set of data by using just a single receiver and without the need of additional microwave connection lines [9]. The second approach is related to tomographic imaging (see Chapter 3) and, for example, it has been proposed for wood material evaluation, including living trees and wood slabs, and to monitor pillars or columns in civil engineering. Other applications discussed in the following sections concern imaging of plastic materials (slabs, pipes, tubes, etc.) as well as various diagnostic investigations in the areas of agriculture, food production, and chemical industry.

It should be mentioned that microwave fields are also applied in the so-called microwave microscopy [13] (not discussed in this book), where a small metallic tip is used as a probe to scan the surface of a given material, usually mounted on a ground plane. An image with fine spatial resolution is obtained by considering the changes in the capacitance between the tip and the ground plane, which of course depend on the local dielectric properties of the sample.

The potentialities of microwave imaging techniques are also very wide because (as already mentioned in the previous chapters) the apparatuses may be quite cheap and safe for the operators. However, as in other applicative scenarios, microwave imaging techniques present some limitations when applied in civil and industrial fields. The main one is related to the requirement of fast and easy-to-use techniques, which ideally operate in real time. Of course, this has an impact on the algorithmic complexity. In particular, while qualitative imaging methods may be quite fast and with a lower complexity since they provide only limited information about the targets of the inspection, quantitative techniques are usually much more complex and computationally expensive. However, with the huge growth in the available computing power, this limitation is being overcome in the very near future. Another limitation is related to the need for designing application-specific probes and antennas, which can vary on a case-by-case basis. Nevertheless, thanks to the constant research in this field, many innovative solutions are continuously proposed, and the number of commercial products and patents applying microwaves for nondestructive testing and evaluations is greatly increasing.

4.2 Electromagnetic Characterization of Some Materials Used in Civil and Industrial Microwave Imaging

There is a plethora of materials and targets that can be inspected by microwave signals and for which microwave imaging techniques have been proposed, including plastic and composite materials, concrete, wood, food and vegetables, and others. Some examples of the related dielectric properties are reported in the following sections, with reference to the general description reported in Chapter 2. Additional data concerning specific materials can be found in the specialized literature.

4.2.1 Dielectric Properties of Concrete

Concrete is a porous and heterogeneous material (with pores partially filled with a ionic solution). Its complex permittivity is frequency dependent and, in general, it also depends on temperature, salinity, porosity, and saturation of the pores. In particular, one of the main factors that influence the dielectric properties of such material is the hydration of cement paste.

In the literature, different dielectric models have been adopted for representing the dielectric properties of concrete [14–19]. In particular, both standard relaxation models, such as Debye and Cole-Cole (see Chapter 2), and mixing formulas such as the complex refraction index model (CRIM) and the self-similarity model [16] have been used. It is worth noting that although Debye models provide less generality and accuracy than Cole-Cole models, they are often considered when performing numerical analyses since they can be more easily implemented in finite-difference time-domain (FDTD) algorithms [20].

Another way of describing the dielectric properties of concrete, which uses the Jonscher model [21], has also been recently proposed [20], [22]. This model, also exploited for the characterization of rocks [23], has been found to be able to represent with good accuracy the measured complex relative permittivity of concrete materials. By using the Jonscher model, the complex dielectric permittivity is approximated by means of the following empirical formula:

$$\varepsilon(\omega) = \varepsilon_\infty + \varepsilon_0 \chi_r \left(\frac{\omega}{\omega_r}\right)^{n-1} \left(1 - j\text{cotg}\left(\frac{n\pi}{2}\right)\right) \tag{4.1}$$

where ω_r is a reference angular frequency (arbitrarily chosen), and ε_∞, χ, and N are parameters to be found on the basis of the measured data [22]. Some examples of the frequency behavior of the dielectric properties for an ordinary Portland cement with different values of the volumetric moisture content (MC) are reported in Figure 4.1 (the Jonscher model in [22] is used).

Another important factor affecting the dielectric properties of cement-based materials is the chloride content. In practical applications, such parameter has a great importance, since high values of its concentration can induce the corrosion of the rebars in reinforced structures (discussed in the following). Consequently, different studies have been performed in order to analyze the changes in the dielectric properties due to the chloride content. In particular, it has been found (for example see [14, 24]) that the effects of the chloride content on the permittivity are higher at the lower frequencies, with the imaginary part showing a greater increase with respect to the real part [24]. Moreover, it has been reported that chloride has very limited effects when dealing with dry cement, especially for frequencies higher than 1 GHz.

In order to improve the robustness characteristics of civil structures, new materials are continuously studied and evaluated. An example is represented by steel fiber reinforced concrete, which is composed of Portland cement and

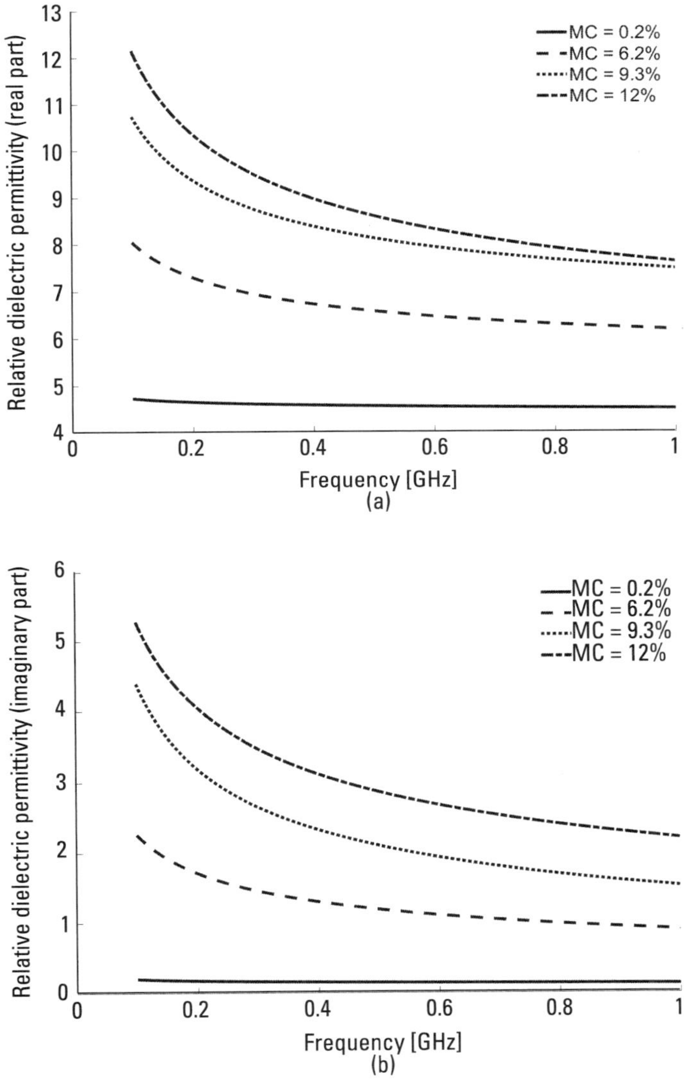

Figure 4.1 Example of dielectric properties of concrete (Jonscher concrete model from [22]). (a) Real and (b) imaginary parts of the relative dielectric permittivity.

coarse aggregates including steel fibers [25, 26]. In order to characterize this material, the inclusions can be considered as electrically small wire scatterers with a given dipole moment. Following this approach and using the Clausius-Mossotti formulation under the hypothesis of low concentration of wires, it results that the effective permittivity of the material is given by [25, 27]

$$\varepsilon(\omega) = \varepsilon_h \frac{1+p}{1-q} \tag{4.2}$$

where ε_h is the dielectric permittivity of the host medium, $q = n\alpha/(3\varepsilon_h\varepsilon_0)$, α and n being the average polarizability and the concentratoin of the wires, and $p = 2q$. Accordingly, the phase and attenuation constants can be calculated by using (2.35) and (2.36).

In addition, civil concrete structures often include the presence of rebars, which are steel bars or meshes of steel wires used for reinforcing and holding the concrete in tension [28]. Rebar may suffer from corrosion problems, that is, part of the steel bar is transformed into rust. This is actually an important problem in civil engineering. In fact, it has been pointed out in [29] that "In the U.S. only, the cost associated with structural repair and maintenance due to such corrosion problems is about $276 billion per year". In particular, when corrosion takes place, the size of the metallic part of the rebar reduces and a dielectric layer made of rust appears around it (as schematically shown in Figure 4.2). Therefore, corrosion in such structures can be evaluated by means of microwave imaging techniques. Consequently, modeling the corrosion of a steel rebar requires a definition of the size of the rebar and of the oxide dielectric layer made of rust.

Assuming the rebar composed by a circular cylinder, it has been reported in [29, 30] that, after corrosion, the radii of the remaining metallic parts, r_b, and of the external ring of oxide, r_o, can be modeled as [30]

$$r_b = r_{b,0}\sqrt{1-P_C}, \; r_o = r_{b,0}\sqrt{1+3P_C} \tag{4.3}$$

where P_C is the percentage of corrosion and $r_{b,0}$ is the radius of the rebar in absence of corrosion. The relative dielectric permittivity of the ring of oxide

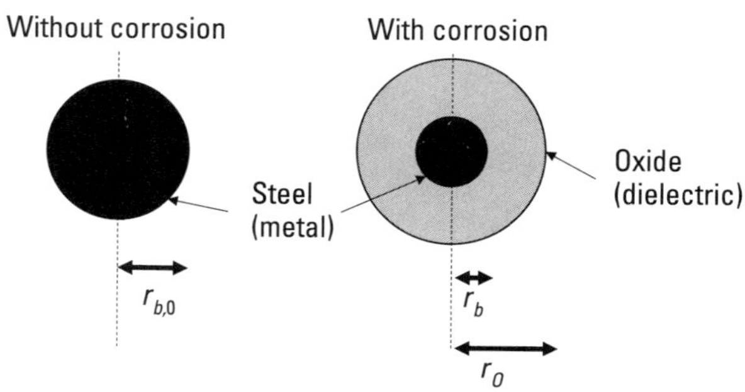

Figure 4.2 Simplified representation of corrosion effect in rebar.

depends of the type of rust considered. For example, in the X band, black rust is characterized by $\varepsilon_r = 12.58 - j2.36$, whereas red rust by $\varepsilon_r = 8.42 - j1.03$ [1].

Figure 4.3 shows an example of the effects of an external oxide layer on the field scattered by a metallic rebar inside a concrete pillar. Such results have

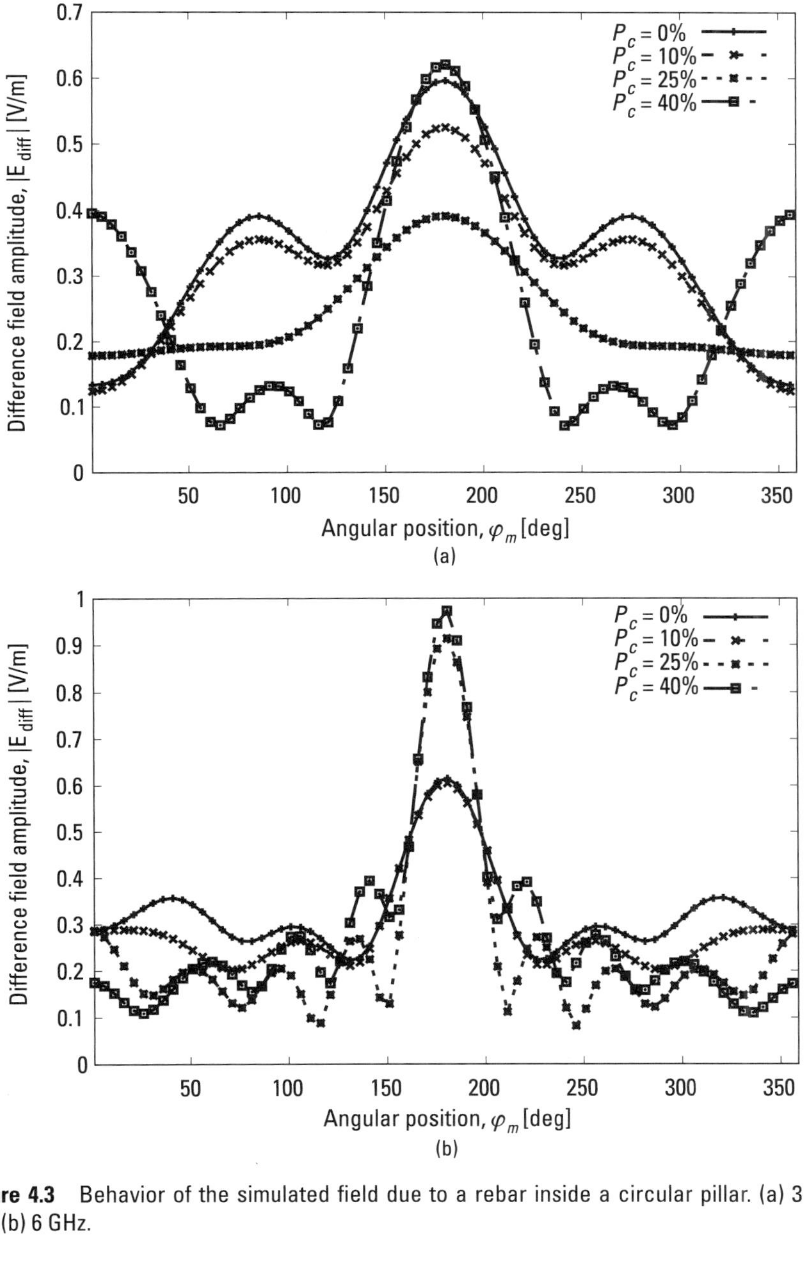

Figure 4.3 Behavior of the simulated field due to a rebar inside a circular pillar. (a) 3 GHz and (b) 6 GHz.

been obtained by using a simplified scenario, which includes an infinite cylindrical pillar with radius r_p = 15 cm and made of dry concrete (the dielectric properties has been set equal to $\varepsilon_r' = 4.4$ and $\sigma = 0.09$ S/m). The rebar is modeled as an infinite PEC cylinder with actual diameter equal to $r_{b,0} = 2.54$ cm (1 inch rebar) located in the center of the pillar, eventually covered with black rust. Under such simplified hypotheses, the electric field values have been computed by using the analytical solution for the scattering from infinite circular cylinders [31] (using 30 modes in the series expansion). In particular, it is supposed that the pillar is illuminated by a line current source with I = 1mA (antenna coupling effects are neglected) in contact with the external surface and the scattered field is computed on a circumference of radius equal to 15 cm (i.e., on the external surface of the pillar). Figure 4.3 reports the differences between the z-components of the electric fields computed with and without the rebar under different percentage of corrosion (ranging from 0%–40%) and for two different values of the working frequency (3 GHz and 6 GHz). As can be seen, for both the considered frequency values, significant differences in the behavior of the scattered electric field can be observed when the external oxide layer is present. These significant differences show that, in principle, the detection of corrosion and rust can be operated with nondestructive techniques based on the measurement of the scattered field outside the structure under test.

Other materials are also used in the constructions of civil structures, for example, different kinds of walls can be adopted. Some of them will be briefly considered in Section 7.2, which is devoted to through-the-wall microwave imaging.

4.2.2 Dielectric Properties of Plastic Materials

Plastic materials have a particular relevance in microwave imaging for several reasons. First, their widespread industrial production and use is demanding automated process control systems, which can be effectively devised using microwave measurement technologies [32–38]. Second, their controlled dielectric properties make them particularly suitable for constructing test targets to be used for the validation phase of microwave imaging techniques. An example about this point is the widely used Frésnel database (developed by the Institute Frésnel, Marseille, France, and already mentioned in Chapter 3), as that even after some years is still considered a standard reference set of targets that allows to test and compare many inversion procedures [39–41]. Last but not least, plastic materials are frequently used in practical microwave imaging systems for building custom antenna assemblies and support structures, or even inside radiating parts, especially now that the low-cost 3-D printing capabilities are freeing the possibilities of designing and prototyping innovative solutions [42].

All these important applications are clearly strongly based on the dielectric characterization of plastic materials at microwave frequencies. The measurement of microwave dielectric properties of plastics and derived materials has attracted the attention of scientists and engineers for a long time [43, 44]. One of the most important aspects of the dielectric behavior of plastic materials at microwaves is the stability of their dielectric properties in a broad range of frequencies and, in case of pure materials, their low loss. The measured values of the dielectric properties of several plastic materials are summarized in Table 4.1 [44]. Moreover, Table 4.2 reports the complex dielectric permittivity of some plastics commonly used in 3-D printing [45]. Furthermore, plastic is often present in composite materials, where it is combined with other particles or substances [46–48]. Those materials have been even used for building phantoms of biological tissues (see also Chapter 5) and the scientific community is unceasingly investigating about the possibilities of innovative compound plastic materials with nanoparticles or nanotubes, which are finding more and more practical applications [49–51].

4.2.3 Dielectric Properties of Food and Vegetables

In the last few years, there has been a significant interest in the characterization of the dielectric properties of food, fruits, and vegetables [52–54]. In fact, the

Table 4.1
Relative Dielectric Permittivity of Some Kinds of Plastics at Room Temperature

Frequency	**100 MHz**		**1 GHz**	
Material	ε_r'	ε_r''	ε_r'	ε_r''
Polymethyl methacrylate (PMMA)	2.7	0.03	2.65	0.02
Polyvinylidene fluoride (PVDF)	3.5	0.53	3	0.3
Polyethilene oxide (PEO)	4.3	0.15	3.8	0.43

Table 4.2
Relative Dielectric Permittivity of Some Common 3-D-Printed Plastic Materials

Frequency	**100 MHz**		**1 GHz**		**10 GHz**	
Material	ε_r'	ε_r''	ε_r'	ε_r''	ε_r'	ε_r''
Acrylonitrile butadiene styrene (ABS)	2.83	0.012	2.80	0.009	2.54	0.038
Polycarbonate (PC)	2.93	0.015	2.89	0.012	2.59	0.013

knowledge of such parameters can be used for the development of diagnostic techniques aimed at, for example, moisture content estimation and quality control [55, 56]. Some examples of measured values of the real and imaginary parts of the dielectric permittivity of some common fruit species are reported in Figure 4.4 [57]. As expected, the values of the permittivities are quite high, due to the high amount of water. It is also worth noting that the dielectric properties of fruits, as can be seen in many other kinds of food, are also significantly influenced by the temperature [52].

4.2.4 Dielectric Properties of Wood

Systematic studies concerning the dielectric properties of wood trunks and derived materials have been performed since 1948 (see, for example, [58, 59]). Because of its fiber structure, wood has different physical properties in the ra-

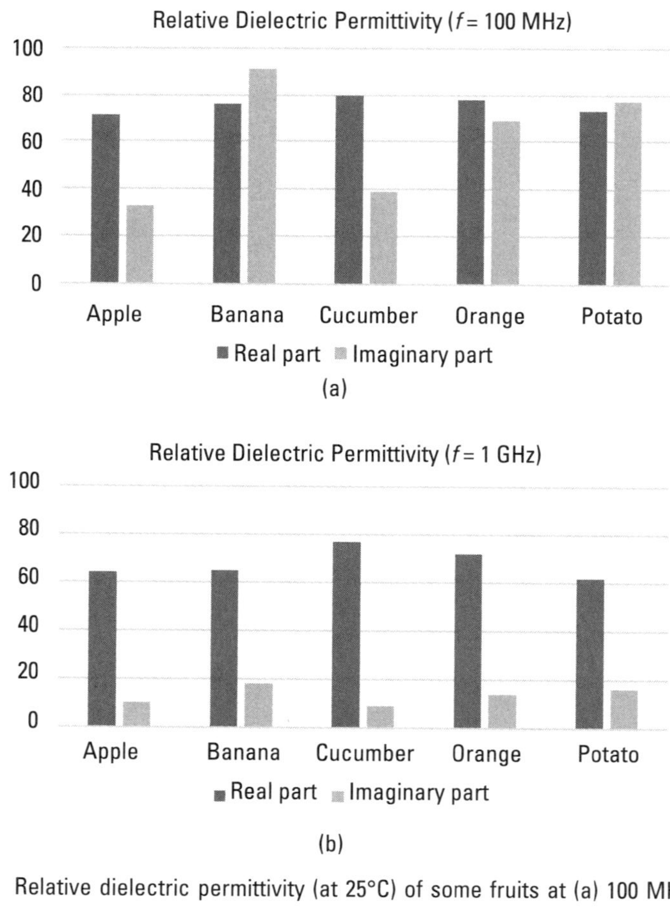

Figure 4.4 Relative dielectric permittivity (at 25°C) of some fruits at (a) 100 MHz and (b) 1 GHz [57].

dial (R), tangential (T), and longitudinal (L) directions (such directions are sketched in Figure 4.5). Consequently, wood is in general an anisotropic material, that is, the relationship between the electric displacement vector **D** and the electric field vector **E** is given by $\mathbf{D}(\mathrm{r}) = \overline{\overline{\varepsilon}}\ (R, \omega) \cdot \mathbf{E}(\mathbf{r})$, $\overline{\overline{\varepsilon}}$ being the dielectric permittivity tensor.

By defining an appropriate cylindrical reference system with unit vectors chosen according to the directions R, T, and L, the dielectric permittivity tensor can be written as

$$\overline{\overline{\varepsilon}} = \begin{bmatrix} \varepsilon_R & 0 & 0 \\ 0 & \varepsilon_T & 0 \\ 0 & 0 & \varepsilon_L \end{bmatrix} \tag{4.4}$$

where ε_R, ε_T, and ε_L are the radial, transversal, and longitudinal complex dielectric permittivity components. However, it has been found that ε_R and ε_T are usually quite similar (for example, see [60] and the reference therein). Consequently, it is often sufficient to consider just the two values $\varepsilon_\perp \cong \varepsilon_R = \varepsilon_T$ and $\varepsilon_\| = \varepsilon_L$. The dielectric permittivity tensor is consequently given by

$$\overline{\overline{\varepsilon}} = \begin{bmatrix} \varepsilon_\perp & 0 & 0 \\ 0 & \varepsilon_\perp & 0 \\ 0 & 0 & \varepsilon_\| \end{bmatrix} \tag{4.5}$$

The real and imaginary parts of the dielectric permittivity terms strongly depend on wood density and moisture content [61, 62]. Generally, both ε' and ε'' increase with the gravimetric moisture content. A monotonic increase can be observed in the real part of the relative permittivity, whereas the imaginary part usually grows up until a maximum value corresponding to fiber saturation is

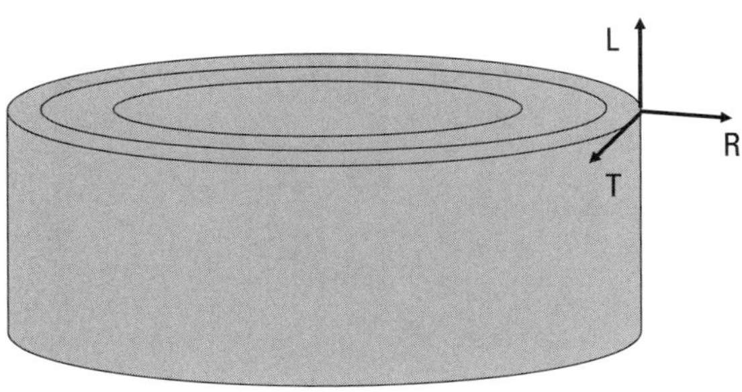

Figure 4.5 Reference directions for the dielectric properties of wood.

reached. After saturation, however, the imaginary part of the permittivity starts to decrease. Some examples of the values of the real and imaginary parts of the dielectric permittivity for different gravimetric moisture contents are reported in Table 4.3, which refers to a Douglas fir tree measured at 3 GHz [63]. Finally, it is worth noting that in general in-vivo wood samples can assume significantly higher values of the dielectric parameters than corresponding cut samples, due to the higher amount of water inside them [64].

An important applicative scenario is related to the detection of defective parts in wood trunks. Usually, two main kinds of defects can be encountered in wood structures: void and rotten parts. In the first case, a significant reduction of the permittivity can be observed. Contrary to the first case, a higher water content is usually present in the second case, leading to an increase in the dielectric permittivity. An example of the measured real and imaginary parts of rotten and healthy chestnut wood is shown in Figure 4.6.

4.3 Imaging of Civil Structures

Microwave imaging techniques can be applied to inspect civil structures, in order to evaluate the condition of buildings, roads, and bridges. To this end, qualitative and quantitative methods can be applied [65]. Such techniques can be based on tomography or on systems scanning a surface outside the structure to be evaluated, as mentioned in Section 4.1.

A significant application in this framework is the inspection of pillars. In this case, if the aim is the retrieval of the presence and positions of inclusions, such as rebar and cracks, qualitative methods can be used. In particular, a possible choice is the linear sampling method (LSM). In fact, such approach has been found to be very effective in retrieving the presence and shape of dielectric discontinuities when full-view tomographic data can be measured [66–69]. It is also worth noting that, although its theoretical formulation would require that

Table 4.3
Relative Dielectric Permittivity (Transverse and Longitudinal Components of the Permittivity Tensor) of a Douglas Fir versus the Gravimetric Moisture Content at 3 GHz

Gravimetric moisture content	**10%**		**70%**	
Permittivity tensor component	ε_r'	ε_r''	ε_r'	ε_r''
Transverse	1.3	0.1	8.7	1.6
Longitudinal	1.8	0.3	14	2.9

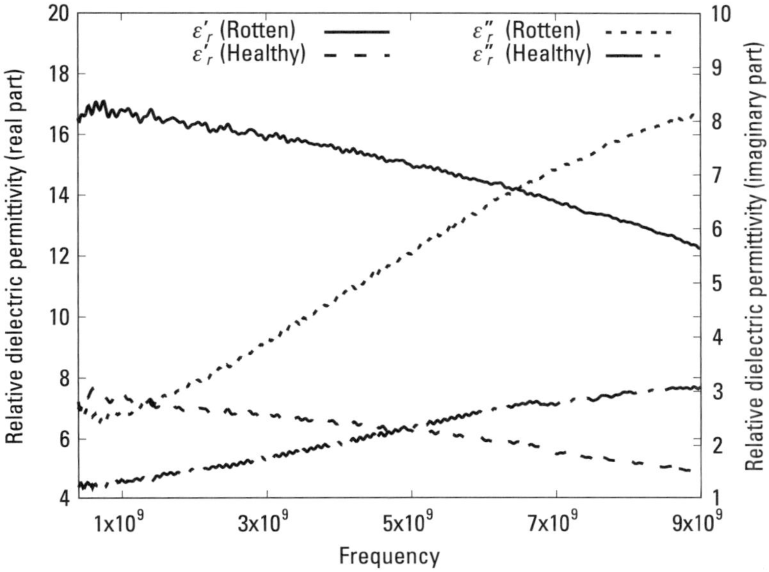

Figure 4.6 Dielectric properties of healthy and rotten chestnut wood versus frequency.

the antennas are located in the far-field region, it has been found that it can be extended to deal with near-field data as well, allowing for its use in civil and industrial applications. Assuming a two-dimensional configuration, the LSM can be formulated as the solution to the following equation

$$\int_0^{2\pi} E_{scatt,z}\left(\mathbf{r}_t',\phi_{inc}\right)\gamma\left(\mathbf{r}_t,\phi_{inc}\right)d\phi_{inc} = g_b\left(\mathbf{r}_t',\mathbf{r}_t\right),\ \mathbf{r}_t \in D, \mathbf{r}_t' \in D_{obs} \quad (4.6)$$

where it is assumed that the target is illuminated by incident fields impinging from directions $\phi_{inc} \in [0, 2\pi]$ and that the corresponding scattered electric field $E_{scatt,z}(\mathbf{r}_t', \phi_{inc})$ is collected at points $\mathbf{r}_t' \in D_{obs}$ located on a circumference surrounding the investigation domain. Moreover, g_b is the Green's function for the considered background. The unknown function γ has the property that its norm blows up near the boundary of the scatterer while remaining large outside it. Consequently, it can be used to visualize the support of the target. In practical implementations, a finite number S of antennas are used to illuminate the target, located at angular positions $\phi_{inc}^s = (2\pi / S)(s-1)$, $s = 1, \ldots, S$, and the scattered field is collected in M points uniformly distributed on a measurement circumference, that is, $\mathbf{r}_{meas}^m = \left(R_{meas}\cos\phi_{meas}^m, R_{meas}\sin\phi_{meas}^m\right)$, with $\phi_{meas}^m = (2\pi / M)(m-1)$, $m = 1, \ldots, M$. Consequently, (4.6) can be written in discrete form as

$$\mathbf{F}\boldsymbol{\gamma}(\mathbf{r}_t) = \mathbf{g}_b(\mathbf{r}_t) \tag{4.7}$$

where $\mathbf{F}$ is a $M \times S$ matrix whose (m, s) element is given by $(2\pi/S)E_{scatt,z}\left(\mathbf{r}_{meas}^{m}, \phi_{inc}^{s}\right)$, $\mathbf{g}_b(\mathbf{r}_t)$ is an array of dimension M whose elements are the values of the Green's function $\mathbf{g}_b\left(\mathbf{r}_{meas}^{m}, \mathbf{r}_t\right)$, and $\gamma(\mathbf{r}_t)$ is an array whose S elements are the values of the unknown function γ at point $\mathbf{r}_t$. The ill-posed linear equation in (4.7) is solved by using the singular value decomposition (SVD) technique [70]. In particular, the square norm of $\boldsymbol{\gamma}(\mathbf{r}_t)$ can be expressed as

$$\left\|\boldsymbol{\gamma}(\mathbf{r}_t)\right\|^2 = \sum_{i=1}^{P} \frac{\sigma_i^2}{\left(\sigma_i^2 + \alpha\right)^2} \left|\left\langle \mathbf{g}_b(\mathbf{r}_t), \mathbf{v}_i \right\rangle\right|^2 \tag{4.8}$$

where σ_i and $\mathbf{v}_i$ are the singular values and the singular vectors of the matrix $\mathbf{F}$. In order to visualize the results, the indicator function $I(\mathbf{r}_t) = (1 + \|\boldsymbol{\gamma}(\mathbf{r}_t)\|^4)^{-1}$ is used in the results reported in this section [71]. It is worth remarking that α is a regularization parameter, and consequently a proper selection rule must be adopted. In the reported example, the generalized discrepancy principle [72] is used for determining its optimal value.

An example of application of the previously described LSM to the detection of rebar inside a concrete pillar is reported in Figure 4.9. The considered pillar has square cross section of side $L = 30$ cm and it contains four metallic rebars located near the four corners at a distance of $d = 5$ cm from the external surface, as shown in Figure 4.7. The radii of the rebars have been set equal to $r_b = 1.2$ cm. In particular, dry concrete (i.e., with volumetric moisture content equal to 0.2%) has been assumed. A tomographic illumination/measurement setup composed by $S = 30$ antennas equally spaced on a circumference of radius

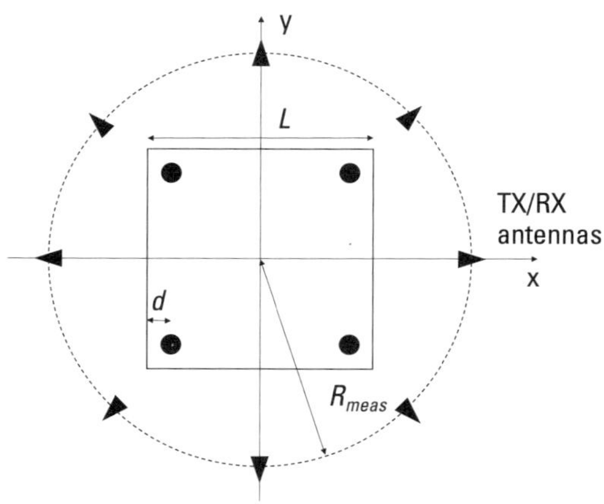

Figure 4.7 Model of a concrete pillar with rebars.

R_{meas} = 30 cm centered at the origin is considered. In particular, each antenna acts in turn as transmitter, whereas all the other anntenas are used to collect the field scattered by the target.

The electric field has been numerically simulated by means of a forward solver based on the FDTD method (the open-source *gprMax* package [73] has been used). Moreover, a white Gaussian noise with zero mean value and variance corresponding to a signal-to-noise ratio (*SNR*) of 25 dB has been added to the computed field values. The dielectric properties of concrete have been obtained by using the Debye model in [20], whereas the rebars have been modeled by using a PEC material. The antennas have been modeled as ideal line-current sources fed by a Ricker pulse (i.e., a negative normalized second derivative of a Gaussian waveform) with unit amplitude and center frequency of 2 GHz. Since LSM (at least in its standard form) is a frequency-domain approach, the computed data have been preprocessed by means of the fast Fourier transform (FFT) in order to extract the data at different frequencies. Moreover, the inhomogeneous Green's function for the configuration without rebars [74] has been used in the LSM. The electric fields due to such reference configuration has been simulated (again, by using the *gprMax* software [73]) and subtracted from the total-field data. Some examples of the behavior of the field due to rebars (i.e., the difference between the total electric field with and without such inclusions) are shown in Figure 4.8. The indicator function provided by the LSM, obtained by using the data at the central frequency (i.e., at 2 GHz), is shown in Figure 4.9. As can be seen, the four rebars have been correctly localized and sized in the reconstructed image.

In the previous example, a direct qualitative approach based on the linear sampling method has been considered. However, it is worth mentioning that other qualitative methods can be adopted as well, for example, those based on radar and beamforming concepts [75–79].

Differently from tomographic techniques, in the so-called near field scanning systems, the target under test is usually in close proximity to the probing antenna and the antennas are moved on a planar surface in order to perform a scan of the target (Figure 4.10). Antennas used for scanning the imaging surface are usually open waveguide antennas (eventually loaded with low-loss dielectric materials). The spatial resolution that can be achieved is directly related to the footprint of the antennas. Two- or three-dimensional images of the targets can be obtained by mapping the measured values of the received signal intensity. To this end, the antenna is usually moved in a raster scan modality on a plane parallel to the surface of the target. The distance between the measurement plane and the target surface is indicated as the standoff distance *h*. In most cases, this distance can be optimized for the considered application. The imaging apparatus can be composed by a coherent reflectometer and a detection system in

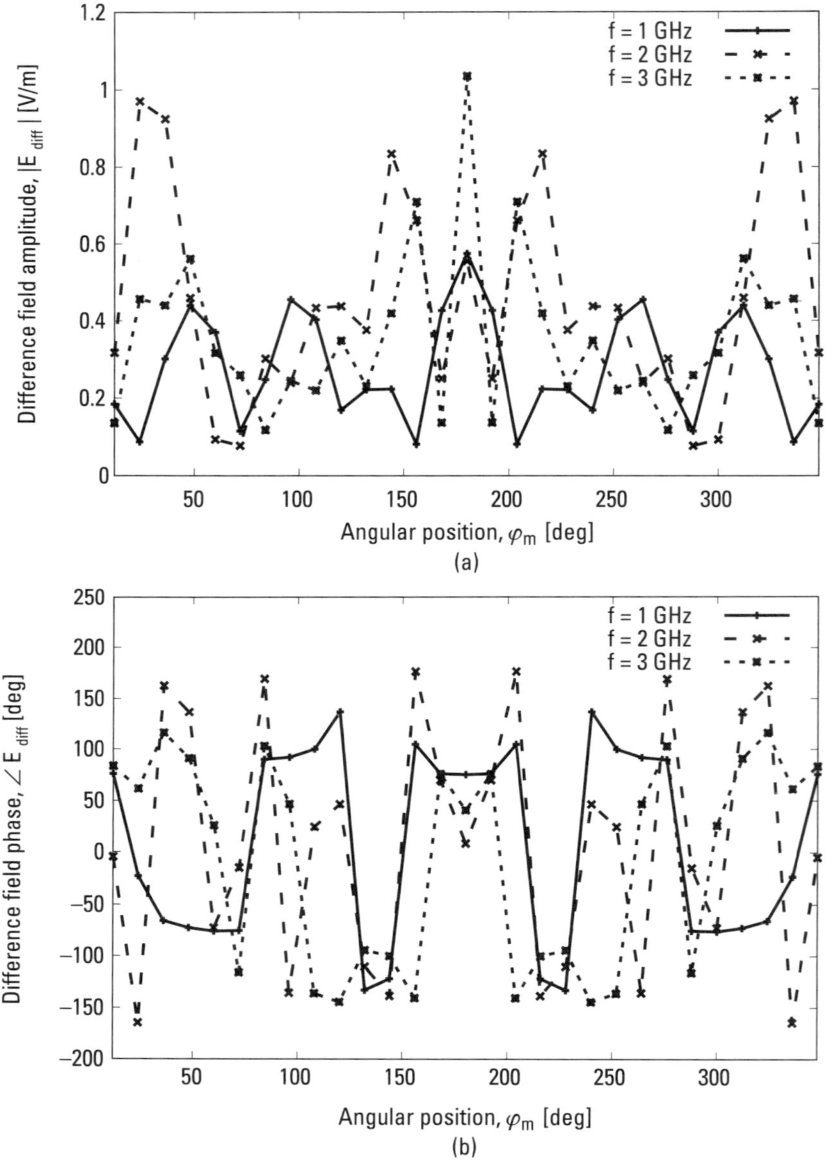

Figure 4.8 Behavior of the simulated field due to the rebars in the measurement points. (a) Amplitude and (b) phase.

order to generate a DC output directly related to the amplitude or phase of the scattered wave (in this case, the main effect is the reflected wave contribution).

SAR techniques, which are not dissimilar to those used in remote sensing applications and subsurface prospection, can be adopted in this short-range

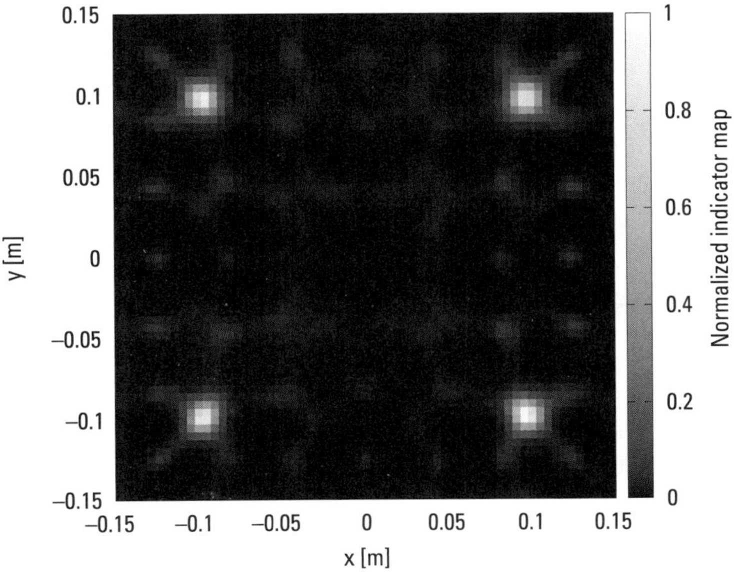

Figure 4.9 Reconstructed image produced by using the linear sampling method. Square pillar with four rebars.

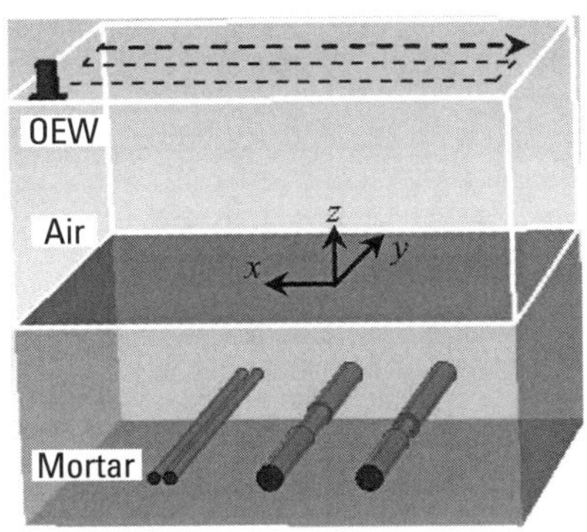

Figure 4.10 Measurement set up for the inspection of corrosion in a mortar sample with rebar [29]. (©2014 IEEE.)

microwave imaging approach. They are based on the idea of migrating or back-propagating the measured field samples in order to focus on specific points inside the investigation area (see also Chapters 6 and 7). This focusing can be performed both in time- and spectral-domains, by adding proper time delays or phase shifts that essentially depend on the propagation velocity of the electromagnetic wave inside the structure. It is evident that certain knowledge of the material composing the structure to be investigated should be a priori known or would have to be estimated. In the field of civil engineering, the possibility of using such microwave imaging techniques for exploring the inner structure of a wall has been considered. Examples are represented by the detection of routed wires or plumbing pipes. For example, a model of a wall composed of different layers (drywall, insulation, and mortar) has been inspected in [29] by using a piecewise version of the frequency-domain SAR technique (referred as PW-SAR), which has been developed in a monostatic arrangement. The basic SAR imaging technique can be summarized by the following relationship

$$I(\mathbf{r}) = \int_{\Omega} \mathcal{F}_{2D}^{-1}\left\{\tilde{S}\left(k_x, k_y, \omega\right) e^{jk_z z}\right\} d\omega, \mathbf{r} \in V \tag{4.9}$$

where $\mathcal{F}_{2D}^{-1}$ denotes the 2-D inverse Fourier transform with respect to the variable k_x and k_y, $\tilde{S} = \mathcal{F}_{2D}(S)$ is the Fourier transform of the scattering parameters $S(x, y, \omega)$ measured on the acquisition surface, and $k_z^2 = 4k_b^2 - k_x^2 - k_y^2$ (k_b being the wavenumber in the material). It is worth noting that such a imaging scheme is equivalent to the frequency-wavenumber migration techniques discussed in Chapter 6 with reference to the case of 2-D subsurface imaging. In [29], the previous SAR imaging scheme has been empirically extended for dealing with a multilayer structure (schematically shown in Figure 4.11) by taking into account the presence of the boundaries among the layers by means of the corresponding Frésnel reflection coefficients. In particular, the PW-SAR

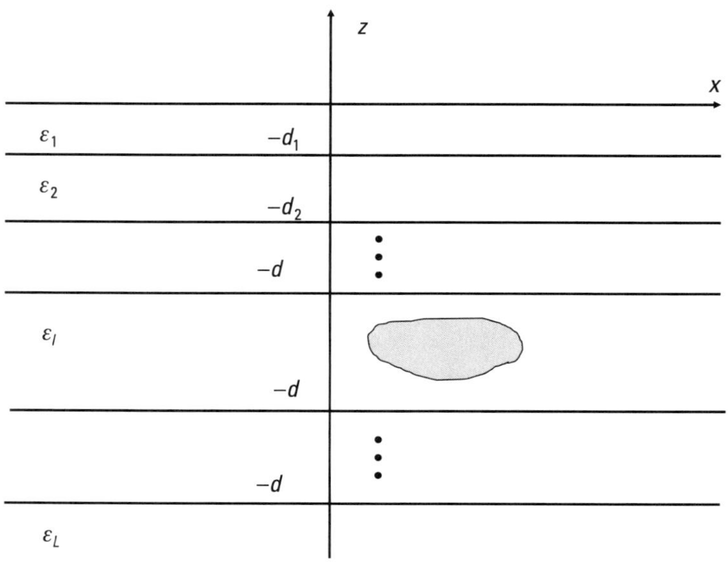

Figure 4.11 Schematic representation of a multilayered medium.

imaging scheme use the following relationship for focusing on an point in the Lth layer [29]

$$I(\mathbf{r}) = \int_{\Omega} \mathcal{F}_{2D}^{-1} \left\{ \frac{\tilde{S}\left(k_x, k_y, \omega\right) \left[\prod_{i=1}^{L-2} e^{jk_{z,i}\left(d_i - d_{i+1}\right)} \right] e^{-jk_{z,1}d_1} e^{jk_{z,L}\left(z + d_{L-1}\right)}}{\prod_{i=1}^{L-1} e^{-j\angle\left(T_{i,i+1}T_{i+1,i}\right)}} \right\} d\omega, \ \mathbf{r} \in V \tag{4.10}$$

where $T_{i,i+1}$ is the Frésnel transmission coefficient between the ith and $(i + 1)$th layers [80] and $k_{z,i}^2 = 4k_i^2 - k_x^2 - k_y^2$, k_i being the wavenumber in the ith layer.

Experimental results obtained by using the previously described procedure for detecting rebar in mortar samples have been reported, for example, in [29]. Three-dimensional images have been obtained by scanning an area of 22 cm × 21 cm at a distance of 13 cm from the mortar sample surface. Figure 4.10 reports the experimental set up, whereas Figure 4.12 provides the 3-D images obtained by using the PW-SAR approach. The results obtained by using another SAR imaging scheme proposed in [29], which is based on the use of a Wiener filter and on the exploitation of the Green's function for layered media (referred as WL-SAR), are also reported in Figure 4.12.

Steel fiber reinforced concrete has been intensively studied as well. For example, in [25] an experimental set up based on time-domain reflectometry has been constructed and several samples with different concentrations of fibers have been successfully inspected.

4.4 Imaging of Plastic Materials

As previously remarked, the nondestructive testing of plastic materials can be useful in different applications and industrial processes. Clearly, both qualitative and quantitative microwave imaging methods can be applied to the inspection of plastic targets. For instance, in [81] a radar-based algorithm in which the phase information is used to reconstruct an image of the unknown target has been applied. In particular, a plastic tube made of polyvinyl chloride (PVC) (external radius of 3 cm, internal radius of 2.5 cm) has been satisfactorily imaged at 3 and 4 GHz by means of a bistatic imaging apparatus in which transmitting and receiving antennas rotate on two separate rings around the target. The adopted antennas are printed patch antennas. The system has also been used for inspecting wood material and is shown in Figure 4.21 [81].

Quantitative inverse scattering techniques can be also exploited in the imaging of plastic materials. An example, in which the previously mentioned

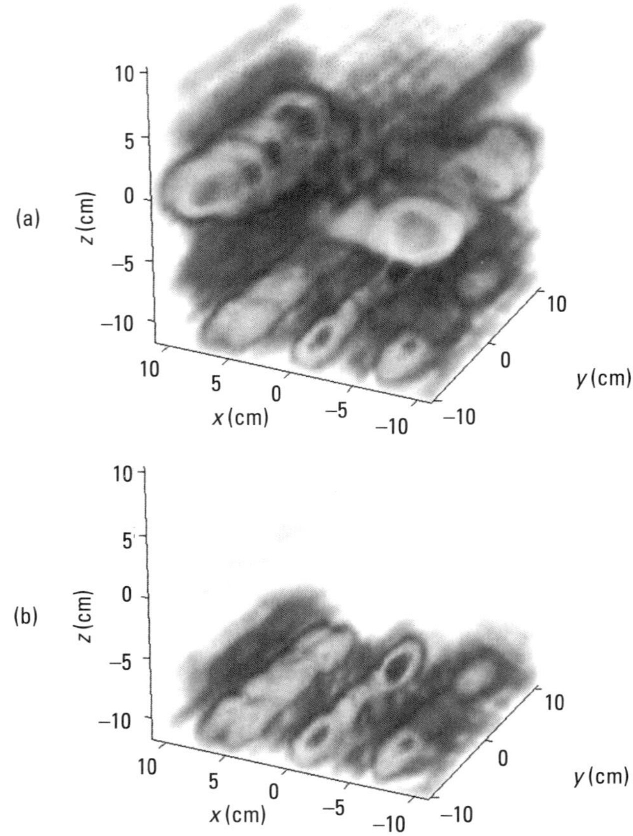

Figure 4.12 Three-dimensional image of the rebars inside a mortar sample obtained by the PW-SAR and WL-SAR approaches developed in [29]. (©2014 IEEE.)

Frésnel experimental database [40] has been considered, is reported in [82]. In particular, the L^p Banach-space inexact Newton/Landweber method described in Section 3.3 has been applied to the reconstruction of two nested cylinders: A plastic circular cylinder with a 31 mm diameter and relative dielectric permittivity $\varepsilon_r = 3$ is surrounded by a foam cylinder with a diameter of 80 mm, characterized by a relative dielectric permittivity $\varepsilon_r = 1.45$. The target under test is illuminated by a transmitting antenna located in 8 different positions, and the scattered electric field is measured in 241 points around it. Electric field measurements have been collected from 2 to 10 GHz with 1 GHz frequency step, and the reconstruction algorithm includes a frequency-hopping procedure, in which data at different frequencies are sequentially exploited starting from the lower frequency values, and using the obtained results as initial guess for the imaging at higher frequencies. The reconstructed distribution of the relative dielectric permittivity of the target for two different numbers of considered frequencies and two values of the Banach-space norm parameter p are shown in Figure 4.13. As can be seen, the quantitative technique allows an accurate reconstruction of the dielectric properties of the sample.

4.5 Imaging of Metallic Structures

It has been shown in Section 4.3 that microwave imaging techniques can be used in civil applications for identifying the presence of metallic objects in dielectric structures. However, other applications involving metallic materials,

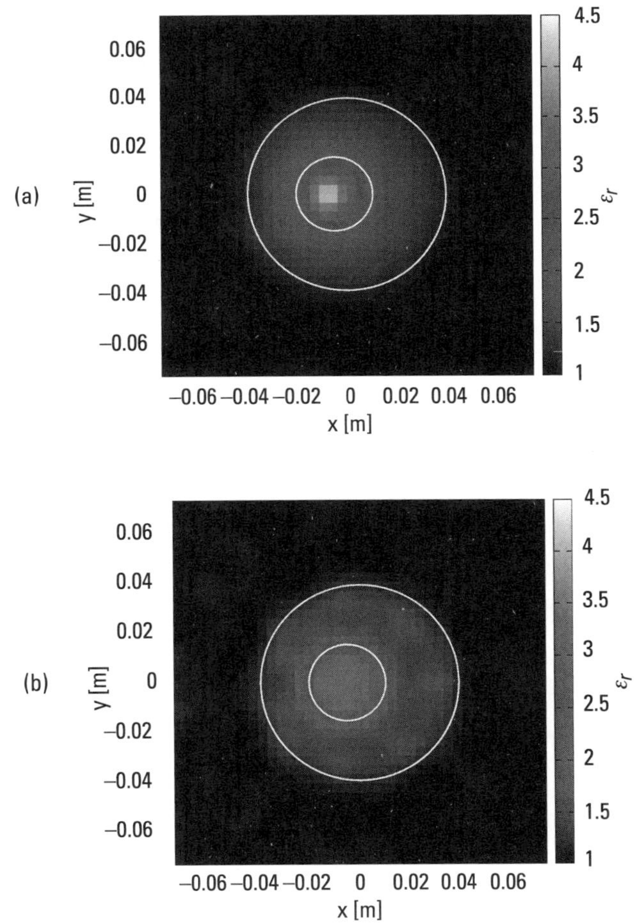

Figure 4.13 Reconstructed distribution of the dielectric permittivity of the *FoamDielIntTM* [40] target using the method presented in [82]. (a) $F=4$ considered frequencies, $p=1.4$, (b) $F=9$ considered frequencies, $p=2.3$. (©2014 IEEE.)

such as the inspection of composite or coated metallic parts, or the detection of surface cracks can be addressed by using microwave imaging.

As an example, the near-field imaging technique mentioned in Section 4.3 can be applied to inspect aluminum substrates. In [83] a system based on this principle and using a circular open-ended waveguide has been used (with a scanning steps of 1 mm and a standoff distance of the same length) to detect flat-bottom holes (with depths ranging from 0.5 mm to 2 mm and a diameter of 2 mm) and 2-mm-wide notches with different lengths and depths in a metallic substrate. It has been shown in [83] that, when radiating in free space, the circular waveguide probe is characterized by a return loss of 14 dB (at 24 GHz), whereas a rectangular probe has a return loss of 12 dB. Moreover, the same system has been successfully applied for corrosion detection under paint. Figure 4.14 shows an example of the reconstructed holes in the metallic structure, whereas Figure 4.15 provides a demonstration of the capabilities of the approach in retrieving the regions affected by the corrosion on a metal structure covered by a painting layer. Corrosion problems in coated mild steels have also been inspected by a microwave imaging system based on a K-band rectangular waveguide and a vector network analyzer (VNA) in [84].

To detect flaws in coated metallic structures, the use of electrically-small microwave ring resonators has also been proposed [85], with good reconstruction results concerning (for example) the corroded regions in aluminum plates coated with Teflon sheets. Reflection-based high-resolution imaging of planar metal-dielectric structures has also been performed by using loaded subwavelength slot aperture antennas and small helix antennas [86].

Other linear techniques can be applied to metallic targets. As an example, the cross sections of several elongated metallic targets provided by the Frésnel database [39] have been reconstructed by using an approach based on the so-called multiple measurement vector (MMV) model by exploiting both TM and TE incident polarizations [87]. Figure 4.16 shows one of the obtained results, that is, the reconstruction of the cross section of a U-shaped metallic cylinder (the *uTM_shaped* target of the database [39]) at a frequency of 8 GHz. The indicator function (already discussed in this chapter) is plotted and comparisons with the reconstructed distributions obtained by using the LSM (in different versions) [88–90] are provided.

4.6 Imaging of Wood Materials

Wood materials are widely used in industry, such as for furniture production and for the preparation of materials used in building construction. The quality of the material can be assessed by using microwaves. In particular, an important problem is the detection of inclusions into trees (e.g., pieces of metal or stones),

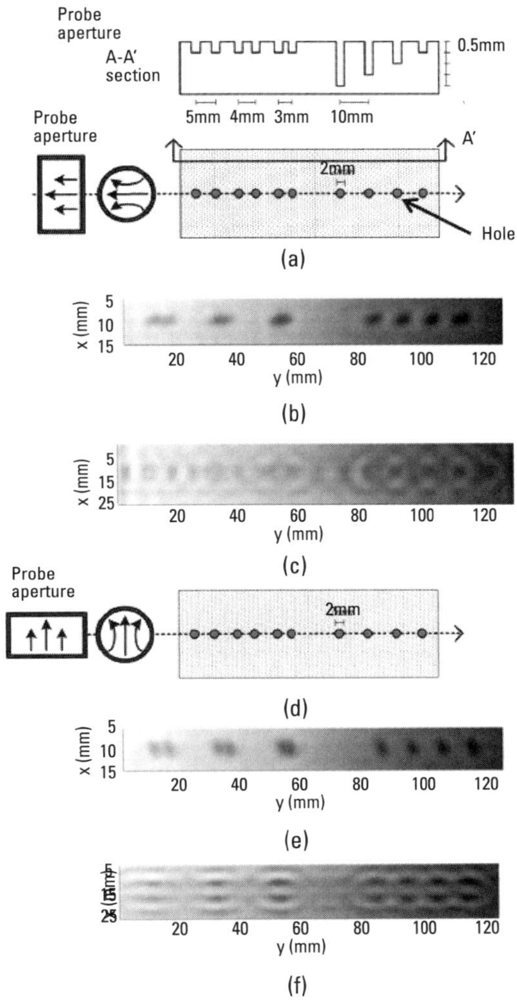

Figure 4.14 (a) Schematic of the flat-bottom hole panel with cross section and Y-scan setup. (b) Y-scan circular probe image. (c) Y-scan rectangular probe image. (d) X-scan setup. (c) X-scan circular probe imaging. (f) X-scan rectangular probe image [83]. (©2017 IEEE.)

which can damage the cutting machinery and consequently produce significant economic losses. In addition to the possible problems related to the wood processing industry, recently there has also been a significant concern about the evaluation of the healthy state of living trees. In fact, the presence of extended rotten parts in the wood or void regions can lead to the fall of the tree and thus cause injuries to people [91].

An example of a tomographic setup for the inspection of wood samples is discussed in this section. It is worth noting that the considered architecture

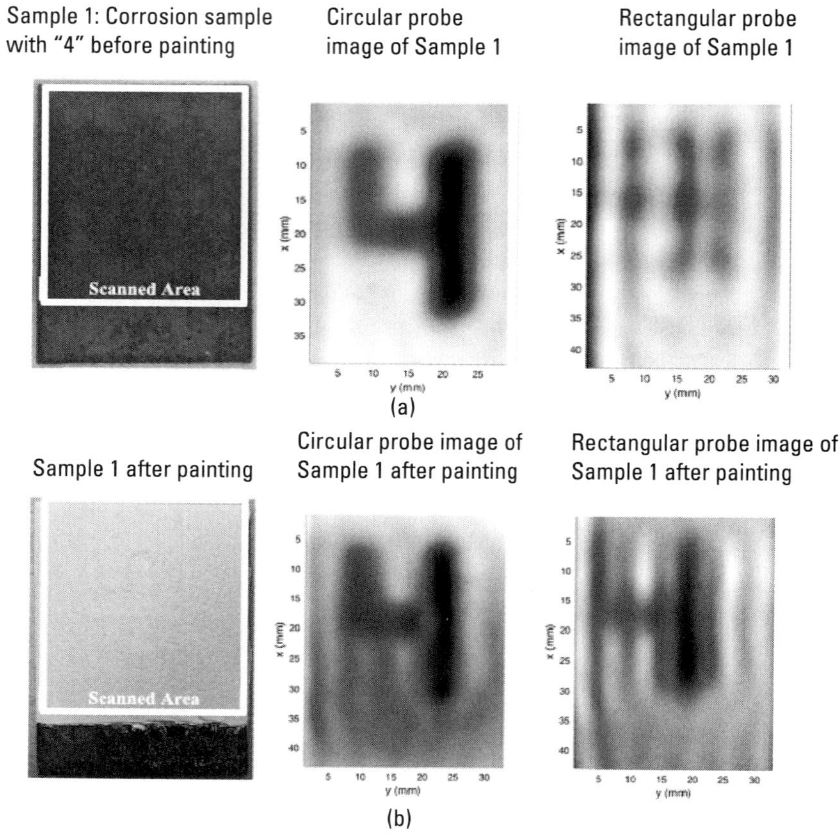

Figure 4.15 (a) Corrosion sample before painting (left) and the corresponding microwave images obtained using the circular probe (middle) and the rectangular probe (right). (b) The same sample after painting (left) and the corresponding microwave images obtained using the circular probe (middle) and the rectangular probe (right) [83]. (©2017 IEEE.)

is also representative of the imaging setups adopted in other applicative fields where tomographic arrangements can be used. The system has been developed by the University of Applied Sciences and Arts of Southern Switzerland, Manno, Switzerland, in cooperation with the University of Genoa, Genoa, Italy [92–94]. Figure 4.17(a) reports a schematic representation of the imaging setup. The system is composed of two antennas that can collect measurements of the field all around the sample under test, which is positioned on a rotating table. The receiving antenna can rotate around the target, whereas the transmitting one is kept at a fixed position. In order to collect multiview multi-illumination data, the table can rotate, too, as schematically shown in Figure 4.17(b). The angular rotation range is 360° and the minimum angular increment is 1°. It is worth noting that, for mechanical reasons, some of the parts of the setup (especially those related to the antenna movements) are made of metallic materials,

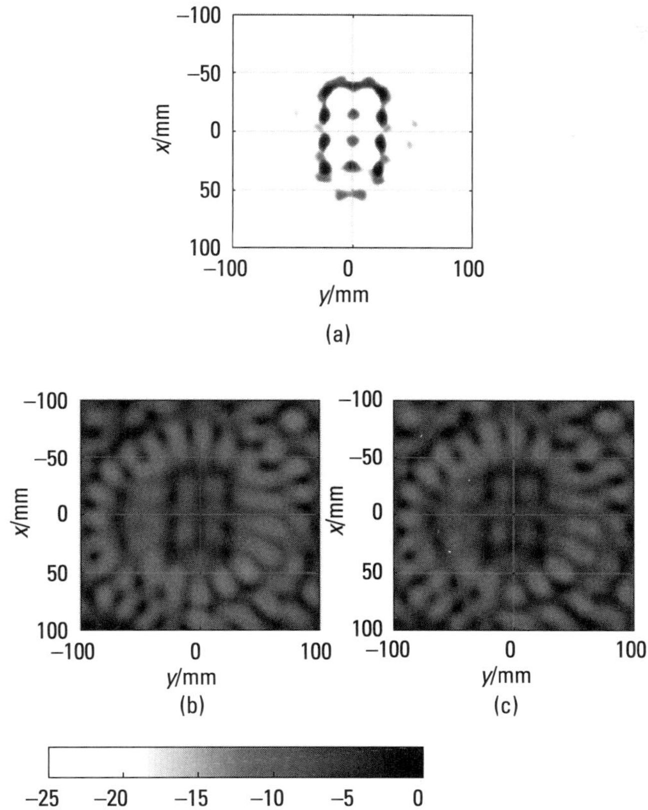

Figure 4.16 Reconstruction of a U-shaped metallic cylinder at 8 GHz [87] (*uTML_shaped* to get from the Frésnel database [39]). Indicator funtion in decibles obtained with (a) MMV, (b) LSM, and (c) improved LSM. A total of 12 transmitter positions and 49 receiver positions for each transmitter are selected for imaging. (©2018 IEEE.)

which can perturb the measurements. Consequently, the rotating platform is covered with absorbing panels with a thickness of 11.4 cm and reflectivity less than –17 dB.

The antennas are connected to the ports of a VNA, which is used for the generation of the signal in input to the transmitting antenna (port 1) and for the measurements of the signal received by the receiving antenna (port 2). The system is designed in order to host different type of antennas, depending on the applicative needs. In particular, the antennas can be located on two vertical arms at different heights ranging from 0 mm to 750 mm from the rotating platform table (with a minimum vertical displacement of 1 mm). The transmitting antenna arm is fixed, whereas the receiving antenna one can rotate around the table (see Figure 4.17[b]) with a minimum angular increment equal to 1°. It is worth noting that for mechanical reasons and in order to avoid excessive direct

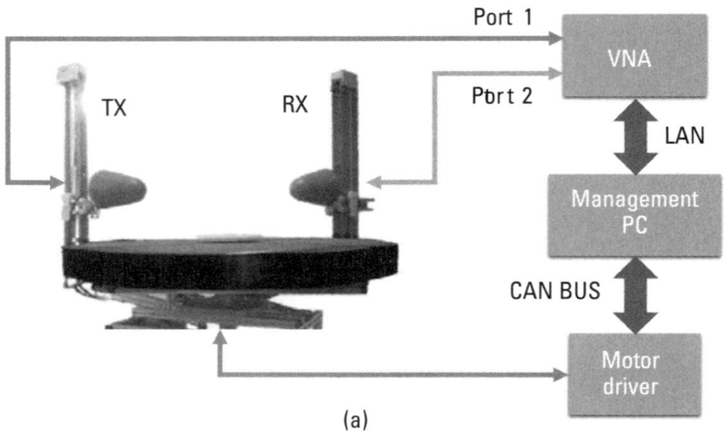

(a)

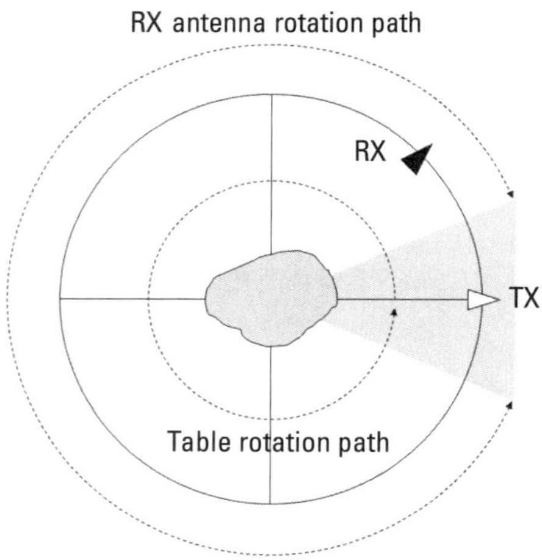

(b)

Figure 4.17 Laboratory tomograph prototype for wood sample inspection [92]. (a) Block scheme of the system and (b) schematic representation of the antenna and table movements.

coupling between the two antenna elements, a blind angular sector of 90°, where the receiving antenna cannot be positioned, is present (shown in grey in Figure 4.17[b]). Moreover, in order to reduce the perturbation of the field, the supporting arms are made of glass resin.

In the specific application considered in this section, linearly-polarized log-periodic antennas are used. The working frequency band is 850 MHz–26.5 GHz, with a voltage standing wave ratio (VSWR) ≤ 2.5. The typical gain is 8.5 dBi. Multifrequency data are collected by using a frequency-stepping strategy.

As shown in Figure 4.17(a), the VNA is controlled by an external personal computer (PC). Such management PC also controls all the movements by means of a set of three motorized actuators connected through a Controller Area Network (CAN) bus. Moreover, it also synchronizes the measurements with the movements of the receiving antenna and of the rotating table hosting the target.

The main drawback of such setup is represented by the acquisition time. In fact, for every view and receiving position, before acquiring the data with the VNA, it is necessary to wait for the antennas to move to the required location and stabilize. Consequently, increasing the number of measurement points produces a rapid increase in the measurement time (for example, 30 seconds are needed with three points per view, whereas ten minutes are necessary when considering 91 antenna positions) [92, 93].

An example of dielectric reconstruction obtained by using real data collected by the previously described system prototype is shown in this section. The considered target is a composition of two adjacent slabs with rectangular cross sections made of different types of wood, as shown in Figure 4.18. The slabs have sides with lengths of 11.5 cm and 7.5 cm. One of the two objects also has a rectangular void of side 5.5 cm × 3.5 cm located in its center. The height of the structure is about 50 cm and measurements are performed at about one-half of the height. The system has been configured in order to acquire data between 1 and 6 GHz. The distance between the antennas and the center of the table hosting the sample under test is such that the corresponding distance of the phase center varies from 49.7 cm to 63 cm. The transmitting antenna has been moved into $S = 16$ angular positions equally spaced on the measurement circumference. For every view, $M = 91$ measurement locations have been considered, uniformly distributed on a 270° arc of circumference (the angular distance between the first receiving position and the transmitting antenna is equal to 45°). An example of the amplitude and phase of the z-component of the scattered electric field vector is shown in Figure 4.19, for the first view and for some of the considered frequencies. It is worth noting that such values have been obtained by performing a preliminary measurement campaign without the target in order to acquire the incident field, which is subsequently subtracted from the measurements of the total electric field. Such incident field is also used for properly scaling the simulated internal incident field used in the inversion strategy. In particular, the internal incident electric field is modeled as the one produced by a line-current source located in the phase center of the transmitting antenna.

The inversion of the field data has been performed by using the L^p Banach-space Newton scheme described in Section 3.3. A frequency-hopping strategy is employed to exploit the multifrequency information. For every frequency, the maximum numbers of iterations of the Newton and Landweber loops have been set equal to $N_{IN} = 30$ and $N_{LW} = 10$, respectively, and the iterations are

(a)

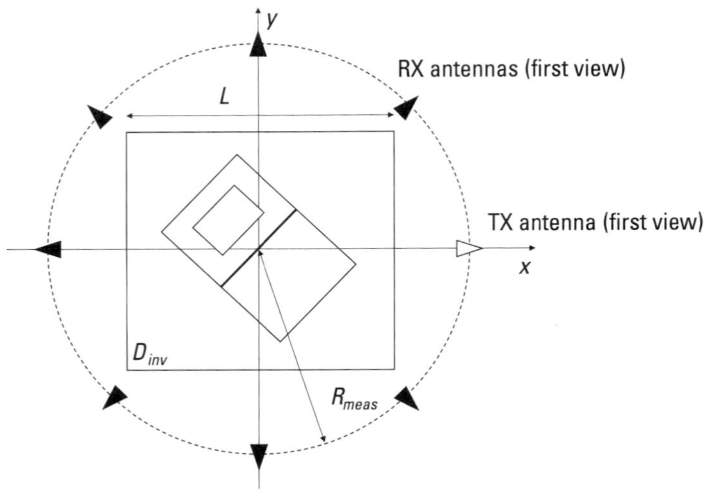

(b)

Figure 4.18 Composition of wood slabs. (a) Schematic representation of the imaging configuration and (b) picture of the actual target.

stopped when the normalized variations of the residual between two consecutive iterations falls below the threshold of 1%. In the inversion, $F = 5$ frequencies (uniformly spaced in the whole frequency range) have been considered. The square investigation area has side 30 cm and it has been discretized into $N = 3969$ square subdomains of side 4.76 mm.

Figure 4.20 shows the distribution of the relative dielectric permittivity reconstructed by using a value of the norm parameter equal to 1.4. As can be seen, it is possible to clearly identify the two slabs. Moreover, the void hole inside the left-up one is visible, too.

In the previous example, an iterative method belonging to the category of quantitative methods has been applied. Such approaches are usually quite heavy

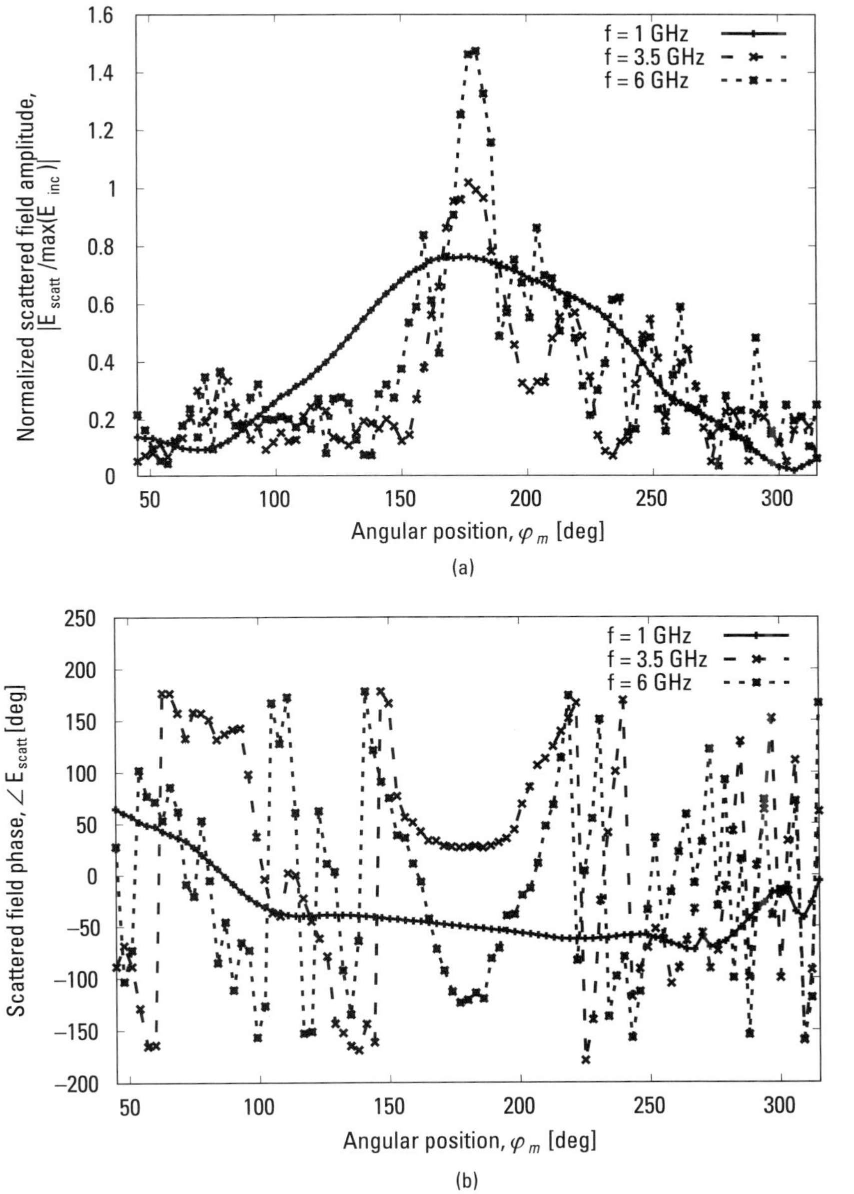

Figure 4.19 Behavior of the (a) amplitude and (b) phase of the measured scattered field due to the composition of wood targets in Figure 4.18.

from a computational point of view. Alternatively, qualitative methods such as those based on radar-based algorithms, can also be adopted for inspecting wood slabs. An example is the imaging procedure used in [81] and previously

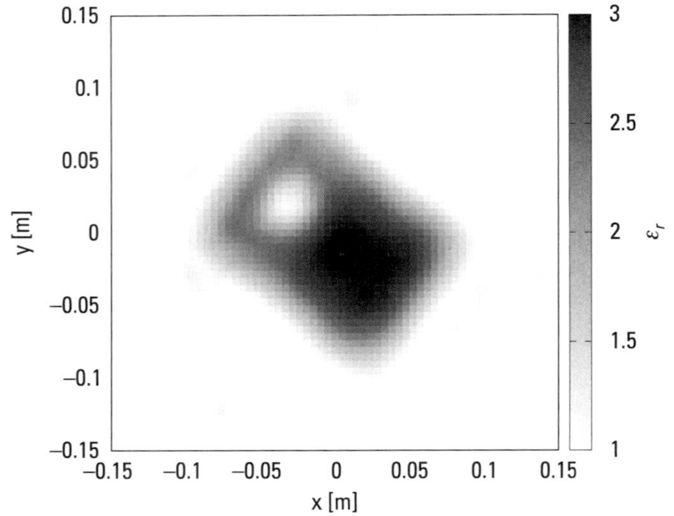

Figure 4.20 Reconstructed distribution of relative dielectric permittivity of the composition of wooden structures shown in Figure 4.18 obtained by using the L^p Banach-space inversion strategy (with $p = p_{opt} = 1.4$).

mentioned for evaluating plastic cylinders. Figure 4.21 shows the bistatic arrangement and the final image of the reconstructed cross section of a square wood object obtained by using this approach.

The LSM [66] can be adopted for inspecting wooden cylinders, too. In [95], for example, this technique has been used in conjunction with an experimental apparatus, in which one of the key aspects is represented by the adoption of a specifically designed corrugated Vivaldi antenna (shown in Figure 4.22), which exhibits a gain ranging from 5.6 to 10.4 dBi in the band 1.96-8.61 GHz. A comparison with other antennas proposed for microwave imaging is also reported in [95] (see Table 4.4).

As previously recalled, in recent years there has been an increasing interest in the evaluation of the state of health of in-vivo trees. Standard systems for performing this task are based on acoustic, resistivity, and thermographic imaging [63, 111]. However, most of them have significant disadvantages. For example, acoustic techniques, which are the most commonly adopted techniques in the field, usually need to insert needles inside the trunk and require high sensitivity and dynamic range in the measurement apparatuses in order to detect the presence of the internal discontinuities. In this framework, microwave imaging is emerging as a possible alternative imaging technique. However, in such applications, the use of the imaging setups previously discussed is not always possible due to the practical limitations related to the transportability and installation of the system and to the acquisition time. To overcome such problems, ground penetrating radar apparatuses can be adopted for collecting the

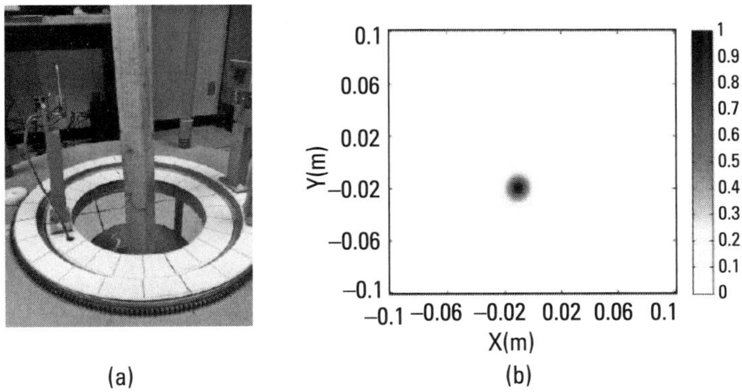

Figure 4.21 Reconstruction of a wood slab. (a) Measurement setup and (b) image obtained by using the single-frequency PCM algorithm [81]. (©2017 IEEE.)

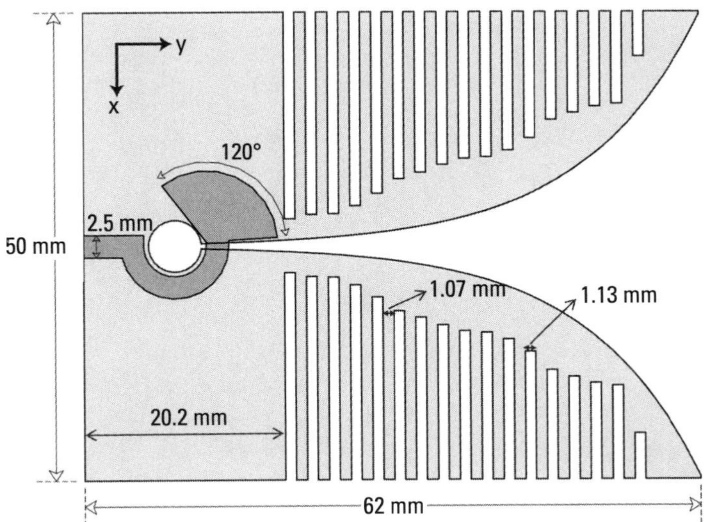

Figure 4.22 Example of antenna used for wood slab imaging in [95]. (©2017 IEEE.)

measurements (see Chapter 6) by using a single antenna or a pair of antennas for scanning the external surface of the trunk. For example, such an approach has been considered in [112–115]. However, it is worth noting that in this case, only monostatic (or quasi-monostatic) measurements can be obtained by using standard apparatuses, thus limiting the quantity of information that is available in the acquired data. Moreover, the imaging procedure is usually limited to the use of synthetic focusing and migration algorithms (see the discussion about such approaches in Chapter 6), which just allow the detection of dielectric

Table 4.4
Comparison of the Performance of Several Antennas Used for Imaging Purposes [95] (©2017 IEEE.)

Reference	Size (W-L-H) [mm^3]	Subs. Rel. Perm. (ε_r)	Bandwidth (min [GHz]–max [GHz]-)	Gain (min [dBi]–max [dBi])
[96]	$40 \times 45 \times 0.8$	4.4	2.9 – 12 – 4.14 : 1	5.2 – 8.2
[97]	$50 \times 78.9 \times 1$	4.5	4 – 16 – 4 : 1	5 – 8.5
[98]	$266 \times 360 \times 0.75$	3	1.5 – 10 – 6.67 : 1	11 – 14.2
[99]	$260 \times 360 \times 1.6$	4.5	1 – 3.4 – 3.4 : 1	7 – 11
[100]	$36.3 \times 59.81 \times 0.64$	6.15	4.73 – 11 – 2.33 : 1	4.1 – 9
[101]	$95 \times 25.4 \times 25.4$	25	1.8 – 12 – 6.67 : 1	NA – NA
[102]	$14 \times 13 \times 1.25$	10.2	4.5 – 10 – 2.22 : 1	NA – 10.2
[103]	$50 \times 50 \times 50$	1	0.7 – 0.97 – 1.39 : 1	3.5 – 5
[104]	$260 \times 180 \times 1.5$	10	0.5 – 2 – 4 : 1	1.0 – 7.0
[105]	$60 \times 43 \times 35$	2.3	3.1 – 40 – 12.90 : 1	4.45 – 9.65
[106]	$40 \times 90 \times 0.508$	3.38	3.4 – 40 – 11.76 : 1	8 – 15
[107]	$15 \times 20 \times 9$	2.33 – 10.2	30.3 – 33.6 – 1.11 : 1	8 – 12
[108]	$150 \times 50 \times 0.5$	2.65	3.1 – 10.6 – 3.42 – 1	5.0 – 16.0
[109]	$150 \times 100 \times 90$	4.4	1.7 – 2.3 – 1.35 : 1	NA – 10
[110]	$89.2 \times 49.2 \times 78.2$	2.33	1.4 – 11 – 7.86 : 1	5 – 10
[95]	$50 \times 62 \times 1.52$	3.5	1.96 – 8.61 – 4.40 : 1	5.6 – 10.4

discontinuities, but do not provide quantitative information about the type of defects. Approaches based on monostatic polarimetric imaging have also been proposed (for an example see [116]), in order to exploit the depolarization introduced by the dielectric inhomogeneities inside the structure.

In order to increase the available information, multistatic setups can be adopted. An example of one such imaging configuration is shown in Figure 4.23. In this case, the antennas are in direct contact with the sample to be inspected, and consequently need to be properly designed in order to maximize the matching with the target. Figure 4.24 reports an example of block scheme for the measurement setup needed to exploit the considered multistatic configuration. As can be seen, with respect to the case shown in Figure 4.17, an additional switch is present in place of the stepped motor, thus allowing the electronic selection of the pair of antennas used to transmit and collect the fields. Clearly, both the switch and the VNA must be synchronized in order to acquire the data.

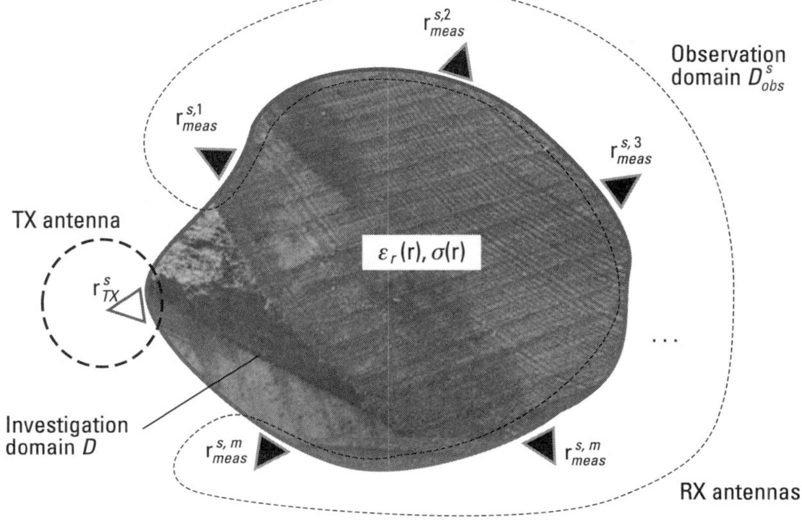

Figure 4.23 Schematic representation of a multistatic wood trunk imaging setup.

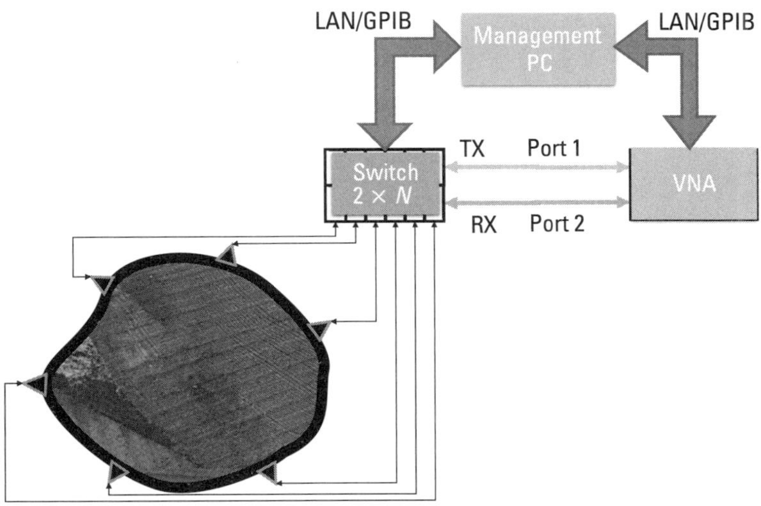

Figure 4.24 Example of block scheme of a multistatic imaging system.

An example of an inversion procedure specifically designed for the present application is discussed below. In this case, it is assumed to have S antennas located at positions $\mathbf{r}_{ant}^{s}$. Each one is used sequentially in transmit mode in order to illuminate the target and the resulting total electric field is measured by the remaining antennas (the set of measurement points for the sth view is thus $\mathbf{r}_{meas}^{s,m} = \mathbf{r}_{ant}^{(s+m)\text{mod}S}$, $m = 1, \ldots, S - 1$). The inversion approach is based on the hybrid scheme in [117], in which a qualitative algorithm is combined with a quantitative procedure as schematically shown in Figure 4.25. The qualitative algorithm has two main goals. The first goal is the estimation of the field contributions due to the inclusions inside the host structure, (i.e., the wood trunk, in the considered example) and the second one is to provide a first qualitative reconstruction of the scenario under test.

The first task is accomplished by means of a filtering procedure in time domain [91], in which the time-domain electric field $e_{tot}^{s}\left(\mathbf{r}_{meas}^{s,m},t\right)$ in the mth measurement point $\mathbf{r}_{meas}^{m,s}$ of the sth view is obtained by applying an inverse Fourier transform to the measured frequency-domain data. The scattered electric field is then approximated as

$$e_{scatt}^{s}\left(\mathbf{r}_{meas}^{s,m},t\right) \approx e_{tot}^{s}\left(\mathbf{r}_{meas}^{s,m},t\right) - \mathbf{w}_{s,m}^{T}\mathbf{q}_{s,m}(t) \tag{4.11}$$

where

$$\mathbf{q}_{s,m}(t) = \begin{bmatrix} \mathbf{w}_{1,m}^{T}\mathbf{e}_{tot}^{1,m}(t) \\ \vdots \\ \mathbf{w}_{s-1,m}^{T}\mathbf{e}_{tot}^{s-1,m}(t) \\ \mathbf{w}_{s+1,m}^{T}\mathbf{e}_{tot}^{s+1,m}(t) \\ \vdots \\ \mathbf{w}_{S,m}^{T}\mathbf{e}_{tot}^{S,m}(t) \end{bmatrix}, \mathbf{e}_{tot}^{s,m}(t) = \begin{bmatrix} e_{tot}^{1}\left(\mathbf{r}_{meas}^{1,m},t\right) \\ \vdots \\ e_{tot}^{s-1}\left(\mathbf{r}_{meas}^{s-1,m},t\right) \\ e_{tot}^{s+1}\left(\mathbf{r}_{meas}^{s+1,m},t\right) \\ \vdots \\ e_{tot}^{S}\left(\mathbf{r}_{meas}^{S,m},t\right) \end{bmatrix} \tag{4.12}$$

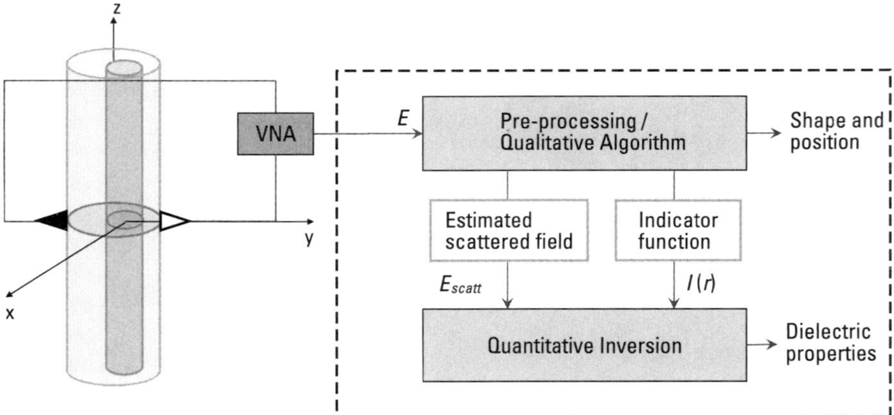

Figure 4.25 Block scheme of the hybrid imaging scheme [117] used for producing the tomographic images in Figure 4.26.

The weights $\mathbf{w}_{s,m}$ are found by minimizing the following functional

$$\Phi\left(\mathbf{w}_{s,m}\right)=\int_{T_l}\left|\mathbf{w}_{s,m}^{T}\mathbf{e}_{tot}^{s,m}\left(t\right)-e_{tot}^{s}\left(\mathbf{r}_{meas}^{s,m},t\right)\right|^{2}dt \tag{4.13}$$

where T_l is the time window in which the main contributions to the scattering are related to the external layers of the trunk. The approximated time-domain scattered fields are then transformed back to frequency domain by using a Fourier transform, and they are used in the second step of the inversion procedure.

The estimated time-domain data are also used to produce a qualitative image $I(\mathbf{r}_t)$ of the inclusions by means of a delay-and-sum (DAS) beamforming approach (see Chapter 5). A quantitative reconstruction is then obtained by applying an inexact-Newton inversion scheme with Landweber inner solver, similar to those described in Chapter 3 (in this case, Hilbert spaces are assumed, that is, the value of p has been set equal to 2 [118]). The inversion procedure has been modified in order to take into account the information obtained in the first step. In particular, the normalized qualitative image $I_n\left(\mathbf{r}_t\right)=I\left(\mathbf{r}_t\right)/\max_{\mathbf{r}_t\in D} I\left(\mathbf{r}_t\right)$ produced by the DAS method is used to weight the iterative updates of the current estimate of the distributions of the dielectric properties, that is, the update step in (3.61) is modified as

$$c_{n+1}\left(\mathbf{r}_t\right)=c_n\left(\mathbf{r}_t\right)+I_n\left(\mathbf{r}_t\right)\overline{h}\left(\mathbf{r}_t\right) \tag{4.14}$$

Such choice allows for focusing the reconstruction mainly on the parts of the trunk that have been found to be possibly affected by defective regions, as identified by the first DAS qualitative procedure [91].

An example of the results that can be obtained by using the previously sketched procedure is reported in Figure 4.26. The scattered field data have been obtained by means of a numerical simulator based on the method of moments [119] with pulse basis functions and Dirac's delta weighting functions. The trunk has been modeled by using a cylindrical structure composed by three layers (i.e., bark [$\varepsilon_r = 7.5$, $\sigma = 0.15$ S/m], sapwood [$\varepsilon_r = 10$, $\sigma = 0.1$ S/m], and heartwood [$\varepsilon_r = 6$, $\sigma = 0.05$ S/m]). The external shape has been randomly generated by deforming a circular structure with a diameter of 0.5m. The distribution of the relative dielectric permittivity of the developed model is shown in Figure 4.26(a). The investigation domain considered in both the qualitative and quantitative procedures corresponds to the area inside the trunk cross section, and it has been discretized into $N_{inv} = 633$ square subdomains of side 0.018m. A total of $S = 30$ antennas (modeled as ideal line-current sources) with equal radial spacing are used. A frequency-stepped measurement setup is simulated by computing the electric field for $F = 64$ frequencies equally spaced between 400

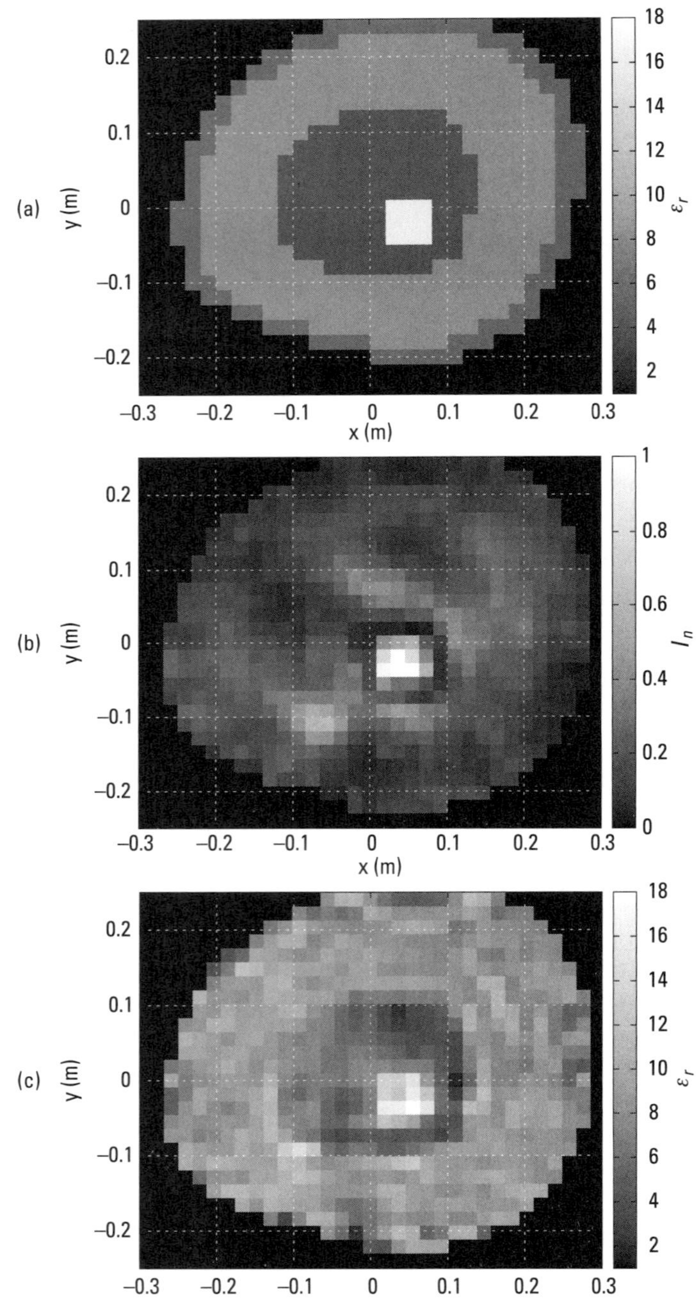

Figure 4.26 Example of reconstruction of a numerical model of a defective wood trunk. (a) Relative dielectric permittivity of the actual model, (b) qualitative indicator map, and (c) reconstructed distribution of the relative dielectric permittivity.

MHz and 1.3 GHz Moreover, the obtained data have been corrupted with an additive white Gaussian noise with zero mean value and $SNR = 15$ dB.

In the considered configuration, a small inclusion is placed in the inner layer. The inclusion has a diameter of approximatively $d_c = 0.06$m and it is characterized by relative dielectric permittivity $\varepsilon_{r,c} = 17$ and electric conductivity $\sigma_c = 0.2$ S/m (rotten wood material). In the quantitative inversion procedure the maximum number of iterations has been set equal to $N_{IN} = N_{LW} = 10$ and the iterations are stopped when the normalized variations of the residuals fall below the fixed threshold of 1%. The qualitative reconstruction result is reported in Figure 4.26(b). As can be seen, the qualitative step is able to produce a spot in the reconstructed image in correspondence to the region in which a significant dielectric contrast with respect to the layered trunk structure is present. However, such information is not sufficient for understanding the type of defect that is present inside the trunk (e.g., a void or a different wood condition). To this end, the subsequent quantitative inversion procedure is applied. The obtained result is shown in Figure 4.26(c), which reports the reconstructed distribution of the relative dielectric permittivity. As can be seen, the approach is able to correctly retrieve the value of the dielectric properties inside the identified region.

4.7 Microwave Imaging in Chemical, Pharmaceutical, Food Industry, and Other Applications

Microwave imaging techniques can also be applied to industrial processes in a wide spectrum of applications connected to chemical, pharmaceutical, agricultural, and food production areas [120]. In such scenarios, imaging methods (e.g., microwave tomography) can be effective for monitoring changing processes, such as multiphase flows, in a noninvasive manner, without introducing apparatuses inside vessels and pipes [121–123]. From microwave measurements, it is also possible to deduce information about the velocity of fluids inside pipes [124] or to perform imaging of granule flows in pipelines [125]. In monitoring industrial processes, it is often important to have a good time-resolution and, possibly, real-time imaging.

As an example, a mixture of dry corn granules (with sizes of 500-2,000 microns) mixed with water flowing in a plastic pipe have been investigated experimentally by using a tomographic system equipped with 16 monopole antennas in a circular arrangement [125]. The working frequency is 1 GHz and the output power in equal to 10 dBm. The inverse scattering solution is obtained by using an iterative Newton-Kantorovich method [126, 127] (in a modified version [125]).

Monitoring the quality of grain stored in silos is another field of great interest in agriculture. This task can be addressed by using microwave imaging techniques [128]. In most cases, in this kind of applications, the target to be inspected can be included in a metallic chamber, which represents a part of the apparatuses for storage or processing the material. This aspect has been studied in [129, 130], in which the use of a metallic enclosure has been proposed also for biomedical applications, where the chamber walls are traditionally constructed by using plastic materials (see also Chapter 5).

For example, in [129], a prototype of air-based metallic resonant chamber has been described (shown in Figure 4.27). Due to the conductive properties of the cylindrical chamber, only the normal components of the electric field have to be measured. This is obtained by using monopole antennas. In particular, in [129], 24 reconfigurable monopole antennas constructed on a double layer PCBs and incorporating five PIN diodes have been used. The idea of using a metallic chamber, instead of a dielectric one, relies on the potentiality of having a better modeling of the boundary conditions, on the possibility of using lossless or low-loss matching media in order to improve the energy delivered on the target under test and the *SNR* of the system, and on the improvements in the shielding properties of the chamber [129].

A metallic enclosure is also an essential part of the microwave measurement system shown in Figure 4.28 [131]. The measurement region is constituted by a circular cylinder of aluminum that is 32 cm in radius. Six rectangular

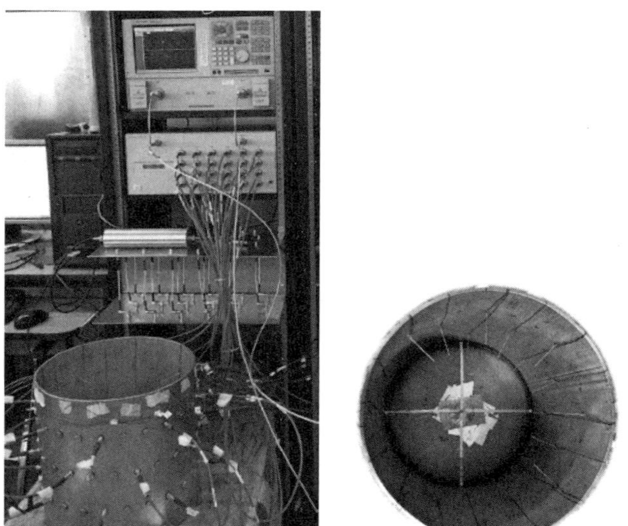

(a) (b)

Figure 4.27 Picture of the imaging prototype presented in [129]. (a) Overall system and (b) metallic chamber. (©2017 IEEE.)

Figure 4.28 Left: microwave measurement system together with the switch and the network analyzer. Right: top-down view of the measurement region of the disassembled microwave system with three acrylic-glass cylinders [131]. (©2018 IEEE.)

waveguides (WR-229) are placed around this region and the investigation domain is a circle of about 11.6 cm. A VNA and a switch allow the automated measurements of the 6×6 scattering matrix in the frequency band from 2.7 to 5.1 GHz. The above system has been used in [131] for the detection and positioning of dielectric objects (acrylic-glass cylinders with circular cross sections with radii of 5.2 mm) inside the metal enclosure. The reconstruction is obtained by using a compressive sensing approach.

Recently, a significant amount of attention concerning the potentialities of microwave imaging techniques has been devoted to assessing the quality of fruits, whose dielectric properties can be correlated to some specific properties (e.g., the moisture content, the maturation, the soluble solids content, and so forth). Moreover, blemishes and defects can also been detected, with great potential improvement for the food industry efficiency [132]. Microwave imaging techniques can be also used to monitor food contamination, that is by localizing foreign objects in food. As an example, the microwave modeling and nondestructive control of apples has been proposed in [56], in which the target has been modeled as an ellipsoid, allowing a sensitivity analysis. Another example concerns the evaluation of the granulation of fruits, which is a problem that can degrade the quality of the fruits. In [133], the granulation in pomelos has been assessed by using a multiview approach based on reflection. The reported results show that changes in the scattering coefficients of the order of 0.033 (amplitude) and 3.9° (phase) can be detected, allowing a possible identification of the granulation phenomenon. Another example, which has been reported in [134], is represented by the determination of the sugar content in fruits by using a chirp pulse microwave tomography technique (with working frequencies in the range 2-3 GHz).

Another challenging problem is represented by the joint determination of the dielectric and velocity distributions of multilayer cylinders moving in the axial direction. In fact, techniques addressing such a problem have potential applications in astrophysics (e.g., for analyzing meteor trails [135]), nuclear and plasma physics (e.g., for tracking of axially moving plasma columns [136–138]), and engineering (e.g., for monitoring jet exhausts [139] and mass flows in pneumatic pipes [140]). Several approaches can be followed for performing this task. As an example, an inverse scattering procedure has been proposed in the already mentioned reference [124] for the case of a multilayer elliptic cylinder, which represent quite an interesting canonical target [141, 142], since it has a sufficiently complex cross section (especially in the multilayer case), which can simulate several relevant structures. Moreover, the circular cylinder represents a particular case in which the two focuses of the ellipses coincide, whereas a strip is a particular case in which the eccentricity tends to zero [143, 144]. The inversion procedure is based on the observation that, when dealing with an object characterized by real valued dielectric permittivity and magnetic permeability (see Chapter 2) in its rest frame, and moving with a velocity v_z in the axial direction (assumed to coincide with the z axis), the constitutive relations can be written as [145, 146]

$$\mathbf{D}_t = \alpha\varepsilon\mathbf{E}_t + \frac{\delta}{c}\hat{\mathbf{z}} \times \mathbf{H}_t \qquad D_z = \varepsilon E_z \tag{4.15}$$

$$\mathbf{B}_t = \alpha\mu\mathbf{H}_t - \frac{\delta}{c}\hat{\mathbf{z}} \times \mathbf{E}_t \qquad B_z = \mu H_z \tag{4.16}$$

where the subscripts t and z denotes the transverse (with respect to the axial direction z) and axial parts of the field vectors, whereas α and δ are given by

$$\alpha = \frac{1-\beta^2}{1-\varepsilon_r\mu_r\beta^2} \qquad \delta = \frac{\beta(\varepsilon_r\mu_r - 1)}{1-\varepsilon_r\mu_r\beta^2} \tag{4.17}$$

being β the normalized axial velocity, that is, $\beta = v_z / v_0$. The moving cylinder can be then modeled as a stationary bianisotropic cylinder, allowing the use of an efficient semi-analytical formula for the computation of the scattered fields [147]. Moreover, in [124], it has been found that for small velocities, the axial movement has negligible effects on the z-component of the electric field, while producing significant changes in the magnetic field.

On the basis of such observations, in order to solve this inverse scattering problem, a two-step inversion procedure has been adopted. In the first step, the permittivity distribution is reconstructed by neglecting all axial movements and by considering the *z*-component of the scattered electric field. In the second part, only the velocity profiles are calculated by using the measured magnetic field. An example of the results obtained in [124] is reported in Figure 4.29. The considered target is a three-layer elliptic cylinder, whose layers are bounded by confocal ellipses of semi-major axes $a_1 = 0.15\lambda_0$, $a_2 = 0.25\lambda_0$, and $a_3 = 0.45\lambda_0$. The half-focal distance is equal to $d = 0.15\lambda_0$. The dielectric permittivities of the three layers are equal to $\varepsilon_{r,1} = 4$, $\varepsilon_{r,2} = 2$, and $\varepsilon_{r,3} = 3$, whereas

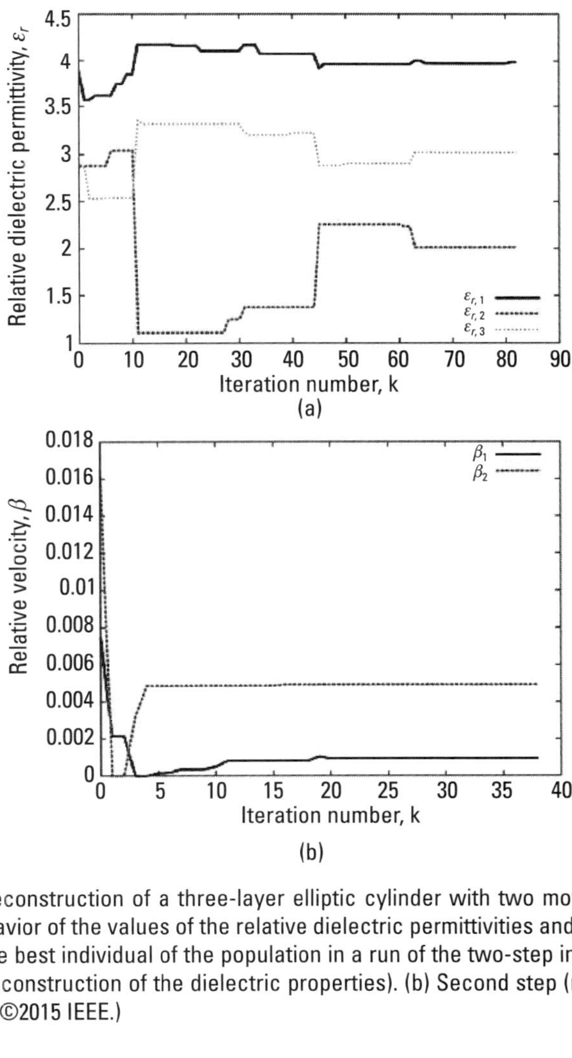

Figure 4.29 Reconstruction of a three-layer elliptic cylinder with two moving layers [124]. Example of behavior of the values of the relative dielectric permittivities and normalized axial velocities for the best individual of the population in a run of the two-step inversion strategy. (a) First step (reconstruction of the dielectric properties). (b) Second step (reconstruction of the velocities). (©2015 IEEE.)

the normalized axial velocities are given by $\beta_1 = 0.001$, $\beta_2 = 0.005$, and $\beta_3 = 0$ (the outer layer is assumed to be at rest in the inversion procedure). A multiview measurement setup is considered. In particular, $S = 4$ transmitting antennas (modeled as ideal line-current sources) are assumed to be uniformly distributed on a circumference of radius $R_{source} = 0.8\lambda_0$ and the z-components of the electric and magnetic fields are collected in $M = 36$ points $\mathbf{r}_{meas}^m$ uniformly distributed on a circumference of radius $R_{meas} = 0.6\lambda_0$. The scattered field data $E_{scatt,z}^{s,meas}\left(\mathbf{r}_{meas}^m\right)$ and $H_{scatt,z}^{s,meas}\left(\mathbf{r}_{meas}^m\right)$ have been numerically computed by using the semi-analytical solution developed in [147] and they have been corrupted with a Gaussian noise having $SNR = 20$ dB. The inverse scattering problem (in each step) is solved by using an algorithm based on the Artificial Bee Colony (ABC) stochastic minimization method [148]. In particular, the following cost functions are minimized in the first and second steps, respectively

$$f_1\left(\mathbf{x}_1\right) = \sum_{s=1}^{S}\sum_{m=1}^{M}\left|E_{scatt,z}^{s,rest,calc}\left(\mathbf{r}_{meas}^m, \mathbf{x}_1\right) - E_{scatt,z}^{s,meas}\left(\mathbf{r}_{meas}^m\right)\right|^2 \tag{4.18}$$

$$f_2\left(\mathbf{x}_2\right) = \sum_{s=1}^{S}\sum_{m=1}^{M}\left|H_{scatt,z}^{s,calc}\left(\mathbf{r}_{meas}^m, \mathbf{x}_1^{opt}, \mathbf{x}_2\right) - H_{scatt,z}^{s,meas}\left(\mathbf{r}_{meas}^m\right)\right|^2 \tag{4.19}$$

where $\mathbf{x}_1 = [\varepsilon_{r,1}, \varepsilon_{r,2}, \varepsilon_{r,3}]^t$ and $\mathbf{x}_2 = [\beta_1, \beta_2]^t$ are the arrays containing the values of the unknown relative dielectric permittivites and normalized axial velocities (the superscript *opt* in [4.19] indicates the optimal solution found in the first step of the procedure). Moreover, $E_{scatt,z}^{s,rest,calc}\left(\mathbf{r}_{meas}^m, \mathbf{x}_1\right)$ is the z-component of the scattered electric field vector due to an elliptic cylinder, which has been computed by assuming that all the layers are at rest, whereas $H_{scatt,z}^{s,calc}$ is the z-components of the magnetic field vector obtained by using the numerical procedure in [147].

In the considered example, the two-step procedure was able to correctly retrieve both the dielectric properties and the velocities of all the layers. In particular, the relative errors (averaged on ten runs) on the relative dielectric permittivities of the three layers found in the first step are 0.18×10^{-2} (for $\varepsilon_{r,1}$), 2.9×10^{-2} (for $\varepsilon_{r,2}$), and 10^{-2} (for $\varepsilon_{r,3}$), whereas the relative errors in the reconstruction of the normalized axial velocities obtained in the second step are 2.7×10^{-2} (for β_1) and 5.7×10^{-2} (for β_2). The mean number of evaluations of the cost functions are 922.3 (step 1) and 583.0 (step 2). An example of the behavior of the reconstructed values (for the best elements in the population of the ABC method) versus the iteration number is shown in Figure 4.29. As can be seen, in the first step (Figure 4.29[a]), the values of the relative dielectric permittivities converge to the actual values after about 80 iterations. Similarly, in the second step, the estimates of the normalized axial velocities are obtained after about 20 iterations.

References

[1] Zoughi, R., *Microwave Non-Destructive Testing and Evaluation*, Dordrecht: Kluwer Academic Publishers, 2000.

[2] Collier, R. J., and A. D. Skinner, *Microwave Measurements, Third Edition*, London: Institute of Engineering & Technology, 2007.

[3] Teppati, V., A. Ferrero, and M. Sayed, (eds.), *Modern RF and Microwave Measurement Techniques*, New York: Cambridge University Press, 2013.

[4] Chen, L. F., et al., *Microwave Electronics: Measurement and Materials Characterization*, Hoboken, NJ: Wiley, 2004.

[5] Nyfors, E., and P. Vainikainen, *Industrial Microwave Sensors*, Norwood, MA: Artech House, 1989.

[6] Ginzton, E. L., *Microwave Measurements*, New York: McGraw-Hill, 1957.

[7] Baker-Jarvis, J., M. D. Janezic, and D. C. Degroot, "High-Frequency Dielectric Measurements," *IEEE Instrum. Meas. Mag.*, Vol. 13, No. 2, April 2010, pp. 24–31.

[8] Kharkovsky, S., and R. Zoughi, "Microwave and Millimeter Wave Nondestructive Testing and Evaluation—Overview and Recent Advances," *IEEE Instrum. Meas. Mag.*, Vol. 10, No. 2, 2007, pp. 26–38.

[9] Bolomey, J. C., and F. E. Gardiol, *Engineering Applications of the Modulated Scatterer Technique*, Boston: Artech House, 2001.

[10] Caorsi, S., M. Donelli, and M. Pastorino, "A Passive Antenna System for Data Acquisition in Scattering Applications," *IEEE Antennas Wireless Propag. Lett.*, Vol. 1, 2002, pp. 203–206.

[11] Hughes, D., and R. Zoughi, "A Novel Method for Determination of Dielectric Properties of Materials Using a Combined Embedded Modulated Scattering and Near-Field Microwave Techniques-Part I: Forward Model," *IEEE Trans. Instrum. Meas.*, Vol. 54, No. 6, December 2005, pp. 2389–2397.

[12] Azaro, R., S. Caorsi, and M. Pastorino, "A 3-GHz Microwave Imaging System Based on a Modulated Scattering Technique and on a Modified Born Approximation," *Int. J. Imaging Syst. Technol.*, Vol. 9, No. 5, January 1998, pp. 395–403.

[13] Wei, Z., et al., "Quantitative Theory for Probe-Sample Interaction with Inhomogeneous Perturbation in Near-Field Scanning Microwave Microscopy," *IEEE Trans. Microw. Theory Techn.*, Vol. 64, No. 5, May 2016, pp. 1402–1408.

[14] Robert, A., "Dielectric Permittivity Of Concrete Between 50 Mhz and 1 Ghz and GPR Measurements for building Materials Evaluation," *J. Appl. Geophys.*, Vol. 40, No. 1–3, October 1998, pp. 89–94.

[15] Tsui, F., and S. L. Matthews, "Analytical Modelling of the Dielectric Properties of Concrete for subsurface Radar Applications," *Constr. Build. Mater.*, Vol. 11, No. 3, April 1997, pp. 149–161.

[16] Sen, P., C. Scala, and M. Cohen, "A Self Similar Model for Sedimentary Rocks with Application to the Dielectric Constant of Fused Glass Beads," *Geophysics*, Vol. 46, No. 5, May 1981, pp. 781–795.

[17] Rhim, H. C., and O. Büyüköztürk, "Electromagnetic Properties of Concrete at Microwave Frequency Range," *ACI Mat. J.*, Vol. 95, No. 3, May 1998, pp. 262–271.

[18] Soutsos, M. N., et al., "Dielectric Properties Of Concrete and Their Influence on Radar Testing," *NDT E Int.*, Vol. 34, No. 6, September 2001, pp. 419–425.

[19] Halabe, U. B., A. Sotoodehnia, K. R. Maser, and E. A. Kausel, "Modeling of the Electromagnetic Properties of Concrete," *Mater. J.*, Vol. 90, No. 6, November 1993, pp. 552–563.

[20] Bourdi, T., J. E. Rhazi, F. Boone, and G. Ballivy, "Modelling Dielectric-Constant Values Of Concrete: An Aid to Shielding Effectiveness Prediction and Ground-Penetrating Radar Wave Technique Interpretation," *J. Phys. Appl. Phys.*, Vol. 45, No. 40, 2012, p. 405401.

[21] Jonscher, A. K., "The 'Universal' Dielectric Response," *Nature*, Vol. 267, No. 5613, June 1977, p. 673.

[22] Bourdi, T., J. E. Rhazi, F. Boone, and G. Ballivy, "Application of Jonscher Model for the Characterization of the Dielectric Permittivity of Concrete," *J. Phys. Appl. Phys.*, Vol. 41, No. 20, October 2008, p. 205410.

[23] Hollender, F., and S. Tillard, "Modeling Ground-Penetrating Radar Wave Propagation and Reflection with the Jonscher Parameterization," *Geophysics*, Vol. 63, No. 6, November 1998, pp. 1933–1942.

[24] Dérobert, X., and G. Villain, "Effect of Water and Chloride Contents and Carbonation on the Electromagnetic Characterization of Concretes on the GPR Frequency Band Through Designs of Experiment," *NDT&E Int.*, Vol. 92, December 2017, pp. 187–198.

[25] Roqueta, G., L. Jofre, J. Romeu, and S. Blanch, "Microwave Time-Domain Reflection Imaging of Steel Fiber Distribution on Reinforced Concrete," *IEEE Trans. Instrum. Meas.*, Vol. 60, No. 12, Decemeber 2011, pp. 3913–3922.

[26] Reinhardt, H. W., C. U. Grosse, and B. Weiler, "Material Characterization of Steel Fiber Reinforced Concrete using Neutron CT, Ultrasound and Quantitative Acoustic Emission techniques," *NDT.net*, Vol. 6, No. 5, May 2001, pp. 1–12.

[27] Sihvola, A. H., *Electromagnetic Mixing Formulas and Applications*, London: Institution of Engineering and Technology, 1999.

[28] Ricketts, J. T., M. K. Loftin, and F. S. Merritt (eds.), *Standard Handbook for Civil Engineers, Fifth Edition,* New York: McGraw-Hill, 2004.

[29] Fallahpour, M., J. T. Case, M. T. Ghasr, and R. Zoughi, "Piecewise and Wiener filter-Based SAR Techniques for Monostatic Microwave Imaging of Layered Structures," *IEEE Trans. Antennas Propag.*, Vol. 62, No. 1, January 2014, pp. 282–294.

[30] Roqueta, G., L. Jofre, and M. Feng, "Microwave Non-Destructive Evaluation of Corrosion in Reinforced Concrete Structures," in *Proc. 5Th European Conference on Antennas and Propagation (EuCAP)*, Rome, Italy, 2011, pp. 787–791.

[31] Bussey H., and J. Richmond, "Scattering by a Lossy Dielectric Circular Cylindrical Multilayer Numerical Values," *IEEE Trans. Antennas Propag.*, Vol. 23, No. 5, September 1975, pp. 723–725.

[32] Karbhari, V. M., *Non-Destructive Evaluation (NDE) of Polymer Matrix Composites*, Cambridge, UK: Woodhead Publishing, June 2013.

[33] Deng, Y., and X. Liu, "Electromagnetic Imaging Methods for Nondestructive Evaluation Applications," *Sensors*, Vol. 11, No. 12, December 2011, pp. 11774–11808.

[34] Diener, L., "Microwave Near-Field Imaging with Open-Ended Waveguide—Comparison with Other Techniques of Nondestructive Testing," *Res. Nondestruct. Eval.*, Vol. 7, No. 2–3, January 1995, pp. 137–152.

[35] Diener, L., "Microwaves for Raster Imaging of Local Anisotropies in Polymer Materials," in *Review of Progress in Quantitative Nondestructive Evaluation*, Boston, MA: Springer, 1995, pp. 615–622.

[36] Bakhtiari, S., S. I. Ganchev, and R. Zoughi, "Open-Ended Rectangular Waveguide for Nondestructive Thickness Measurement and Variation Detection of Lossy Dielectric Slabs Backed by a Conducting Plate," *IEEE Trans. Instrum. Meas.*, Vol. 42, No. 1, February 1993, pp. 19–24.

[37] Osaki, S., "A New Method for Quick Determination of Molecular Orientation in Poly(ethylene terephthalate) Films by Use of Polarized Microwaves," *Polym. J.*, Vol. 19, No. 7, July 1987, p. 821.

[38] Mukherjee, S., A. Tamburrino, L. Udpa, and S. Udpa, "NDE of Composite Structures Using Microwave Time Reversal Imaging," *AIP Conf. Proc.*, Vol. 1706, No. 1, February 2016, p. 100002.

[39] Belkebir, K., and M. Saillard, "Special Section: Testing Inversion Algorithms Against Experimental Data," *Inverse Probl.*, Vol. 17, No. 6, December 2001, pp. 1565–1571.

[40] Geffrin, J.-M., P. Sabouroux, and C. Eyraud, "Free Space Experimental Scattering Database Continuation: Experimental Set-Up and Measurement Precision," *Inverse Probl.*, Vol. 21, No. 6, December 2005, pp. S117–S130.

[41] Geffrin, J.-M., and P. Sabouroux, "Continuing with the Frésnel Database: Experimental Setup and Improvements In 3D Scattering Measurements," *Inverse Probl.*, Vol. 25, No. 2, February 2009, p. 024001.

[42] Yurduseven O., et al., "Computational Microwave Imaging Using 3D Printed Conductive Polymer Frequency-Diverse Metasurface Antennas," *IET Microw. Antennas Amp Propag.*, Vol. 11, No. 14, September 2017, pp. 1962–1969.

[43] Von Hippel, A. R., *Dielectric Materials and Applications*. Boston, MA: Artech House, 1995.

[44] Bur, A. J., "Dielectric Properties of Polymers at Microwave Frequencies: A Review," *Polymer*, Vol. 26, No. 7, July 1985, pp. 963–977.

[45] Deffenbaugh, P. I., R. C. Rumpf, and K. H. Church, "Broadband Microwave Frequency Characterization of 3-D Printed Materials," *IEEE Trans. Compon. Packag. Manuf. Technol.*, Vol. 3, No. 12, December 2013, pp. 2147–2155.

[46] Al-Ghamdi, AA, et al., "Dielectric and Microwave Properties of Polyvinyl Chloride/ Graphite/Nickel Composites and Its Applications," *J. Thermoplast. Compos. Mater.*, Vol. 27, No. 4, April 2014, pp. 528–540.

[47] Saybolt, M., S. Mukherjee, L. Udpa, and P. Chahal, "A Low Cost Microwave Imaging System Using A 6-Port Reflectometer for NDE of Composites," in *Proc. 31st Annual Technical Conference of the American Society for Composites, (ASC)* 2016, Williamsburg VA.

[48] Xie, S.-H., et al., "Polyimide/BaTiO3 Composites with Controllable Dielectric Properties," *Compos. Part Appl. Sci. Manuf.*, Vol. 36, No. 8, August 2005, pp. 1152–1157.

[49] Sharma, B. K., N. Khare, S. K. Dhawan, and H. C. Gupta, "Dielectric Properties of Nano ZnO-Polyaniline Composite in the Microwave Frequency Range," *J. Alloys Compd.*, Vol. 477, No. 1, May 2009, pp. 370–373.

[50] Zhao, D.-L., X. Li, and Z.-M. Shen, "Microwave Absorbing Property and Complex Permittivity and Permeability of Epoxy Composites Containing Ni-Coated and Ag Filled Carbon Nanotubes," *Compos. Sci. Technol.*, Vol. 68, No. 14, November 2008, pp. 2902–2908.

[51] Mo, Z., et al., "Preparation and Characterization of a PMMA/Ce(OH)3, Pr2O3/Graphite Nanosheet Composite," *Polymer*, Vol. 46, No. 26, December 2005, pp. 12670–12676.

[52] Nelson, S. O., "Dielectric Spectroscopy in Agriculture," *J. Non-Cryst. Solids*, Vol. 351, No. 33, Sep. 2005, pp. 2940–2944.

[53] Trabelsi, S., and S. O. Nelson, "Nondestructive Sensing of Physical Properties of Granular Materials by Microwave Permittivity Measurement," *IEEE Trans. Instrum. Meas.*, Vol. 55, No. 3, June 2006, pp. 953–963.

[54] Nelson, S. O., "Dielectric Spectroscopy of Fresh Fruit and Vegetable Tissues from 10 to 1800 MHz," *J. Microw. Power Electromagn. Energy*, Vol. 40, No. 1, 2005, pp. 31–47.

[55] Khaled, D. E., et al., "Fruit and Vegetable Quality Assessment via Dielectric Sensing," *Sensors*, Vol. 15, No. 7, June 2015, pp. 15363–15397.

[56] Jawadm, H., et al., "Microwave Modeling and Experiments for Non Destructive Control Improved Quality of fruit," in *Proc. 2017 IEEE Conference On Antenna Measurements & Applications (CAMA)*, Tsukuba, Japan, 2017, pp. 124–127.

[57] Nelson, S. O., "Dielectric Spectroscopy of Fresh Fruits and Vegetables," in *Proc. 2005 IEEE Instrumentation And Measurement Technology Conference Proceedings*, Ottawa, Canada, 2005, Vol. 1, pp. 360–364.

[58] Skaar, C., *The Dielectric Properties of Wood at Several Radio Frequencies*, New York State College of Forestry at Syracuse University, 1948.

[59] Brown, J. H., R. W. Davidson, and C. Skaar, "Mechanism of Electrical Conduction of Wood," *For. Prod. J.*, October 1963, pp. 455–459.

[60] Torgovnikov, G. I., *Dielectric Properties of Wood and Wood-Based Materials*. Berlin, Heidelberg: Springer Berlin Heidelberg, 1993.

[61] James, W. L., "Dielectric Properties of Wood and Hardboard: Variation with Temperature, Frequency, Moisture Content, and Grain Orientation.," U.S. Department of Agriculture, Madison, WI, FPL 245, 1975.

[62] Wert, C. A., M. Weller, and D. Caulfield, "Dynamic loss properties of wood," *J. Appl. Phys.*, Vol. 56, No. 9, November 1984, pp. 2453–2458.

[63] Bucur, V., *Nondestructive Characterization and Imaging of Wood*, Berlin: Springer, 2003.

[64] Franchois, A., Y. Pineiro, and R. H. Lang, "Microwave permittivity measurements of two conifers," *IEEE Trans. Geosci. Remote Sens.*, Vol. 36, No. 5, 1998, pp. 1384–1395.

[65] Catapano, I., A. Randazzo, E. Slob, and R. Solimene, "GPR Imaging Via Qualitative and Quantitative Approaches," in *Civil Engineering Applications of Ground Penetrating Radar*, Benedetto, A., and L. Pajewski (eds.), Cham, Switzerland: Springer International Publishing AG, 2015, pp. 239–280.

[66] Colton, D., H. Haddar, and M. Piana, "The Linear Sampling Method in Inverse Electromagnetic Scattering Theory," *Inverse Probl.*, Vol. 19, No. 6, December 2003, pp. S105–S137.

[67] Catapano, I., L. Crocco, and T. Isernia, "On Simple Methods for Shape Reconstruction of Unknown Scatterers," *IEEE Trans. Antennas Propag.*, Vol. 55, No. 5, May 2007, pp. 1431–1436.

[68] Brignone, M., et al., "A Hybrid Approach to 3D Microwave Imaging by Using Linear Sampling and ACO," *IEEE Trans. Antennas Propag.*, Vol. 56, No. 10, October 2008, pp. 3224–3232.

[69] Crocco, L., I. Catapano, L. Di Donato, and T. Isernia, "The Linear Sampling Method as a Way to Quantitative Inverse Scattering," *IEEE Trans. Antennas Propag.*, Vol. 60, No. 4, April 2012, pp. 1844–1853.

[70] Bertero, M., and P. Boccacci, *Introduction to Inverse Problems in Imaging*, Bristol, UK: Institute of Physics Publishing, 1998.

[71] Bozza, G., M. Brignone, and M. Pastorino, "Application of the No-Sampling Linear Sampling Method to Breast Cancer Detection," *IEEE Trans. Biomed. Eng.*, Vol. 57, No. 10, October 2010, pp. 2525–2534.

[72] Tikhonov, A. N., A. Goncharsky, V. V. Stepanov, and A. G. Yagola, *Numerical Methods for the Solution of Ill-Posed Problems*, Berlin: Springer, 1995.

[73] Warren, C., A. Giannopoulos, and I. Giannakis, "gprMax: Open Source Software to Simulate Electromagnetic Wave Propagation for Ground Penetrating Radar," *Comput. Phys. Commun.*, Vol. 209, December 2016, pp. 163–170.

[74] Bozza, G., et al.,"Crack Detection in Dielectric Structures by a Linear Sampling Approach," *Int. J. Signal Imaging Syst. Eng.*, Vol. 3, No. 2, 2010, p. 73.

[75] D'Errico, M. S., B. L. Douglas, and H. Lee, "Subsurface Microwave Imaging for Nondestructive Evaluation of Civil Structures," Vol.5, 1993, pp. 453–456.

[76] Shaw, M. R., et al.,"Location of Steel Reinforcement in Concrete Using Ground Penetrating Radar and Neural Networks," *NDT E Int.*, Vol. 38, No. 3, April 2005, pp. 203–212.

[77] Maierhofer, C., "Nondestructive Evaluation of Concrete Infrastructure with Ground Penetrating Radar," *J. Mater. Civ. Eng.*, Vol. 15, No. 3, June 2003, pp. 287–297.

[78] Mast, J. E., H. Lee, W. C. Chew, and J. P. Murtha, "Pulse Echo Holographic Techniques for Microwave Subsurface NDE," in *Proceedings of Nondestructive Evaluation of Civil Structures and Materials*, 1990, pp. 177–191.

[79] Lim, M. K., and C. A. Olson, "Use of Nondestructive Impulse Radar in Evaluating Civil Engineering Structures," in *Proceedings of Nondestructive Evaluation of Civil Structures and Materials*, 1990, pp. 167–176.

[80] Chew, W. C., *Waves and Fields in Inhomogeneous Media*. Piscataway, NY: IEEE Press, 1995.

[81] Shao, W., A. Edalati, T. R. McCollough, and W. J. McCollough, "A Phase Confocal Method for Near-Field Microwave Imaging," *IEEE Trans. Microw. Theory Techn.*, Vol. 65, No. 7, July 2017, pp. 2508–2515.

[82] Fedeli, A., M. Pastorino, and A. Randazzo, "Experimental Validation of a Novel Gauss-Newton Inversion Method for Microwave Tomographic Imaging," in *Proc. 2014 XXXITH URSI General Assembly and Scientific Symposium (URSI GASS)*, Beijing, China, 2014.

[83] Ramzi, M. R., M. Abou-Khousa, and I. Prayudi, "Near-Field Microwave Imaging Using Open-Ended Circular Waveguide Probes," *IEEE Sens. J.*, Vol. 17, No. 8, April 2017, pp. 2359–2366.

[84] Zhang, H., et al., "Evaluation of Atmospheric Corrosion on Coated Steel Using K-Band Sweep Frequency Microwave Imaging," *IEEE Sens. J.*, Vol. 16, No. 9, May 2016, pp. 3025–3033.

[85] Ali, A., M. E. Badawe, and O. M. Ramahi, "Microwave Imaging of Subsurface Flaws in Coated Metallic Structures Using Complementary Split-Ring Resonators," *IEEE Sens. J.*, Vol. 16, No. 18, September 2016, pp. 6890–6898.

[86] Malyuskin, O., and V. F. Fusco, "High-Resolution Microwave Near-Field Surface Imaging Using Resonance Probes," *IEEE Trans. Instrum. Meas.*, Vol. 65, No. 1, January 2016, pp. 189–200.

[87] Sun, S., B. J. Kooij, and A. G. Yarovoy, "A Linear Model for Microwave Imaging of Highly Conductive Scatterers," *IEEE Trans. Microw. Theory Techn.*, Vol. 66, No. 3, March 2018, pp. 1149–1164.

[88] Colton, D., and A. Kirsch, "A Simple Method for Solving Inverse Scattering Problems in the Resonance Region," *Inverse Probl.*, Vol. 12, No. 4, 199, p. 3836.

[89] Colton, D., M. Piana, and R. Potthast, "A Simple Method Using Morozov's Discrepancy Principle for solving Inverse Scattering Problems," *Inverse Probl.*, Vol. 13, No. 6, December 1997, pp. 1477–1493.

[90] Crocco, L., L. Di Donato, I. Catapano, and T. Isernia, "An Improved Simple Method for Imaging the Shape of Complex Targets," *IEEE Trans. Antennas Propag.*, Vol. 61, No. 2, February 2013, pp. 843–851.

[91] Boero, F., et al., "Microwave Tomography for the Inspection of Wood Materials: Imaging System and Experimental Results," *IEEE Trans. Microw. Theory Techn.*, In print.

[92] Salvade, A., et al., "A New Microwave Axial Tomograph for the Inspection of Dielectric Materials," *IEEE Trans. Instrum. Meas.*, Vol. 58, No. 7, July 2009, pp. 2072–2079.

[93] Monleone, R. D., et al., "Impact of Background Noise on Dielectric Reconstructions Obtained by a Prototype of Microwave Axial Tomograph," *IEEE Trans. Instrum. Meas.*, Vol. 61, No. 1, January 2012, pp. 140–148.

[94] Pastorino, M., et al., "A Microwave Tomographic System for Wood Characterization in the Forest Products Industry," *Wood Mater. Sci. Eng.*, Vol. 10, No. 1, 2015, pp. 75–85.

[95] Abbak, M., M. N. Akinci, M. Cayoren, and I. Akduman, "Experimental Microwave Imaging with a Novel Corrugated Vivaldi Antenna," *IEEE Trans. Antennas Propag.*, Vol. 65, No. 6, June 2017, pp. 3302–3307.

[96] Kumar Pandey, G., and M. Kumar Meshram, "A Printed High Gain UWB Vivaldi Antenna Design Using Tapered Corrugation and Grating Elements: A Printed High Gain UWB Vivaldi Antenna Design," *Int. J. RF Microw. Comput.-Aided Eng.*, Vol. 25, No. 7, September 2015, pp. 610–618.

[97] Teni, G., N. Zhang, J. Qiu, and P. Zhang, "Research on a Novel Miniaturized Antipodal Vivaldi Antenna With Improved Radiation," *IEEE Antennas Wireless Propag. Lett.*, Vol. 12, 2013, pp. 417–420.

[98] Ascama, H. D. O., C. D. Pantaleon, and S. T. Kofuji, "Effect of the Substrate on the Performance of Corrugated UWB Antennas," in *Proc. IEEE Latin-American Conference on Communications*, Bogota, Columbia, 2010.

[99] Wang, Z., and H. Zhang, "Improvements in a High Gain UWB Antenna with Corrugated Edges," *Prog. Electromagn. Res. C*, Vol. 6, 2009, pp. 159–166.

[100] De Oliveira, A. M., M. B. Perotoni, S. T. Kofuji, and J. F. Justo, "A Palm Tree Antipodal Vivaldi Antenna with Exponential Slot Edge for Improved Radiation Pattern," *IEEE Antennas Wireless Propag. Lett.*, Vol. 14, 2015, pp. 1334–1337.

[101] Bourqui, J., and E. C. Fear, "Shielded UWB Sensor for Biomedical Applications," *IEEE Antennas Wireless Propag. Lett.*, Vol. 11, 2012, pp. 1614–1617.

[102] Gibbins, D., et al., "A Comparison of a Wide-Slot and a Stacked Patch Antenna for the Purpose of Breast Cancer Detection," *IEEE Trans. Antennas Propag.*, Vol. 58, No. 3, March 2010, pp. 665–674.

[103] Ahdi Rezaeieh, S., A. Zamani, K. S. Bialkowski, and A. M. Abbosh, "Foam Embedded Wideband Antenna Array for Early Congestive Heart Failure Detection with Tests Using Artificial Phantom with Animal Organs," *IEEE Trans. Antennas Propag.*, Vol. 63, No. 11, November 2015, pp. 5138–5143.

[104] Fioranelli, F., S. Salous, I. Ndip, and X. Raimundo, "Through-The-Wall Detection With Gated FMCW Signals Using Optimized Patch-Like and Vivaldi Antennas," *IEEE Trans. Antennas Propag.*, Vol. 63, No. 3, March 2015, pp. 1106–1117.

[105] Yurduseven, O., D. Smith, and M. Elsdon, "Printed Slot Loaded Bow-Tie Antenna with Super Wideband Radiation Characteristics for Imaging Applications," *IEEE Trans. Antennas Propag.*, Vol. 61, No. 12, December 2013, pp. 6206–6210.

[106] Moosazadeh, M., and S. Kharkovsky, "A Compact High-Gain and Front-to-Back Ratio Elliptically Tapered Antipodal Vivaldi Antenna with Trapezoid-Shaped Dielectric Lens," *IEEE Antennas Wireless Propag. Lett.*, Vol. 15, 2016, pp. 552–555.

[107] Elboushi, A., and A. Sebak, "High-Gain Hybrid Microstrip/Conical Horn Antenna for MMW Applications," *IEEE Antennas Wireless Propag. Lett.*, Vol. 11, 2012, pp. 129–132.

[108] Wang, Y.-W., et al., "Ultra-Wideband E-Plane Monopulse Antenna Using Vivaldi Antenna," *IEEE Trans. Antennas Propag.*, Vol. 62, No. 10, October 2014, pp. 4961–4969.

[109] Kang, W., S. Lee, and K. Kim, "A Ground-Folded Slot Antenna for Imaging Radar Applications," *IEEE Antennas Wireless Propag. Lett.*, Vol. 10, 2011, pp. 155–158.

[110] Oloumi, D., P. Mousavi, M. I. Pettersson, and D. G. Elliott, "A Modified TEM Horn Antenna Customized for Oil Well Monitoring Applications," *IEEE Trans. Antennas Propag.*, Vol. 61, No. 12, December 2013, pp. 5902–5909.

[111] Bucur, V., "Techniques for High Resolution Imaging of Wood Structure: A Review," *Meas. Sci. Technol.*, Vol. 14, No. 12, December 2003, p. R91.

[112] Jezova, J., and S. Lambot, "Reflection Waveforms Occurring in GPR Tree Trunk Testing," in *Proc. 16th International Conference On Ground Penetrating Radar (GPR)*, Hong Kong, China, 2016, pp. 1–5.

[113] Jezova, J., S. Lambot, A. Fedeli, and A. Randazzo, "Ground-Penetrating Radar for Tree Trunk Investigation," in *Proc. 9th International Workshop On Advanced Ground Penetrating Radar (IWAGPR)*, Edinburgh, UK, 2017, pp. 1–6.

[114] Yilmaz, B., C. Ozdemir, and S. Gokkan, "A Focusing Algorithm for Tree-Penetrating Radar Imaging: An Experimental Study and Concept Evaluation," in *Proc. International Radar Symposium (IRS)*, Krakow, Poland, 2016, pp. 1–4.

[115] Yılmaz, B., S. Gökkan, and C. Özdemir, "An Experimental Study and Concept Evaluation on Tree-Interior Imaging Radar Using Sinusoidal Template-Based Focusing Algorithm," *J. Electr. Comput. Eng.*, Vol. 2017, 2017, pp. 1–7.

[116] Kaestner, A. P., and L. B. Baath, "Microwave Polarimetry Tomography of Wood," *IEEE Sens. J.*, Vol. 5, No. 2, April 2005, pp. 209–215.

[117] Fedeli, A., et al., "Experimental Analysis of Dielectric Structures with a Two-Step Electromagnetic Imaging Method," in *Proc. URSI International Symposium On Electromagnetic Theory (EMTS)*, Espoo, Finland, 2016, pp. 13–15.

[118] Bozza, G., C. Estatico, M. Pastorino, and A. Randazzo, "An Inexact Newton Method for Microwave Reconstruction of Strong Scatterers," *IEEE Antennas Wireless Propag. Lett.*, Vol. 5, No. 1, December 2006, pp. 61–64.

[119] Harrington, R. F., *Field Computation by Moment Methods*, Piscataway, NJ: IEEE, 1993.

[120] Manickavasagan, A., and H. Jayasuriya (eds.), *Imaging with Electromagnetic Spectrum Applications in Food and Agriculture*, Berlin: Springer, 2014.

[121] Mallach, M., P. Gebhardt, and T. Musch, "2-D Microwave Tomography System for Imaging of Multiphase Flows in Metal Pipes," *Flow Meas. Instrum.*, Vol. 53, March 2017, pp. 80–88.

[122] Wu, Z., et al., "Microwave-Tomographic System for Oil- and Gas-Multiphase-Flow Imaging," *Meas. Sci. Technol.*, Vol. 20, No. 10, 2009, p. 104026.

[123] Wu, Z., "Developing a Microwave Tomographic System for Multiphase Flow Imaging: Advances and Challenges," *Trans. Inst. Meas. Control*, Vol. 37, No. 6, July 2015, pp. 760–768.

[124] Pastorino, M., M. Raffetto, and A. Randazzo, "Electromagnetic Inverse Scattering of Axially Moving Cylindrical Targets," *IEEE Trans. Geosci. Remote Sens.*, Vol. 53, No. 3, March 2015, pp. 1452–1462.

[125] Wu, Z., and H. Wang, "Microwave Tomography for Industrial Process Imaging: Example Applications and Experimental Results.," *IEEE Antennas Propag. Mag.*, Vol. 59, No. 5, October 2017, pp. 61–71.

[126] Roger, A., "Newton-Kantorovitch Algorithm Applied to an Electromagnetic Inverse Problem," *IEEE Trans. Antennas Propag.*, Vol. 29, No. 2, March 1981, pp. 232–238.

[127] Joachimowicz, N., C. Pichot, and J. P. Hugonin, "Inverse Scattering: An Iterative Numerical Method for electromagnetic Imaging," *IEEE Trans. Antennas Propag.*, Vol. 39, No. 12, December 1991, pp. 1742–1753.

[128] Abbak, M., et al., "Wideband Compact Dipole Antenna for Microwave Imaging Applications," *IET Microw. Antennas Propag.*, Vol. 11, No. 2, January 2017, pp. 265–270.

[129] Asefi, M., A. Zakaria, and J. LoVetri, "Microwave Imaging Using Normal Electric-Field Components Inside Metallic Resonant Chambers," *IEEE Trans. Microw. Theory Techn.*, Vol. 65, No. 3, March 2017, pp. 923–933.

[130] Mojabi, P., and J. LoVetri, "A Novel Microwave Tomography System Using a Rotatable Conductive Enclosure," *IEEE Trans. Antennas Propag.*, Vol. 59, No. 5, May 2011, pp. 1597–1605.

[131] Winges, J., et al., "Compressed Sensing for the Detection and Positioning of Dielectric Objects Inside Metal Enclosures by Means of Microwave Measurements," *IEEE Trans. Microw. Theory Techn.*, Vol. 66, No. 1, January 2018, pp. 462–476.

[132] Butz, P., C. Hofmann, and B. Tauscher, "Recent Developments in Noninvasive Techniques for Fresh Fruit and Vegetable Internal Quality Analysis," *J. Food Sci.*, Vol. 70, No. 9, pp. R131–R141, May 2006.

[133] C. Kittiyanpunya, C. Phongcharoenpanich, and M. Krairiksh, "Detecting Granulated Fruit by Measuring Dual-Frequency Back Scattered Waves," in *Proc. IEEE Conference on Antenna Measurements & Applications (CAMA)*, Tsukuba, Japan, 2017, pp. 128–129.

[134] Watanabe, M., et al., "Nondestructive Imaging of the Sugar Distribution Inside a Fruit by Use of CP-MCT," in *Proc. 2005 European Microwave Conference*, 2005, Paris, France.

[135] Nickel, J., J. Parker, and R. Gould, "Resonance Oscillations in a hot nonuniform plasma column," *Phys. Rev. Lett.*, Vol. 11, No. 5, September 1963, pp. 183–185.

[136] Messiaen, A., and P. Vandenplas, "High-Frequency Effect Due to the Axial Drift Velocity of a Plasma Column," *Phys. Rev.*, Vol. 149, No. 1, September 1966, pp. 131–140.

[137] Yeh, C., "Scattering Obliquely Incident Microwaves by a Moving Plasma Column," *J. Appl. Phys.*, Vol. 40, No. 13, December 1969, pp. 5066–5075.

[138] Shiozawa, T., and S. Seikai, "Scattering of Electromagnetic Waves from an Inhomogeneous Magnetoplasma Column Moving in the Axial Direction," *IEEE Trans. Antennas Propag.*, Vol. 20, No. 4, July 1972, pp. 455–463.

[139] Censor, D., "Scattering of Electromagnetic Waves by a Cylinder Moving Along its Axis," *IEEE Trans. Microw. Theory Techn.*, Vol. 17, No. 3, March 1969, pp. 154–158.

[140] Yan, Y., "Mass Flow Measurement of Bulk Solids in Pneumatic Pipelines," *Meas. Sci. Technol.*, Vol. 7, No. 12, December 1996, p. 1687.

[141] Bowman, J. J., T. B. A. Senior, P. L. E. Uslenghi, and J. S. Asvestas, *Electromagnetic And Acoustic Scattering by Simple Shapes*, New York, NY: Hemisphere Pub. Corp., 1987.

[142] Pastorino, M., *Microwave Imaging*, Hoboken, NJ: Wiley, 2010.

[143] Ragheb, H. A., and L. Shafai, "Electromagnetic Scattering from a Dielectric-Coated Elliptic Cylinder," *Can. J. Phys.*, Vol. 66, No. 12, December 1988, pp. 1115–1122.

[144] Caorsi, S., M. Pastorino, and M. Raffetto, "Electromagnetic Scattering by a Conducting Strip with a Multilayer Elliptic Dielectric Coating," *IEEE Trans. Electromagn. Compat.*, Vol. 41, No. 4, November 1999, pp. 335–343.

[145] Tai, C., "The Dyadic Green's Function for a Moving Isotropic Medium," *IEEE Trans. Antennas Propag.*, Vol. 13, No. 2, March 1965, pp. 322–323.

[146] Kong, J. A., and D. K. Cheng, "On Guided Waves in Moving Anisotropic Media," *IEEE Trans. Microw. Theory Techn.*, Vol. 16, No. 2, February 1968, pp. 99–103.

[147] Pastorino, M., and M. Raffetto, "Scattering of Electromagnetic Waves from a Multilayer Elliptic Cylinder Moving in the Axial Direction," *IEEE Trans. Antennas Propag.*, Vol. 61, No. 9, September 2013, pp. 4741–4753.

[148] Karaboga, D., and B. Basturk, "Artificial Bee Colony (ABC) Optimization Algorithm for Solving Constrained Optimization Problems," in *Foundations Of Fuzzy Logic and Soft Computing*, Melin, P., et al. (eds.), Berlin, Germany: Springer, 2007, pp. 789–798.

5

Microwave Imaging for Biomedical Applications

This chapter discusses microwave imaging techniques for some relevant diagnostic applications in the biomedical field. The chapter begins with a brief discussion about the dielectric properties of biological tissues, and then an overview of some of the more recent techniques for breast cancer detection, brain stroke imaging, and other proposed applications is reported.

5.1 Potentialities and Limitation of Microwave Imaging in Biomedical Applications

From a historical point of view, the first research concerning the present applications dates back to the final twenty years of the last century. The fundamental book by Larsen and Jacoby [1] summarized the preliminary proposals. That book included a review of studies concerning the characterization of biological tissues in terms of dispersion and attenuation, but also proposed preliminary methods for imaging, both based on radar concepts and inverse scattering. In most cases, very simplified models of the biological bodies were considered. Successively, some of these pioneering techniques reveled themselves to be unable to produce realistic images of the bodies to be inspected. Nevertheless, they pose the basis for successive studies. In particular, the work of Larsen and Jacobi concerning the imaging of perfused kidney is worth mentioning, since it clearly proved the need for the use of suitable matching media in order to allow the electromagnetic energy to better penetrate inside the body. The work by Bolomey et al. [2] introduced the use of tomographic techniques in the biomedical microwave imaging arena, as an extension to diffracted wavefield of the projection methods commonly applied in computerized tomography (CT), based on X-ray radiation. Being essentially based on the first order Born approximation, for which the scattered electric field is expressed in terms of the internal incident field only (linearization of the scattering problem), these techniques have been successively modified in order to better take into account the strong scattering nature of biological bodies, even if surrounded by proper coupling media.

The potential advantages of using microwaves in biomedical diagnostics are basically related to the following aspects. First of all, microwave imaging is potentially able to directly retrieve the dielectric properties of the tissues, which represents information that cannot be directly obtained by other, more consolidated diagnostic techniques. The values of these parameters can be associated with the health state of the biological tissues, the presence of tumors, and so forth.

In addition, the low levels of electromagnetic power needed for imaging makes the technique healthy for both patients and practitioners, as the microwave radiation is nonionizing in nature. Finally, the apparatuses used are similar to those adopted for telecommunication purposes, and are of course not as expensive as those used for X-ray CT, nuclear magnetic resonance (NMR), positron emission tomography (PET), and so forth.

The main limitations essentially concern (as it is well known) spatial resolution, information content of scattered data, and power levels of the scattered signals with respect to the power of the incident radiation. These aspects, which are common to other applications of microwave imaging, will be discussed in the following paragraphs.

It must be stressed that microwave imaging techniques can be applied mainly in the cases in which there are significant changes in the dielectric properties of the biological tissues due to diseases. This is essentially the main reason for which the most attention has been focused on breast cancer detection [3]. More recently, however, a large number of studies have been concentrated on the detection and imaging of brain strokes as well.

5.2 Electromagnetic Characterization of Biological Materials

Biological tissues exhibit dielectric properties that are strongly dependent upon their water contents. In the past years, there has been a wide research activity concerning the measurement of these properties, as well as concerning the modeling of their dispersive nature, which is particularly relevant at microwave frequencies. In this frequency range, most of the biological tissues are also very lossy, resulting in significant attenuation of the signal inside tissues. The mechanism of interaction between electromagnetic fields and biological tissues at these frequencies poses severe limitations in the possibility of associating microwave imaging techniques with cancer, diseases, and other aspects of medical diagnostics.

In particular, it has been found (for an example, see [4]) that biological tissues are characterized by three main dispersion regions, which are then characterized by significant variations in the values of the dielectric properties versus frequencies. The first region (α dispersion) happens between 0.1 kHz and 100 kHz and it is mainly related to the diffusion mechanism in the outer cellular membranes. The β dispersion is between approximatively 1 MHz and 20 MHz, and it is related to the presence of bound water in macromolecules (e.g., proteins). Finally, the γ dispersion is located around 20 GHz and it is mainly due to the polarization of the water molecules (see Figure 2.2). It is worth noting that sometimes a fourth transition region (named δ relaxation) can also be present between the α and β dispersion regions.

The first systematic works devoted to the characterization of the dielectric properties of biological tissues date back to the middle of the last century. More recently, several other measurement campaigns have been performed in order to extend and validate the available data [4–7]. Among others, a significant set of measurements have been performed by Gabriel et al. [8]. In particular, different types of tissues have been considered, both from animals and humans. Most of these measurements have been performed post mortem or ex-vivo, except in some particular cases (e.g., for the skin). Consequently, in some cases, there could be some differences in the dielectric properties, as discussed in [9]. In particular, the equivalent conductivity could be higher for in-vivo tissues.

An important feature of this study is the development of Cole-Cole models (as defined in Chapter 2) of the measured data [10]. In particular, a four-pole model has been fitted to the measurements in the range 10 Hz–100 GHz. Such models are widely used in the scientific literature for modeling the biological tissues (and they are also provided by the U.S. Federal Communications Commission (FCC) [11]). An alternative set of Cole-Cole models have also been adopted in [12]. They have been derived using a fitting algorithm specifically designed for devising the optimal parameters of multiple dispersion models [13].

Some examples of the behavior of the (equivalent) dielectric parameters of some tissues versus the frequency (in the range of 300 MHz–10 GHz) are shown in Figure 5.1, together with the measured data provided in [8].

As discussed in Section 2.2, the presence of nonnegligible values of the equivalent conductivity (either due to real conduction phenomena or to dispersion) produces attenuations of the electromagnetic waves propagating inside the material. Such behavior is particularly important in the case of biological tissues, since the attenuation can assume significant values in the microwave

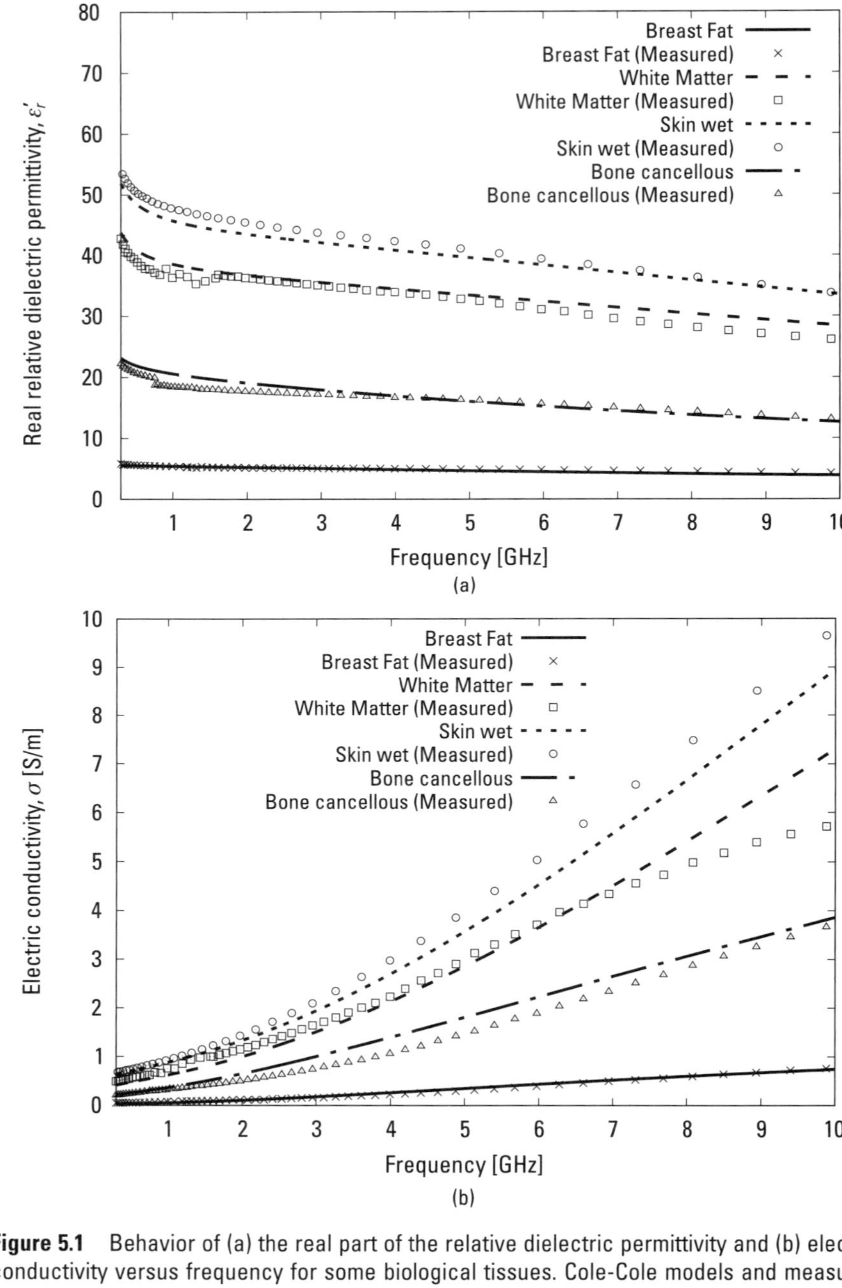

Figure 5.1 Behavior of (a) the real part of the relative dielectric permittivity and (b) electric conductivity versus frequency for some biological tissues. Cole-Cole models and measured data are obtained from [10] and [8], respectively

frequency band. Some examples of the penetration depth (defined as in [2.37]) versus the frequency are shown in Figure 5.2 for the tissues considered in Figure

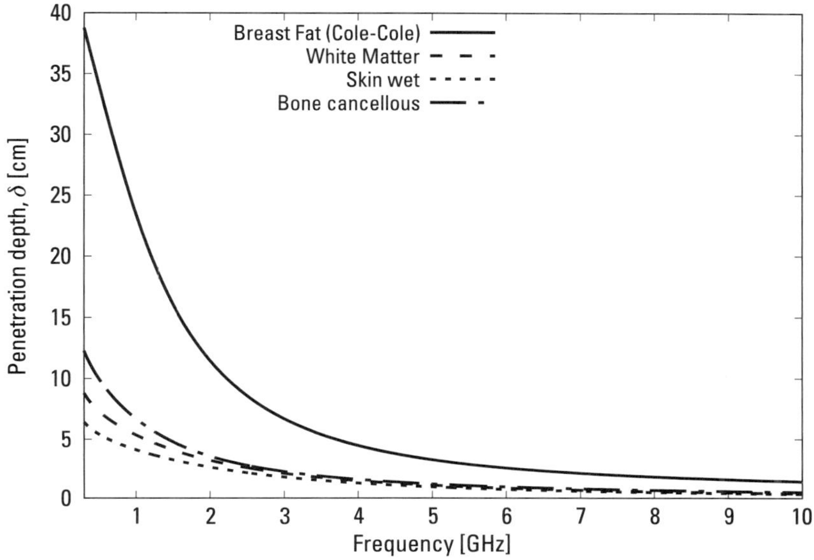

Figure 5.2 Behavior of the penetration depth versus frequency for some of the biological tissues of Figure 5.1.

5.1. As can be seen, in most cases the electromagnetic waves penetrate for just few centimeters, and the penetration depth strongly reduces when the frequency increases. Such behavior must be properly taken into account in designing biomedical imaging systems working with microwaves.

The knowledge of the dielectric properties of the tissues is important for several reasons. Among others, it allows one to simulate the response of the body (or a part of it) to an electromagnetic illumination. However, it is worth noting that the integration of Cole-Cole models into numerical simulators, especially those working in the time domain, is not straightforward. For example, in the case of the finite-difference time-domain (FDTD) method, specific implementations with reduced computational demands have been proposed [14–16].

Alternatively, representations based on Debye-type models on limited frequency bands can be used. In fact, the Debye relations can be more easily integrated inside FDTD-based forward solvers [17, 18]. For example, in [19] a three-pole Debye model has been used to describe a subset of the tissues reported in [10], in a smaller frequency range. In [20], four-pole models have been developed for accurately describing the tissues of the head in the range of 0.1 GHz–3 GHz. In both these cases, the models were fitted on the data provided by Gabriel et al. [8].

It should be also mentioned that the dielectric properties of living tissues may exhibit notable changes depending on the temperature of the body [21–

24]. Such a situation may hold, for example, during some medical treatments such as microwave ablation [25, 26].

5.2.1 Dielectric Properties of Breast Tissues

Significant efforts have been devoted in the few last years to the accurate characterization of the dielectric properties of breast tissues. In fact, the development of microwave imaging systems able to detect (possibly in early stage) breast tumors has been one of the main pursued applications in the biomedical field.

Such interest was mainly motivated by the initial studies on the dielectric properties of normal and malignant breast tissues (e.g., those reported in [27], concerning the frequencies between 3 MHz and 3 GHz), which indicated that significant differences were present. Similar results were also obtained by other subsequent experimental investigations. It is worth noting that the dielectric properties of benign and malignant tumor samples were also found to be very similar (for example, see [28]), thus excluding the possibility of discriminating between these two types of tumors, at least at the considered frequencies.

More recently, a large scale study of the dielectric properties of both normal and tumor tissues has been performed. The results were published in [29] (where the tissues were obtained from reduction surgery) and [30] (where the samples were obtained from cancer surgery). In particular, the normal breast samples were subdivided into three groups, depending on the percentage of adipose tissues (group 1: adipose tissue < 30%; group 2: adipose tissue between 31%, and 84%; group 3: adipose tissue > 84%). For each group, the median values of the dielectric properties were extracted and used for creating a single-pole Cole-Cole model. Figure 5.3 shows a plot of the values of the real and imaginary parts of the complex relative dielectric permittivity provided by such models.

According to [29] and [30], it has been found that a significant variability exists for the dielectric properties of the normal breast tissues, ranging from low (for mostly adipose breast tissue) to high values (for breast with significant glandular and fibroconnective tissues). Such behavior is in accordance with other studies of the dielectric properties of the breast (e.g., this fact was also previously suggested in [28], although in this study the frequency was limited to 3.2 GHz).

Moreover, in [30] it has been found that the contrast between malignant and normal tissues can be far less than previously expected. In fact, for high-density fibroconnective/glandular tissues it is less than about 10% in the frequency range of interest. For mostly adipose tissues, the contrast is instead quite high (the malignant relative permittivity is about ten times that of normal tissues). Following such findings, the use of agents able to increase the contrast between malignant and dense healthy tissues has been recently proposed. In

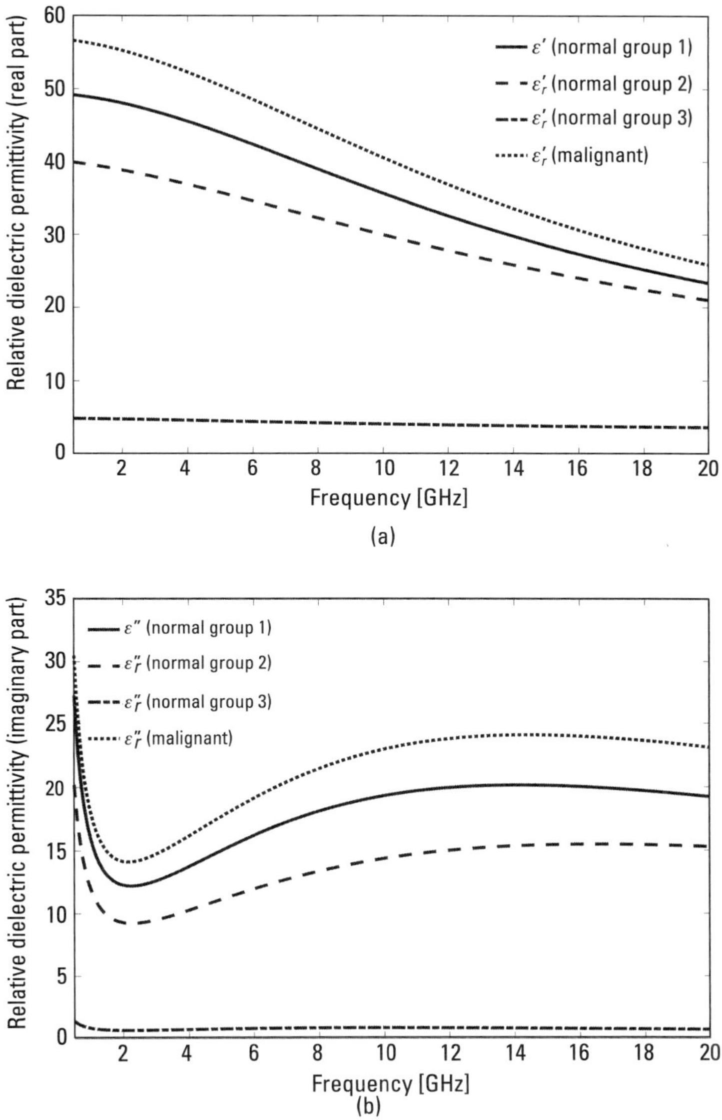

Figure 5.3 Dielectric properties—(a) real and (b) imaginary part of the relative dielectric permittivity—of the normal and malignant breast tissue derived in [29, 30].

particular, magnetic nanoparticles have been found to be a promising choice [31–34] (see also Chapter 8).

It is worth noting that the research in this framework is still going on. For example, the work in [35] includes the results of a set of measurements up to 50 GHz and proper parametric models from such data are reported. However, up to now, the models developed by Lazebnik et al. are still considered a reference in most of the works related to microwave breast imaging.

5.2.2 Dielectric Properties of Brain Tissues

Microwave imaging is becoming a powerful tool in brain imaging, especially for what concerns the detection and monitoring of strokes. Consequently, in the last few years, there has been an increasing interest in the development and use of dielectric models of brain tissues. It is also worth noting that a proper electromagnetic characterization of the head is of great importance in other fields, such as the prediction of the specific absorption rate (SAR) due to the exposition to mobile terminals or other radiating elements (for example, see [36]).

Concerning the head tissues, the dielectric parameters usually employed are based on the measurements reported by Gabriel et al. [8]. However, especially in the field of brain stroke monitoring, it has been pointed out (for example, see [20]) that for the frequency range of interest (i.e., around 1 GHz, as discussed in Section 5.5), the Cole-Cole models could exhibit some disagreements with respect to the measured values. Moreover, as previously recalled, the direct use of Cole-Cole models in time-domain numerical simulators is not straightforward. Consequently, as previously mentioned, alternative models of the head tissues, limited to the frequency range of interest, have been developed (for example, in [19, 20]).

In brain stroke imaging, it is also necessary to have a proper model of the stroke properties. Actually, two possible types of strokes can be present: hemorrhagic and ischemic strokes (see Section 5.5). A hemorrhagic stroke is due to a rupture of a cerebral blood vessel, which causes a hemorrhage in the brain. In this case, the dielectric properties of the region affected by the stroke can be assumed to be equal to that of the blood [37, 38]. Another possible approach is to use an average of the dielectric properties of the blood and of the tissues in which the stroke occurs [39]. On the contrary, ischemic strokes are due to a vascular occlusion caused by atherosclerotic thrombosis or embolism, which leads, if not properly treated, to the tissue necrosis. In this case, a common way to model the dielectric properties of the part of the head affected by the stroke is to reduce the dielectric properties of the involved tissues by a factor equal to 10% [40, 41]. The behaviors of the dielectric properties of healthy tissues (white and grey matter) and of the corresponding values for the stroke regions are shown in Figure 5.4 for the frequency range of interest for such application.

5.2.3 Dielectric Properties of Matching Media

From the discussions reported in the previous sections, it is clear that biological tissues present quite high values of the (complex) dielectric permittivity. Such high values may produce undesirable, strong reflections from the outer

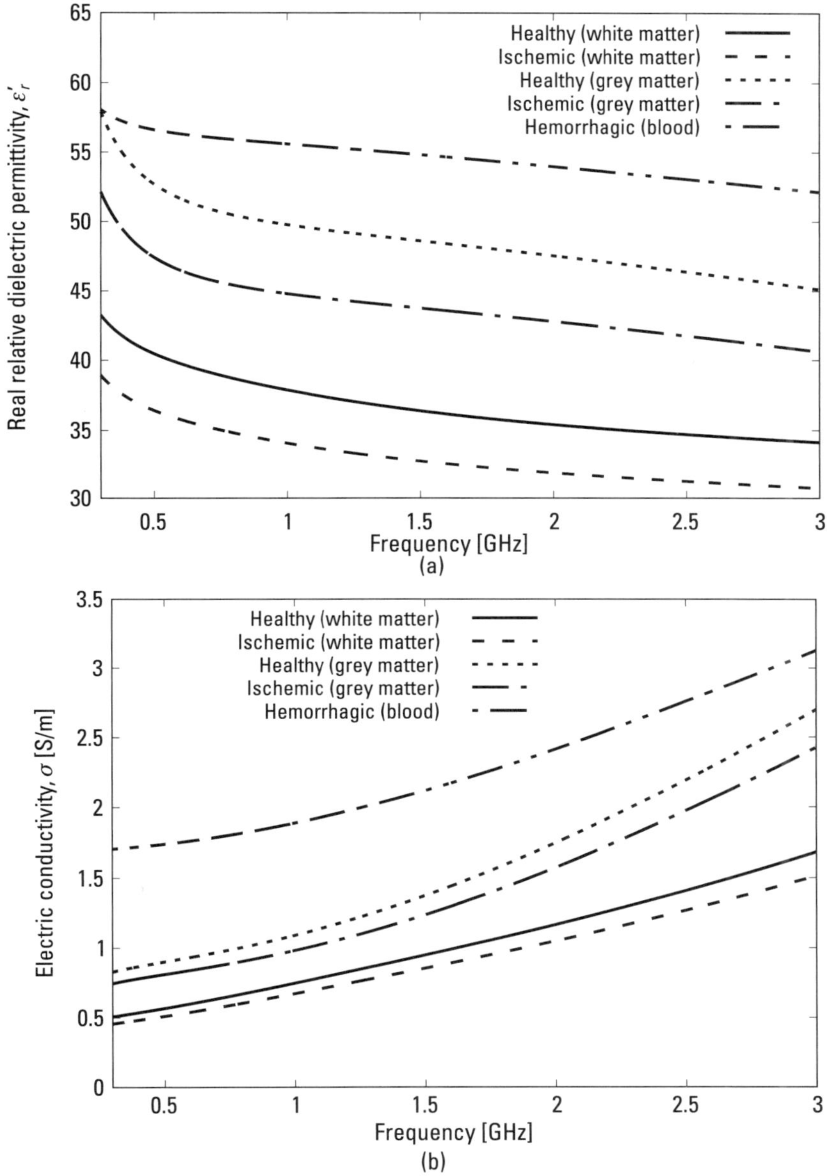

Figure 5.4 Comparison among the dielectric properties of healthy, ischemic, and hemorrhagic brain tissues between 300 MHz and 3 GHz. The fourth-order Debye models developed in [20] are used.

boundaries of the body, which can mask the scattering contributions from the searched inclusions [42]. Moreover, a high dielectric contrast can also be difficult to be retrieved by inverse scattering algorithms. Consequently, coupling media that try to match the dielectric properties of the inspected tissues are usually adopted. To this end, the dielectric properties of the matching medium are chosen in order to maximize the power radiated inside the biological part under investigation.

Materials commonly used in biomedical applications ranges from oils to water and saline mixtures. The liquids with lower permittivity values are usually adopted for tissues with low water contents (e.g., the breast tissues), whereas saline solutions are often used for tissues with higher dielectric contrast (e.g., the brain tissues). In particular, for breast and head imaging, some typically used solutions are corn syrup [43], Triton X-100 [44], and water/glycerin mixtures [45]. Some examples of the dielectric properties of such materials are shown in Figure 5.5. It is worth noting that matching media are usually lossy (as can be seen from Figure 5.5[b]). Consequently, they also introduce attenuation in the received signal. However, it has been found (for example, see [42]) that such unwanted attenuation can be useful in order to reduce the mutual coupling between adjacent antennas in multistatic setups.

5.3 Numerical and Experimental Phantoms

In the design of microwave imaging systems, an important phase is represented by the simulation of the response of the targets to be investigated. In the specific case of biomedical imaging, it is thus necessary to be able to model in a realistic way the human body. Apart the knowledge of the dielectric properties of the tissues (discussed in the previous Section 5.2), it is also important to create specific numerical models mimicking the real anatomy.

In this framework, several phantoms describing the whole body or just limited specific parts have been made available in the scientific community. Table 5.1 reports some examples of numerical phantoms that have been used in microwave simulations. It is worth noting that often such phantoms are obtained starting from MRI or CT images, which has been properly segmented in order to identify the various tissues and rescaled in order to obtain resolutions suitable for electromagnetic simulations in the microwave range. Some models have also been obtained starting from the photographic data collected in the framework of the Visible Human Project [46].

5.3.1 Breast Models

In order to aid the evaluation of the feasibility of microwave imaging systems and techniques for breast cancer detection, specific numerical models have been made available by the University of Wisconsin-Madison [47, 48]. In particular, several types of breasts, differing in the tissue densities, have been implemented starting from MRI images. In order to use such models in numerical simula-

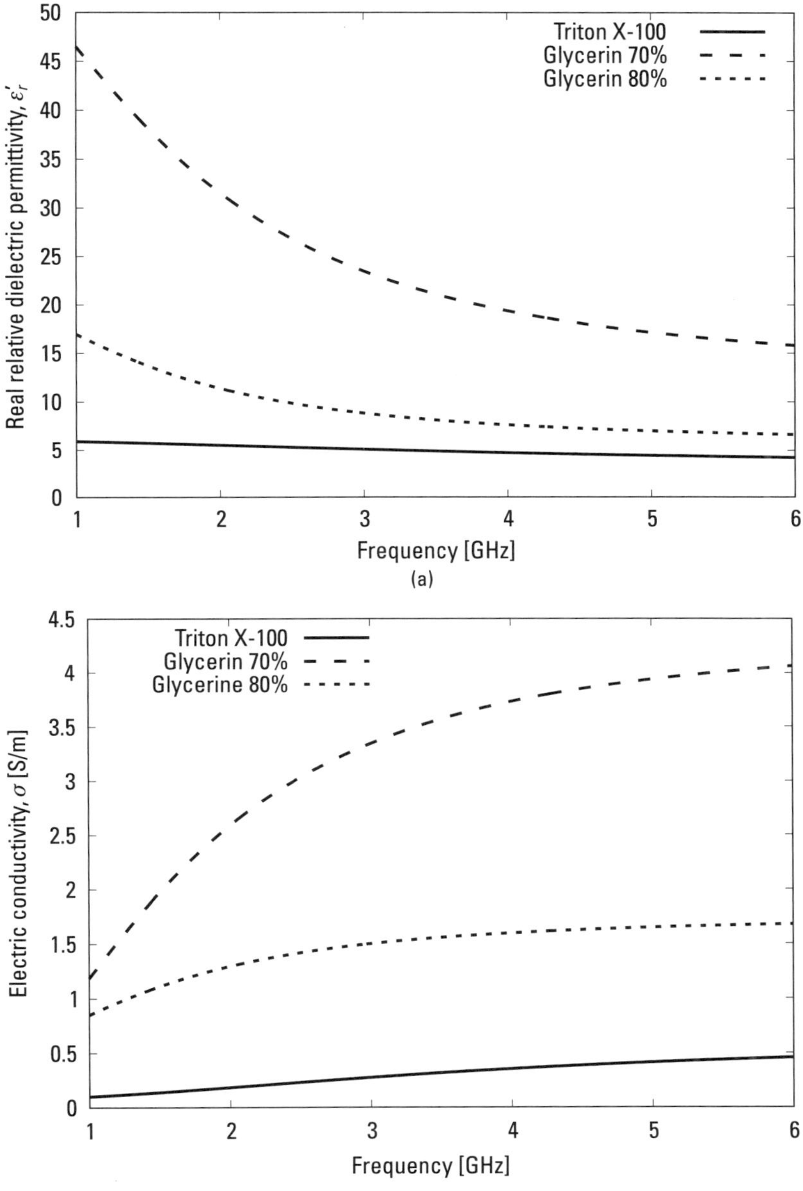

Figure 5.5 Dielectric properties of some mixture solutions usually employed as matching media in medical microwave imaging. (a) Real part of the relative dielectric permittivity and (b) electric conductivity.

tors in time domain (e.g., FDTD-based methods), a set of Debye models of the dielectric properties measured by Lazebnik et al. are also provided [49].

Table 5.1
Some Examples of Numerical Phantoms Available in the Scientific Literature

Name	Source	Resolution [mm³]	Type	Year
VoxelMan [50]	CT	3.6×3.6×3.6	Whole body (Male)	1992
	MRI	1.1×1.1×1.4	Head (Male)	
Norman [51]	MRI	2.04×2.04×1.95	Whole body (Male)	1997
Golem [52]	CT	2.08×2.08×8	Whole body (Male)	1998
VIP-man [52]	Photos	0.33×0.33×1	Whole body (Male)	2000
UWCEM Numerical Breast Phantom [53]	MRI	0.5×0.5×´0.5	Breast (Female)	2008
Virtual family [47]	MRI	0.5×0.5´0.5	Whole body (Male, Female, Child)	2010
AustinMan/ AustinWoman [54]	Photos	1×1×1	Whole body (Male, Female)	2012
CMODEL [55]	Photos	0.16×0.16×0.25	Head and shoulder	2016

An example of a scattered fibroglandular breast phantom obtained from such a database is shown in Figure 5.6. Concerning the eventually present tumors, in many cases they are simulated by assuming simple geometrical shapes (e.g., spheres). Recently, more realistic and sophisticated models incorporating the knowledge arising from clinical data have also been proposed [56] (e.g., they are represented by polygonal volumes with approximately elliptical shapes and with spicules on the surface).

Beside the numerical phantoms, experimental phantoms are also available in the scientific community, mainly with the aim of providing a common set of standard targets on which to test the imaging systems and techniques. One of the first models made available is that provided by Burfeindt et al. in [57]. Such a model, which is shown in Figure 5.7, has been derived from one of the heterogeneously dense numerical breast phantoms in the repository of the University of Wisconsin-Madison. The adipose tissues are modeled by the solid plastic internal parts, whereas a filling liquid (with proper dielectric properties mimicking the real dielectric properties) is used for the fibroglandular tissues.

Another phantom has been proposed in [58] (see Figure 5.8). In this case, the various parts of the breast are modeled by using carbon/rubber mixtures, which have the advantages of being flexible, reasonably robust, and with stable dielectric properties.

The *GeePs-L2S* phantom shown in Figure 5.9 has been developed by research groups of the CentraleSupélec (France) in the framework of the MiMed European COST Action, and it is being proposed as a standard benchmark for testing microwave imaging systems and reconstruction algorithms [59]. The structure is made of plastic material (ABS) and has been realized by using a 3-D printer. It is composed of four parts: an outer shell with a thickness of about 1.5

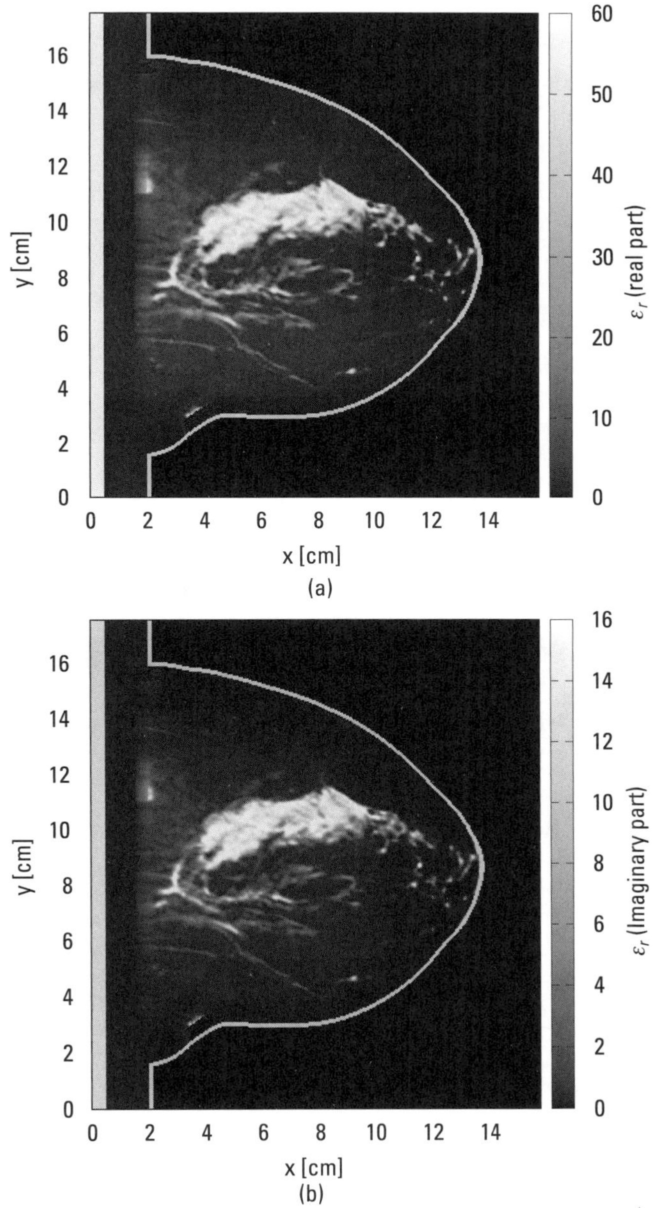

Figure 5.6 Example of a numerical model of the scattered fibroglandular breast obtained from the data available in the University of Wisconsin-Madison database [47] (sagittal slice). (a) Real and (b) imaginary parts of the relative dielectric permittivity at 3 GHz.

mm, a tank containing material mimicking the fibroglandular tissue, a removable inclusion to simulate the tumor, and a support plate on which the previous

Figure 5.7 Experimental breast phantoms from [57]. (©2012 IEEE.)

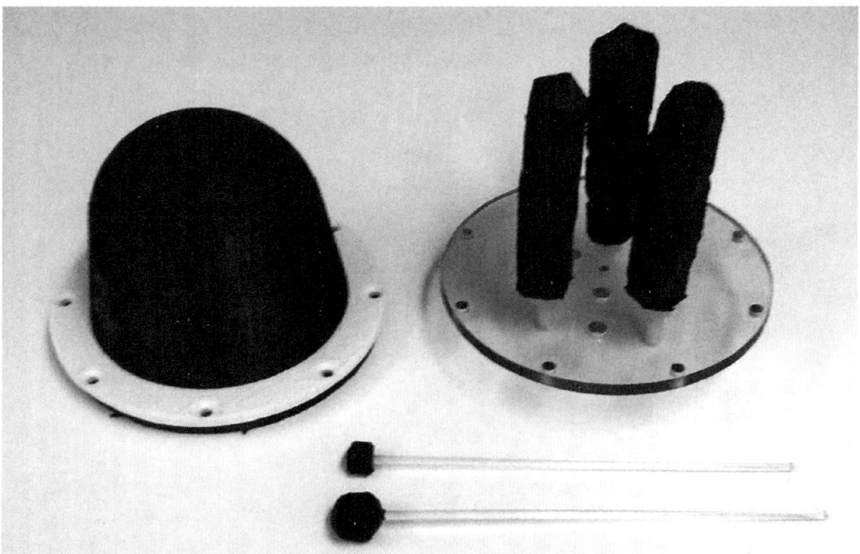

Figure 5.8 Experimental breast phantoms from [58]. (©2015 IEEE.)

parts can be fixed. The breast tissues are simulated by solutions of Triton X-100 and salted water. Such a choice has been taken on the basis of the results reported in [60], where "it has been shown that liquid mixtures based upon Triton X-100 and salted water solutions are able to mimic the various breast tissues and tumors in a large frequency range." Moreover, it has been found that the dielectric properties of such solutions can be accurately predicted by using binary fluid mixture model [61] in conjunction with the Debye models of Triton X-100 and salted water.

Finally, it is worth reporting that in [62] a realistic breast phantom for microwave imaging and hyperthermia applications has been developed. In this

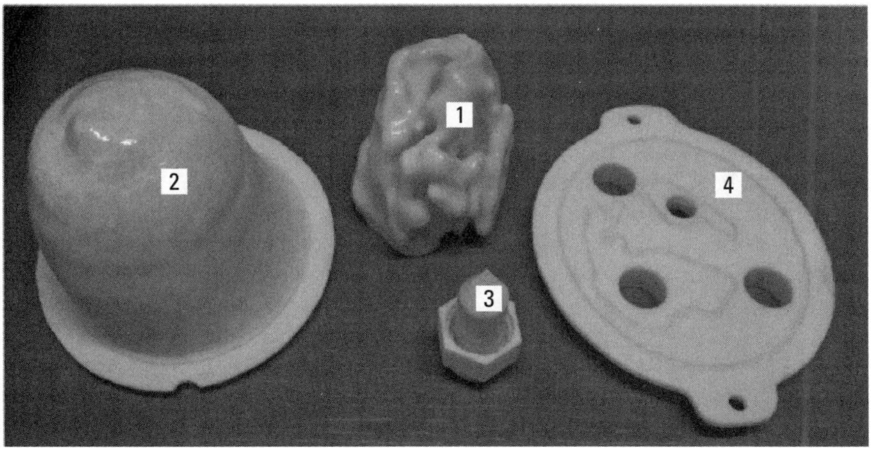

Figure 5.9 Experimental GeePs-L2S breast phantom [59]: (1) tank containing the fibroglandular part, (2) outer shell, (3) removable tumor, and (4) support plate. (©2017 IEEE.)

case, an important issue is related to the realization of materials that properly mimic both the dielectric and thermal properties of the tissues. In the proposed phantom, this is accomplished by using plastic molds for skin and glands (created with 3-D printers) and mixtures poured into the plastic parts for the fat tissues.

5.3.2 Head Models

As indicated in the beginning of Section 5.3, there are several numerical anatomical phantoms derived from photos or MRI images that comprise the head as well. In the framework of microwave imaging, the mainly-used ones are those based on the VoxelMan [50] and AustinMan/AustinWoman datasets [55]. The VoxelMan dataset provides a segmented model of the head derived from MRI images, which is composed by 128 slices (along the vertical direction) discretized into 256 × 256 pixels of dimensions 1.1 mm × 1.1 mm. The distance between the slices is 1.4 mm. In such a model, 60 tissue types have been identified and associated with the various pixels. A 3-D view of the phantom is shown in Figure 5.10(a).

The AustinMan/AustinWoman models are based on the photos available from the Visible Human Project [63]. They provide full body voxel models of a man and of a woman, both with a resolution of 1 mm^3. A view of the head of the phantom is shown in Figure 5.10(b).

Concerning experimental head phantoms, it should be mentioned that many efforts have been devoted to the development of anatomically realistic structures for the evaluation of the SAR due to the exposure to electromagnetic

(a) (b)

Figure 5.10 Examples of numerical models of head. (a) VoxelMan and (b) AustinMan numerical phantoms.

waves, mainly caused by mobile terminals. A reference phantom is defined in the IEEE 1528-2013 standard [64], whose shape and dimensions have been devised from the 90th percentile of male anthropometric data of the U.S. Army. Such a phantom is basically constituted by an external shape made of a low-loss dielectric material, which can be filled with tissue-mimicking liquid mixtures. It is worth noting that a homogeneous phantom is assumed, with dielectric properties selected in order to obtain a suitable estimate of the peak spatial-average SAR values.

Such a phantom is, however, not exactly suitable for microwave imaging purposes, since it neglects the interface effects of all the layers of different tissues. Consequently, more realistic experimental phantoms have been realized for the specific application. In particular, in [65] an MRI-derived human head phantom comprised of materials mimicking the frequency-dependent dielectric properties of the tissues is used (Figure 5.11). The exterior part of the phantom, which is realized by using a 3-D printer, represents the combined effects of fat, skin, muscular parts, and skull. Such a shell is filled with tissue-mimicking materials modeling the dielectric properties of the main brain tissues (cerebrospinal fluid [CSF], gray and white matters), which are obtained by mixing water, corn flour, agar, gelatin, sodium azide, and propylene glycol [66]. The brain injury (bleeding) is again modeled by placing blood mimicking materials inside the head. It is also worth noting that the liquid mixtures developed in [60] (based on Triton X-100 or glycerin with salted water) have been found to be able to mimic the frequency behavior of different head tissues as well.

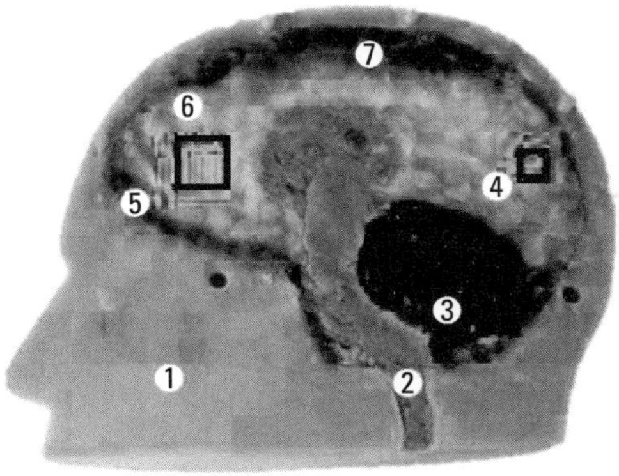

Figure 5.11 Internal structure of a realistic head phantom [65]. (1) Exterior section, (2) spinal cord (3) cerebellum, (4) white matter, (5) Dura, (6) gray matter, (7) CSF. Black squares show the locations of bleeding. (©2017 IEEE.)

5.4 Breast Imaging

Breast imaging for cancer detection is one of the main applications of microwave imaging in the biomedical field [67–72]. In this framework, several techniques and systems have been developed. The working frequencies are usually comprised of frequencies between 1 and 4 GHz, although UWB radar setups can employ also higher frequencies (up to about 10 GHz). However, it is worth mentioning that very recently the possibility of using millimeter-wave illuminations (e.g., around 30 GHz) has also been considered, since it has been claimed that "in several cases, a penetration depth of a few centimeters is possible, while maintaining reasonable safety margins" [73]. Similarly to other applicative fields, the target (the breast in this case) is illuminated by an incident field generated by antennas located all around the breast. Multistatic setups are often adopted and, in many cases like those discussed next, (ultra)wideband pulsed signals (eventually synthesized by using a frequency-stepped system) are employed. The scattered signal is measured and recorded by a set of receiving antennas located around the breast (in the multistatic configuration).

Two main types of antenna configurations can be used. The first configuration (schematized in Figure 5.12[a]) refers to the antennas located on the surface of a cylinder containing the breast. In particular, the probes are positioned on circumferences with different heights. In order to minimize the mutual coupling among the elements, an angular shift between adjacent probing lines in the vertical direction is usually assumed. In this case, the radiating elements have usually vertical polarizations (e.g., dipole or monopole antennas can be used). This configuration is often adopted for gathering data for inverse scattering-based techniques. The second type of configuration involves radiating elements positioned on a hemispherical surface (as schematically shown in

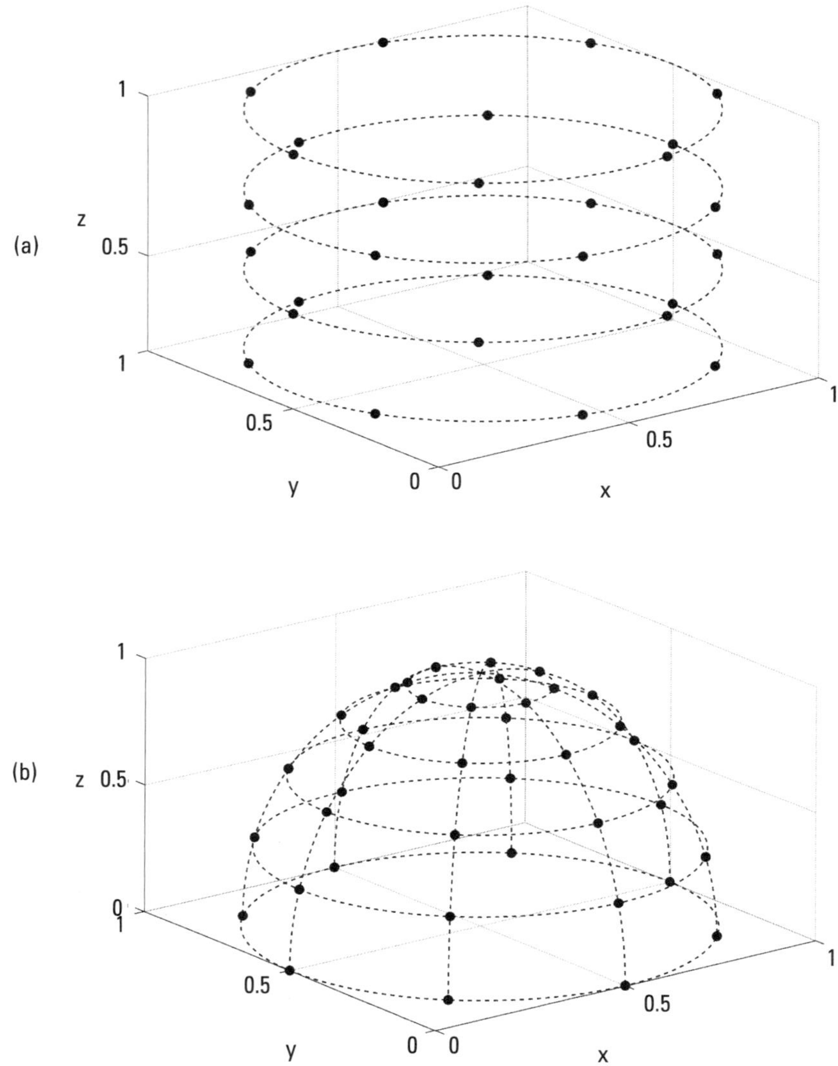

Figure 5.12 Schematic representation of common measurement setups employed in breast imaging. (a) Cylindrical and (b) hemispherical configurations.

Figure 5.12[b]). In both cases, a matching medium between the antennas and the breast surface should be inserted, in order to reduce the dielectric contrast. Some examples of practical implementations of such measurement arrangements are shown next.

Concerning the imaging approaches, they can be classified into two main categories: Beamforming-based and inverse scattering-based techniques. The aim of approaches belonging to the beamforming-based techniques is simply to identify strong discontinuities in the inspected region. As discussed in Section 5.2.1, in the early studies concerning the dielectric properties of breast tissues, high values of contrast between malignant and normal tissues have been reported. Consequently, approaches of this kind have been generally followed and found to be quite robust when dealing with homogeneous or low-density breasts.

Imaging methods based on inverse scattering techniques aim at reconstructing the full distributions of the dielectric properties, and consequently these methods are, in principle, able to provide more comprehensive information about the inhomogeneous internal structure of the breast.

Finally, it is worth mentioning that the use of differential imaging methods, working off the differences between the reconstructed images of the two breasts, have also been recently proposed [74].

5.4.1 Beamforming-Based Imaging Techniques

In this section, some of the main basic concepts concerning beamforming approaches are discussed. Detailed descriptions can be found in the scientific literature, including review papers and books (for example, see [71, 72]). Moreover, comparative assessments of the performance achieved by different beamforming techniques and analyses about the assumptions behind the use of such approaches for near-field imaging can be found in [75–77].

The delay-and-sum (DAS) beamforming method proposed in [76, 78] is the first implementation of an approach of this kind for breast cancer detection. In this case, the basic idea is to apply proper time shifts to the measured signals in order to align the backscattered pulses due to the reflection from the eventually present dielectric discontinuities. In particular, assuming that the measurement setup is composed by M antenna located at positions $\mathbf{r}_m$, an indicator function $I(\mathbf{r})$, $\mathbf{r} \in V$, is created as

$$I(\mathbf{r}) = \begin{cases} \displaystyle\int_0^T \left(\sum_{m=1}^{M} w_{mm}(\mathbf{r}) s_{tot}^{(m)}\left(\mathbf{r}_m, t - \tau_{mm}(\mathbf{r})\right) \right)^2 dt & \textit{Monostatic} \\ \displaystyle\int_0^T \left(\sum_{m=1}^{M} \sum_{n=1}^{M} w_{mn}(\mathbf{r}) s_{tot}^{(m)}\left(\mathbf{r}_n, t - \tau_{mn}(\mathbf{r})\right) \right)^2 dt & \textit{Multistatic} \end{cases} \tag{5.1}$$

where $s_{tot}^{(m)}\left(\mathbf{r}_n, t\right)$ is the time-domain signal (e.g., the voltage) at the output of the nth receiving antenna due to the incident field produced by the mth transmitting antenna, T denotes the time window considered for the integration, τ_{mn} is the applied time delay given by

$$\tau_{mn}(\mathbf{r}) = \frac{d_m(\mathbf{r}) + d_n(\mathbf{r})}{v} \tag{5.2}$$

being $d_p(\mathbf{r}) = |\mathbf{r} - \mathbf{r}_p|$, $p = m, n$, the distance between the pth antenna and the current scan point $\mathbf{r}$, and v is the average speed of the electromagnetic waves inside the breast. Equation (5.1) includes weighting factors $w_{mn}(\mathbf{r})$ that may be added in order to compensate for the attenuation effect. In the monostatic case discussed in [78], for example, these weights have been set equal to $w_{mn}(\mathbf{r}) = 1/|\mathbf{r} - \mathbf{r}_m|$. It is worth noting that in this approach, the heterogeneity of the breast is usually neglected and the propagation velocity, as mentioned, is assumed to be constant and equal to $v = v_0 / \sqrt{\varepsilon'_{r,breast}}$, with $\varepsilon'_{r,breast}$ being the average value of the real parts of the dielectric permittivities of the breast tissues. Moreover, since the dispersive nature of the tissues is not considered, some degradation in the reconstructed images (e.g., the presence of clutters or artifacts) should be expected in practical applications.

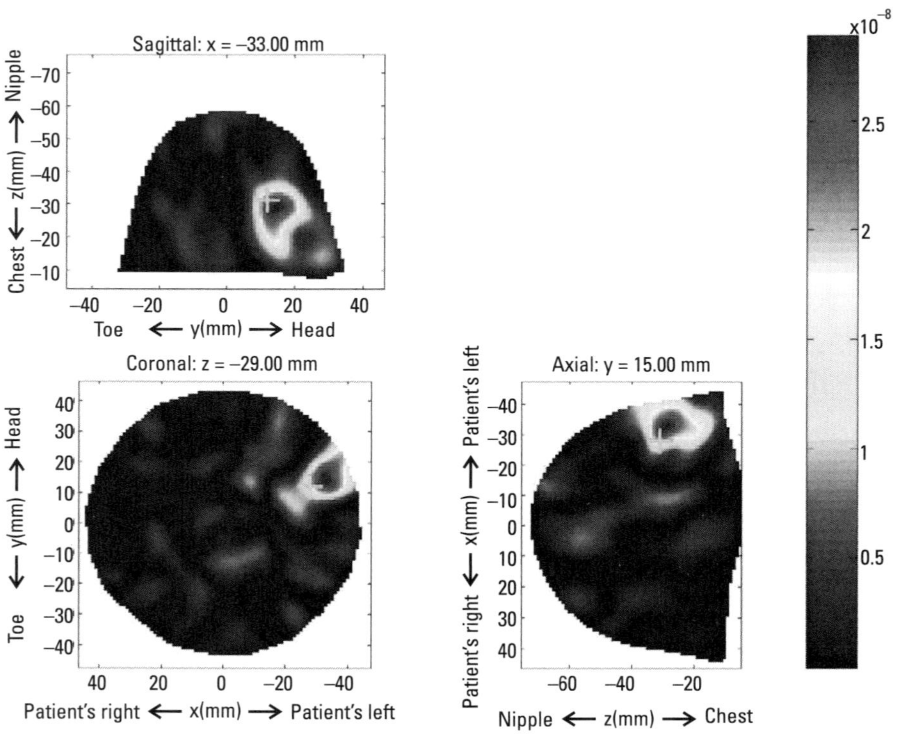

Figure 5.13 Examples of slices of qualitative images provided by the DAS beamforming technique [75]. (©2017 IEEE.)

An example of application of the DAS algorithm is shown in Figure 5.13, which reports some reconstructions obtained in [75] by using the experimental breast phantoms proposed in [58] (in the case of presence of a 16-mm tumor). A UWB Gaussian pulse with center frequency of 4 GHz and a bandwidth of 5 GHz is used to drive the transmitting antennas. The scattered electric field is collected at 140 points located on seven equally spaced rings surrounding the breast model. As can be seen in this case, the distribution of the indicator function provided by the DAS exhibits peak values in the region corresponding to the tumor location (indicated by the + sign).

Modified versions of the basic DAS method, which maintain the basic principle of focusing the received signals in the investigation domain, can be developed. For example, in the delay-multiply-and-sum (DMAS) beamforming approach [79] the shifted signals are multiplied in pairs before summation. For example, for the monostatic case, the indicator function can be defined as

$$I(\mathbf{r}) = \int_0^T \left(\sum_{m=1}^{M-1} \sum_{n=m+1}^{M} s_{tot}^{(m)}\left(\mathbf{r}_m, t - \tau_{mm}(\mathbf{r})\right) s_{tot}^{(n)}\left(\mathbf{r}_n, t - \tau_{nn}(\mathbf{r})\right) \right)^2 dt \tag{5.3}$$

Such an approach has been found to provide a good clutter rejection [79]. Other proposed variants of the DAS algorithm introduce additional multiplicative weighting factors (e.g., the quality factor in the improved delay-and-sum (IDAS) method [80] and the coherence factor in [81]).

A more sophisticated technique, which also aims at compensating the frequency-dependent effects on the wave propagation inside the breast, is the microwave imaging via space-time (MIST) beamforming [82]. In this method, before summation, the time-shifted backscattered signals are modified by a set of finite impulse response (FIR) filters, specifically designed for compensating the frequency-dependent propagation effects due to the material dispersion. It is worth noting that the original MIST algorithm reported in [82] was designed for a monostatic measurement setup. However, the extension to the multistatic case has been also proposed in the literature [83, 84].

All the previous approaches are usually referred to as data-independent beamforming methods, since they are only based on (usually simplified) propagation models. On the contrary, data-dependent methods change the beamforming parameters in order to adapt them to the measured data. Two of the main methods belonging to this class that can be used for breast imaging are the robust capon beamforming (RCB) and the multistatic adaptive microwave imaging (MAMI) [85] algorithms. A wide band adaptive strategy has been also presented in [86]. Such approaches have been found to be able to attain very good resolutions and significant clutter rejections.

5.4.2 Inverse Scattering Techniques

Although beamforming approaches are widely employed in microwave breast imaging applications, they still exhibit some limitations. First of all, when dealing with very complex anatomical structures (e.g., in the presence of strongly inhomogeneous fibroglandular breasts) the clutters in the reconstructed images may be significant, especially when inclusions (the tumors) with relatively small dielectric contrasts are to be detected [87]. Secondly, the indicator function that they provide contains only information about the presence of dielectric discontinuities, but does not give detailed information about the dielectric properties of the identified targets.

To address such limitations, inverse scattering approaches, which are in principle able to retrieve the whole dielectric structure, can be applied. Clearly, there are some drawbacks in such approaches as well. In fact, since the anatomical structure of the breast does not fully allow using simplified assumptions, the full vector 3-D inverse scattering equations must be used (see Chapter 3). Moreover, the spatial discretization that is needed to correctly solve the scattering problem and to obtain a suitable resolution in the final image is quite high (in the order of millimeters) [88]. These two facts pose significant computational issues, both concerning memory requirements and the times needed to perform the inversion (although nowadays it is possible to use high-performance computers and graphical processing unit resources). Moreover, it is necessary to carefully address the problems related to the nonlinearity and ill-posed nature of the inverse scattering problem.

In the framework of microwave breast imaging, a common choice for addressing the inverse scattering problem is to use methods based on the distorted wave Born approximation (DWBA) [89, 90]. This approximation assumes that a reference distribution of the dielectric properties, $\varepsilon_{r,ref}(\mathbf{r})$, is available (e.g., the distributions of the dielectric parameters of a healthy or a homogeneous breast). Moreover, as previously mentioned, in breast imaging applications a multistatic measurement/illumination setup is usually adopted. For the sake of simplicity, in this section, a single-view formulation is discussed (the extension to the multiview/multistatic case is straightforward, however). Let us assume that the actual inspected configuration be constituted by a dielectric contrast, with respect to the reference profile, that can be described by the following differential contrast function

$$c_{diff}(\mathbf{r}) = \frac{\varepsilon_r(\mathbf{r}) - \varepsilon_{r,ref}(\mathbf{r})}{\varepsilon_{r,b}} = c(\mathbf{r}) - c_{ref}(\mathbf{r}) \tag{5.4}$$

where c and c_{ref} are the contrast function (defined as in Chapter 3) of the actual and reference dielectric distributions. In this case, the scattering problem can be defined in terms of the following equation (see [3.5]) [89]

$$\begin{aligned}\mathbf{E}_{diff}(\mathbf{r}) &= \mathbf{E}_{tot}(\mathbf{r}) - \mathbf{E}_{ref}(\mathbf{r}) \\ &= -k_b^2 \int_V c_{diff}(\mathbf{r}')\mathbf{E}_{tot}(\mathbf{r}') \bullet \overline{\overline{\mathbf{G}}}_{ref}(\mathbf{r},\mathbf{r}')d\mathbf{r}',\ \mathbf{r} \in D_{obs}\end{aligned} \tag{5.5}$$

where $\varepsilon_{r,b}$ is the (eventually complex) relative dielectric permittivity of the matching medium, $\mathbf{E}_{tot}$ and $\mathbf{E}_{ref}$ are the total electric fields due to the actual and reference configurations, respectively, and $\overline{\overline{\mathbf{G}}}_{ref}$ is the dyadic Green's function for the inhomogeneous dielectric distribution $\varepsilon_{r,ref}$. If the differences between the actual and reference profiles are sufficiently small, (5.5) can be approximated as

$$\mathbf{E}_{diff}(\mathbf{r}) \cong -k_b^2 \int_V c_{diff}(\mathbf{r}')\mathbf{E}_{ref}(\mathbf{r}') \bullet \overline{\overline{\mathbf{G}}}_{ref}(\mathbf{r},\mathbf{r}')d\mathbf{r}',\ \mathbf{r} \in D_{obs} \tag{5.6}$$

In this way, given the reference configuration $\varepsilon_{r,ref}$, the problem can be written in terms of the following linear equation

$$\mathbf{E}_{diff}(\mathbf{r}) = \mathbf{G}_{DB}\left(c_{diff}\right)(\mathbf{r}),\ \mathbf{r} \in D_{obs} \tag{5.7}$$

where $\mathbf{G}_{DB}$ is a linear vector operator whose kernel is the dyadic Green's function for the reference configuration, that is,

$$\mathbf{G}_{DB}(\bullet)(\mathbf{r}) = -k_b^2 \int_V (\bullet)(\mathbf{r}')\mathbf{E}_{ref}(\mathbf{r}') \bullet \overline{\overline{\mathbf{G}}}_{ref}(\mathbf{r},\mathbf{r}')d\mathbf{r}',\ \mathbf{r} \in D_{obs} \tag{5.8}$$

Such a linear relationship can be solved by using standard linear inversion schemes. Since the problem is still ill-posed, regularization schemes must be employed in this case as well.

The above formulation can be iteratively applied by performing subsequent reconstructions and updating the reference dielectric profile. The obtained method is usually referred to as the distorted Born iterative method (DBIM) [89]. In particular, the DBIM in its basic form can be summarized by the following steps:

1. Set the iteration number to $n = 0$. Initialize the inversion procedure with an initial reference profile corresponding to a contrast function $c_{ref}^{(0)}$ (which can eventually include the available a priori information about the target).

2. Calculate the internal electric fields $\mathbf{E}_{ref}^{(n)}(\mathbf{r})$ and the dyadic Green's functions $\overline{\overline{\mathbf{G}}}_{ref}^{(n)}(\mathbf{r},\mathbf{r}')$ due to the distribution of dielectric parameters described by $c_{ref}^{(n)}$ by solving a set of forward scattering problems.
3. Solve (5.7) by means of a linear regularization strategy (e.g., by using a conjugate gradient [CG] method or a singular value decomposition [SVD] algorithm) in order to find a regularized solution $\tilde{c}_{diff}^{(n)}$.
4. Update the contrast function of the reference profile with

$$c_{ref}^{(n+1)}(\mathbf{r}) = c_{ref}^{(n)}(\mathbf{r}) + \tilde{c}_{diff}^{(n)}(\mathbf{r}) \tag{5.9}$$

5. Check if a proper stopping criterion is fulfilled. If not, repeat steps 2–4, otherwise stop. In the latter case, the updated reference profile is the estimated contrast function, that is, $\tilde{c} = c_{ref}^{N_{opt}}$ (N_{opt} being the iteration at which the algorithm is stopped).

It is worth remarking that the DBIM is equivalent to a Newton scheme applied to the solution of the combined inverse scattering equation (see [3.15]) [91]. Inversion approaches based on DBIM (or equivalent Newton schemes) have been proposed in several studies (for example, see [92–95]). It is worth noting that DBIM minimizes a least-square residual $\mathcal{R}_{LS}$ (similar to the one reported in [3.62] or [3.63]), subject to some regularization constraints. The regularization term can be forced directly in the outer linearization step (by adding an additional additive or multiplicative penalty term) or during the solution of the linearized problem [95]. The second option is sometimes preferred, since it allows employing standard regularization procedures for linear problems, which also permit to use well-known regulation parameter strategies. For example, in [93] a truncated CG is applied, where the regularization parameter is the number of performed iterations. It is worth noting that standard linear regularization procedures (e.g., those based on Tikhonov-type penalty terms) search for the solution with minimum norm. Consequently, the obtained solutions are typically affected by a certain low-pass filtering effect [96], resulting in a slight underestimation of the actual values and/or in the possible presence of ringing artifacts. Specific regularization strategies can be adopted for mitigating such drawbacks. For example, in [92] a weighted-norm total variation multiplicative regularization is applied to the contrast source formulation (Chapter 3), which has been found to be quite useful in preserving edges. In this approach, the following residual function is minimized

$$\mathcal{R}(c) = \mathcal{R}_{LS}(c)\mathcal{R}_{MR}(c) \tag{5.10}$$

where $\mathcal{R}_{LS}$ denotes a least square cost function measuring the misfit between the measured data and the scattered field computed from the dielectric distribution described by c (as in [3.63]), whereas the multiplicative term $\mathcal{R}_{MR}$ introduces a constraint on the variations of the contrast function, that is,

$$\mathcal{R}_{MR}\left(c_n\right) = \frac{1}{\Delta V} \int_V \frac{\left|\nabla c_n\left(\mathbf{r}'\right)\right|^2 + \delta_n^2}{\left|\nabla c_{n-1}\left(\mathbf{r}'\right)\right|^2 + \delta_n} d\mathbf{r}' \tag{5.11}$$

with ΔV being the volume of the investigation domain V, ∇ is the gradient operator with respect to the spatial coordinates, c_n is the contrast function reconstructed at the nth iteration, and $\delta_n^2 = \mathcal{R}_{state}\left(c_{n-1}\right) / \Delta^2$ (where $\mathcal{R}_{state}$ is the term of the least square cost function which is related to the state equation, and Δ is the linear dimension of the subdomains in which the investigation domain is discretized). Other similar multiplicative terms can be used in conjunction with the electric-field integral-equation (EFIE) formulation [97].

It is worth noting that the choice of the initial guess is quite important for the convergence of the DBIM approach (as well as for other deterministic inversion procedures). When no a priori information is available, a common choice is to use a homogeneous profile (e.g., with dielectric properties equal to that of the matching medium). Another possible approach is to use the average dielectric properties of the breast and, eventually, the information on the external surface of the breast can be included in the reconstruction process. Such information can be obtained, for example, by means of radar-based techniques [98–100]. An interesting study about the effects of initial solution in breast imaging is reported in [101], where it has been pointed out that "obtaining good results with DBIM requires patient specific estimates of the average properties" of the tissues inside the breast. In order to reduce the effects of the initial guess in the reconstructed images, alternative formulations can be used. For example, in [93, 102], it has been suggested that a log transformation of the complex data (in which the inverse scattering equation (5.7) is rewritten in terms of the logarithmic magnitude and of the unwrapped phase of the measured fields) can help in mitigating the issues related to the choice of the initial guess.

Other possible strategies to pursue the same objective rely on the use of a priori information. For example, constraints on the spatial variations in specific regions (e.g., those composed by tissues of the same type) can be imposed by adding specific regularization terms to the residual function minimized by the inversion procedure [103, 104]. For example, it can be assumed that, although the biological body is composed by different tissues, inside the same type of tissue the variations of the dielectric properties are small [104]. Consequently, during the iterative process, the updates of the unknowns (e.g., the contrast function) for points belonging to the same tissue (or located in the same region) are forced to be similar. Such approach has been found to "perform substantially better than the conventional DBIM" [104]. Methods based on the use of wavelet-based representations of the unknown dielectric profile have also been proposed [105].

In order to increase the available information and to enhance the robustness of the inversion schemes, multifrequency data can be exploited. In this case, it is assumed to have at its disposal the electric field data collected at several angular frequencies ω_1, ω_2, ..., ω_F, which can be directly obtained by using a stepped-frequency measurement scheme or can be extracted by the time-domain response with a Fourier transform. The simplest way to take into account all the available data is the use of a frequency hopping strategy [94, 106, 107], which can be summarized as follows (the dependence from the frequency is explicitly reported in the following relationships):

1. Set the frequency step to $f = 1$. Initialize the frequency-hopping scheme with an initial contrast function $c_{ref}^{(0)}(\mathbf{r}, \omega_1)$.
2. Solve a single-frequency inverse scattering problem to find an estimated distribution of the dielectric contrast $\tilde{c}\left(\mathbf{r}, \omega_f\right)$, at frequency ω_f
3. Set the new starting guess to $c_{ref}^{(0)}\left(\mathbf{r}, \omega_{f+1}\right) = \mathcal{M}\left(\tilde{c}\left(\mathbf{r}, \omega_f\right)\right)$, where $\mathcal{M}$ is an operator that modifies the values of the dielectric properties on the basis of a suitable dispersion model to take into account the frequency change.
4. Repeat steps 2–3 until all the available frequencies are processed.

A different strategy concerns the processing of all the available data (collected at different frequencies) at the same time [108]. However, since the dielectric parameters, in the frequency range considered for breast imaging applications, exhibit a significant variability with respect to the frequency, it is necessary to reformulate the inverse scattering problem by including a proper model of the dielectric properties based on frequency-independent parameters, that is, by assuming that

$$c_{diff}(\mathbf{r}, \omega) = T_\omega\left(\mathbf{x}_{diff}\right)(\mathbf{r}, \omega) \tag{5.12}$$

where $\mathbf{x}_{diff}$ is an array of functions containing the relevant parameters of the considered model (which are independent of the frequency), and T_ω is an operator expressing the mapping between the model parameters and the differential contrast function. In the field of breast imaging, a popular choice is the use of a single-pole Debye model (Chapter 3), which has been found to be able to cover the range of frequencies 1.0–3.5 GHz, typically assumed in this framework. It this case, the operator T_ω is nonlinear, and, consequently, the linearization of the scattering equations should take into account this term, too. However, a simplified version of the approach can be developed by assuming a constant value for the relaxation parameter τ of the Debye model. Such an assumption is motivated by the fact that, at least for the considered tissues and in the assumed frequency band, the relaxation parameter has a limited variability [108]. In this case, the operator T_ω can be expressed as

$$T_\omega\left(\mathbf{x}_{diff}\right)(\mathbf{r},\omega) = \frac{1}{\varepsilon_{r,b}}\left(\underbrace{\left(\varepsilon_\infty(\mathbf{r}) + \frac{\Delta\varepsilon(\mathbf{r})}{1+j\omega\tau} - j\frac{\sigma_s(\mathbf{r})}{\omega\varepsilon_0}\right)}_{\varepsilon_r(\mathbf{r})} - \underbrace{\left(\varepsilon_{\infty,ref}(\mathbf{r}) + \frac{\Delta\varepsilon_{ref}(\mathbf{r})}{1+j\omega\tau} - j\frac{\sigma_{s,ref}(\mathbf{r})}{\omega\varepsilon_0}\right)}_{\varepsilon_{r,ref}(\mathbf{r})}\right) \quad (5.13)$$

with

$$\mathbf{x}_{diff}(\mathbf{r}) = \mathbf{x}(\mathbf{r}) - \mathbf{x}_{ref}(\mathbf{r}) = \begin{bmatrix} \varepsilon_\infty(\mathbf{r}) - \varepsilon_{\infty,ref}(\mathbf{r}) \\ \Delta\varepsilon(\mathbf{r}) - \Delta\varepsilon_{ref}(\mathbf{r}) \\ \sigma_s(\mathbf{r}) - \sigma_{s,ref}(\mathbf{r}) \end{bmatrix} \quad (5.14)$$

where $\mathbf{x}$ and $\mathbf{x}_{ref}$ are the arrays containing the space-dependent Debye parameters for the actual and reference configurations.

Under the above assumptions, it is possible to straightforwardly apply the linearized relationship (5.7) and to directly reuse the DBIM scheme previously discussed [108]. In particular, the DWBA-based equation (5.7) can be rewritten as

$$\mathbf{E}_{diff}(\mathbf{r},\omega) = \mathbf{G}_{DB}\left(T_\omega\left(\mathbf{x}_{diff}\right)\right)(\mathbf{r},\omega) = \mathbf{G}_{DB,\omega}\left(\mathbf{x}_{diff}\right)(\mathbf{r},\omega),\ \mathbf{r}\in D_{obs} \quad (5.15)$$

where $\mathbf{G}_{DB,\omega}(\bullet) = \mathbf{G}_{DB}(T_\omega(\bullet))$ is a linear operator acting on the frequency-independent unknown vector $\mathbf{x}_{diff}$. By taking into account all the available frequencies, the linear inverse scattering problem that must be solved in step 3 of the previously described iterative procedure, can be then formulated as

$$\begin{bmatrix} \mathbf{E}_d(\mathbf{r},\omega_1) \\ \vdots \\ \mathbf{E}_d(\mathbf{r},\omega_F) \end{bmatrix} = \begin{bmatrix} \mathbf{G}_{DB,\omega_1}\left(\mathbf{x}_{diff}\right)(\mathbf{r},\omega_1) \\ \vdots \\ \mathbf{G}_{DB,\omega_F}\left(\mathbf{x}_{diff}\right)(\mathbf{r},\omega_F) \end{bmatrix} = \mathbf{G}_{DB,MF}\left(\mathbf{x}_{diff}\right)(\mathbf{r}),\ \mathbf{r}\in D_{obs} \quad (5.16)$$

where $\mathbf{G}_{DB,MF}$ is the overall operator obtained by combining all the single-frequency operators $\mathbf{G}_{DB,\omega_f}$, $f = 1, \ldots, F$, and $\mathbf{x}_{diff}$ is now the quantity to be estimated.

An example of a quantitative reconstruction of a heterogeneously dense breast model is shown in Figure 5.14 [104]. The measurement setup is composed by 40 dipoles uniformly distributed on five rings located on a cylindrical surface surrounding the target (with minimum distance from the skin equal to 1 cm) and excited with a Gaussian pulse. A lossless coupling medium with ε_r

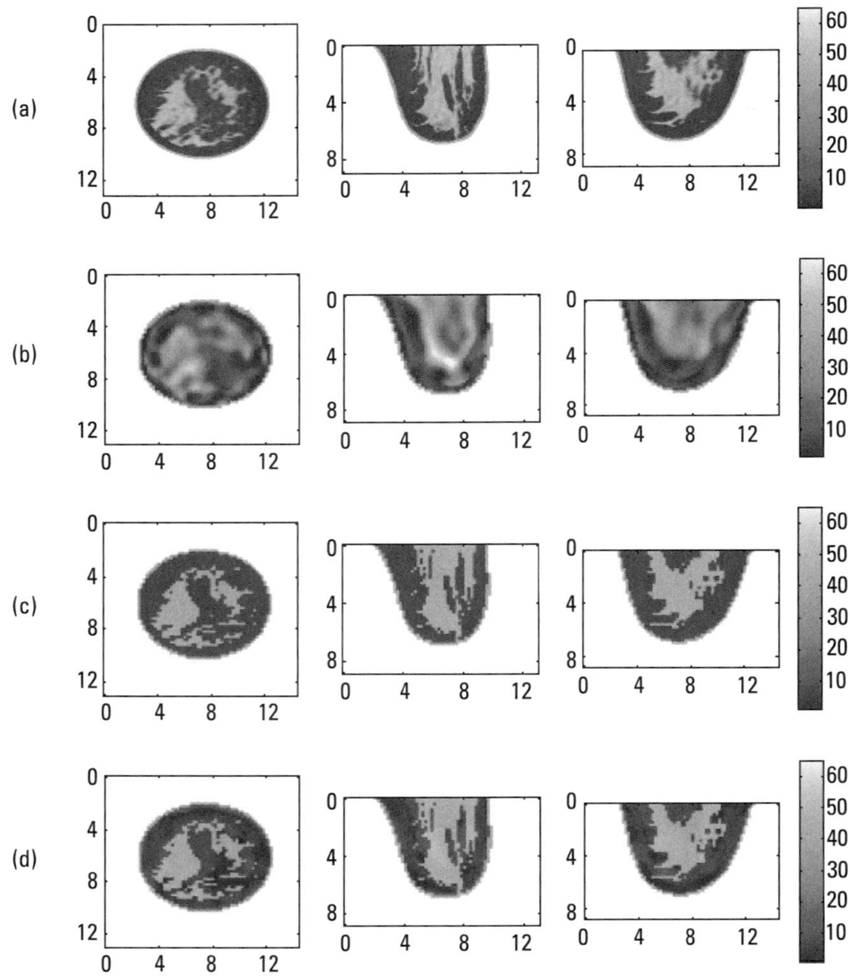

Figure 5.14 Examples of quantitative reconstructions of a heterogeneously dense breast phantom [104]. (a) Actual distributions of the dielectric properties at 2 GHz. (b) Reconstruction with the standard DBIM. (c) and (d) Reconstructions with the DBIM including spatial prior information [104]. (©2017 IEEE.)

= 2.6 is considered. The scattered field data has been simulated by using the FDTD method, and four frequencies (i.e., 1, 1.5, 2, and 2.5 GHz) are extracted by means of the Fourier transform for the application of the DBIM method. As can be seen from the reconstructed images (Figure 5.14 [b]), "the conventional DBIM shows the evidence of correctly identifying the general regions of adipose and fibroglandular tissue, but it suffers from blurred boundaries between the different tissues and inaccurate properties estimation" [104]. As discussed in [104], the a priori information is useful for enhancing the reconstruction quality. This is confirmed by Figure 5.14(c) and (d), which are related to two different choices of the regularization parameter.

As indicated in the introduction to this section, one of the main limitations of the use of inverse scattering approaches is related to the high dimensionality that arises when the related equations are discretized, especially when simple discretization schemes (such as the one based on pulse basis functions and described in Section 3.2) are applied. Two-dimensional approximations have been widely used in the literature for reducing the computational load. However, in real applications, it would be important to address the full vector 3-D problem. It is worth noting that in several systems proposed for breast imaging, the employed antennas are linearly polarized and directed along the same direction. Consequently, the measurement setup is only able to acquire information about one of the components of the electric field (in the following, assumed to be parallel to the z axis). In this case, it is possible to reduce the computational burden, mainly for what concerns the memory consumption. For example, the single-frequency inverse scattering equation (5.6) can be simplified as

$$\begin{aligned} E_{diff,z}(\mathbf{r}) &= E_{tot,z}(\mathbf{r}) - E_{ref,z}(\mathbf{r}) \\ &= -k_b^2 \int_V c_{diff}(\mathbf{r}')\mathbf{E}_{ref}(\mathbf{r}') \bullet \mathbf{G}_{ref,z}(\mathbf{r},\mathbf{r}')d\mathbf{r}',\ \mathbf{r} \in D_{obs} \end{aligned} \tag{5.17}$$

where only the z-component of the field vectors in the measurement points are considered and $\mathbf{G}_{ref,z} = \overline{\overline{\mathbf{G}}}_{ref} \bullet \hat{\mathbf{z}}$. Moreover, under the approximated assumption that only the z-directed component of the electric field of the reference profile is nonzero, the previous equation can be rewritten as a scalar equation [88–108].

Another possible way to approach the problem is to use smarter representations for the discretization of the inspected region. For example, in [88], a specific set of custom smooth basis functions has been derived for the interior of the breast, with the aim of reducing the dimensionality of the problem.

Other specific modifications of the DBIM inversion scheme have also been proposed. In [109], a beamforming strategy has been hybridized with the

DBIM. In particular, a set of modified equations is obtained by processing the measured data in order to synthetically focus the reconstruction on a series of focal locations in the imaging region. The obtained set of equations is solved by using a DBIM-like scheme. Such an approach turned out to be quite robust against the noise on the data. In [94], the linearized problems arising in the DBIM are solved by using a two-step iterative shrinkage/thresholding method, in which the updates of the iterative solution are obtained by considering not only the current solution, but also the one at the previous iteration (in this case, the weighting terms represent parameters to be properly tuned).

As already mentioned, conventional regularization techniques usually produce smoothed images. This fact can of course constitute a practical limitation in biomedical imaging. Consequently, other research directions have been explored. Compressive sensing (CS) techniques and sparsity-based inversion approaches have been considered, not only in the biomedical field, as a possible alternative approaches [110–112]. It is well known that, especially when dealing with inhomogeneous breast configurations, the contrast function is not in general a sparse function; however, it seems to be possible to represent it, at least in an approximate way, by using some basis set able to promote the sparsity. To this aim, in order to describe the dielectric profile, it is possible to introduce a sparse function χ such that

$$c(\mathbf{r}) = \Psi(\chi)(\mathbf{r}) \tag{5.18}$$

where Ψ is an operator mapping the new unknown χ with the contrast fucntion c unknowns. In this framework, wavelets have been proposed as a way to produce sparse representations of the contrast function [111, 113].

Moreover, it has been pointed out in [111] that using pure L^1-based CS approaches within the DBIM iterations can lead to some numerical difficulties. Consequently, several different strategies re-adding some kind of L^2 regularization have been studied. Basically, when applied to the DWBA, the following optimization problem is usually solved (with different possible implementations [111, 112, 114, 115])

$$\min_{c_{diff}} \left\| \mathbf{E}_{diff} - \mathbf{G}_{DB}\left(c_{diff}\right) \right\|_2^2 + \gamma_1 \left\| c_{diff} \right\|_1 + \gamma_2 \left\| c_{diff} \right\|_2^2 \tag{5.19}$$

where γ_1 and γ_2 are weighting parameters.

As previously pointed out, if c_{diff} is not sparse, a suitable sparsifying operator should be applied (which is not included in [5.19]). Clearly, when $\gamma_2 \to 0$, the approach becomes similar to the one involved in the standard CS formulation (see Chapter 3), whereas for $\gamma_1 \to 0$ a Tikhonov-type regularization is attained.

Other inversion schemes can be applied to solve the inverse scattering problem arising in breast imaging. Concerning quantitative approaches (which have been proposed for long time for solving diagnostic problems in biomedical applications [116–118]), the use of the contrast source formulation (see [3.19]) has been, for example, proposed in [119, 120]. Qualitative methods have been adopted as well. In particular, the multiple signal classification (MUSIC) method [121], the linear sampling method (LSM) [122], and the level set method [123] have been successfully applied. Since they provide only qualitative reconstructions, hybridized implementations involving also quantitative methods can be very effective (e.g., with the previously described DBIM). As an example, in [120], the LSM is used to create a suitable initial guess for the DBIM, whereas in [123] the DBIM is used to initialize a technique based on the level set method in order to identify the boundaries among different tissues with high resolution.

Finally, time-domain inverse scattering approaches can be used, too. In particular, procedures based on the Born approximation and on the DWBA implemented in time domain have been applied for breast imaging. For example, in [124] the outer linearization/approximation steps are performed in time domain, whereas the solution of the linearized problem is solved in the frequency domain (the Fourier transform is used to obtain the related input data). This solution is obtained by applying a CG method with a Tikhonov regularization. Forward-backward time stepping schemes [125, 126], and time-reversal approaches [127–130] have also been exploited.

5.4.3 Overview of Experimental Apparatuses

In the last few years, several experimental systems and apparatuses have been specifically designed and realized for breast microwave imaging. A selection of them is listed in Table 5.2. As for other applicative diagnostic scenarios considered in the previous chapters, there are essentially two possible ways of designing a measurement system for breast imaging, that is, by considering frequency-domain or time-domain apparatuses and techniques. In the first case, the scattered field data are collected by means of a frequency-stepped strategy, that is, several single frequency measurements (which, for example, can be obtained by using a vector network analyzer) are sequentially performed.

The prototype developed in [131] represents an example of a stepped system. In this system, a multistatic imaging setup with 16 monopole antennas that can act in transmitting and receiving modalities has been realized. The antennas, which are located on a circumference, can also move vertically, in order to obtain measurements at different heights. The working frequency changes in the range 500 MHz–3 GHz. The breast is inserted in a tank (which is located below a table on which the patient can lie). This tank is filled with a coupling

Table 5.2
Selected Examples of Microwave Breast Imaging Systems

	Antennas	Coupling media	Frequency	Notes
Dartmouth College [131]	16 monopole antennas located on a circumference (multistatic, vertical position can be changed)	Water/glycerin mixture	500 MHz–3 GHz	Frequency stepped. A laser system is used for obtaining the outer breast surface.
University of Bristol [132]	60 cavity/backed UWB slot antennas located on a hemispherical cup (multistatic)	Water/oil mixture	4–8 GHz	Frequency stepped. A pulse is synthesized from frequency domain data.
University of Calgary [133]	1 balanced antipodal Vivaldi antenna (mechanically moved to scan a cylindrical surface, monostatic)	Canola oil	50 MHz–15 GHz	Frequency stepped. A pulse is synthesized from frequency domain data. A laser system is used for obtaining the outer breast surface.
McGill University [134]	16 traveling-wave tapered and loaded transmission-line antennas located on a hemispherical cup (multistatic)	Ultrasound gel	2–4 GHz	Time domain system, Pulses with duration of 70 ps and repetition rate of 25 MHz.

medium made by a water/glycerin mixture. Moreover, in the most recent revision of this imaging setup, a laser displacement measuring system has been added, in order to reconstruct the external profile of the breast to be used as a priori information for the inversion algorithms.

In [133], an imaging setup with a single antenna that can be moved in order to collect monostatic data in several positions on a cylindrical surface (as in the previous described system) has been described and tested. For any scan, a set of about 200 measurements is performed by using a vector network analyzer. The measurement samples are acquired for 1,601 different frequencies in the range 50 MHz–15 GHz. Since the applied inversion procedure is based on a beamforming technique, the measured data are processed in order to synthesize a Gaussian-derivative pulse with a half-power bandwith between 1.3 GHz and 7.6 GHz. A laser system for obtaining the information about the external shape of the breast is adopted in this case as well.

The system proposed in [132] is based on a multistatic setup, in which 60 antennas are located on a hemispherical cup in which the breast can be inserted. The antennas are UWB slot antennas. They are connected, through a switch, to a vector network analyzer that collects frequency-stepped data between 4 GHz and 8 GHz. As for the system developed in [133], the measured frequency-domain data are used to synthesize the pulses needed for the application of beamforming algorithms.

In [135], a custom transreceiver module, based on CMOS technology, has been designed for medical imaging applications, in order to avoid the use of the vector network analyzer, which represents quite expensive laboratory equipment. The possibility of using low-cost off-the-shelf components is also discussed in [136]. In this case, low-cost printed antennas are immersed in a coupling liquid and are applied to collect multistatic data at frequencies between 1.4 GHz and 1.6 GHz. The experimental results reported in that work seem to indicate that the "detection is possible with accuracy similar to what can be achieved using standard costly microwave equipment" [136].

The use of metallic tanks has also been recently investigated [137], since they can provide some advantages with respect to the dielectric enclosures usually employed in the imaging setups, such as more accurate system modeling, ability to use low-loss matching medium, less ambiguity on the field near the walls, and shielding of the imaging scenario.

Another interesting example is represented by a planar microwave camera [138]. Such a system, differently from the previous ones, is based on the use of two horn antennas with an aperture of 30 × 30 cm, which are positioned beside a water tank containing the target. The modulated scattering technique (MST) (see Chapter 4) is employed for acquiring the data on a retina composed by 32 × 32 dipoles loaded by PIN diodes.

In the second class of systems, a pulse (or a train of pulses) with a short duration is transmitted and the time-domain received signal is recorded. Such an approach is potentially able to reduce the acquisition time, which in biomedical applications can be a serious limiting factor, since the patients can move during the measurement phase.

An example of a time-domain multistatic radar system for breast imaging has been developed in [134]. Such a system has also been used for performing some clinical tests on volunteer patients, whose results have been reported in [139]. The apparatus includes 16 resistively loaded traveling-wave antennas [140] mounted on a hemispherical dielectric structure (four elements are located in any quadrant), as shown in Figure 5.15. In particular, the antennas are positioned in order to collect both co- and cross-polarized signals. The transmitting/receiving antenna pair is selected by using a 16 × 2 electromechanical switching matrix. Pulsed incident fields with duration of 70 ps are generated. The pulses are shaped by using a passive microstrip line in order to obtain a working frequency in the band 2 GHz–4 GHz. Finally, the repetition frequency is 25 MHz.

A flexible measurement system has also been proposed [141, 142]. A set of 16 antennas (arranged in 4 equally spaced linear arrays) are mounted on a flexible structure (see Figure 5.16[a]) to be put directly in contact with the

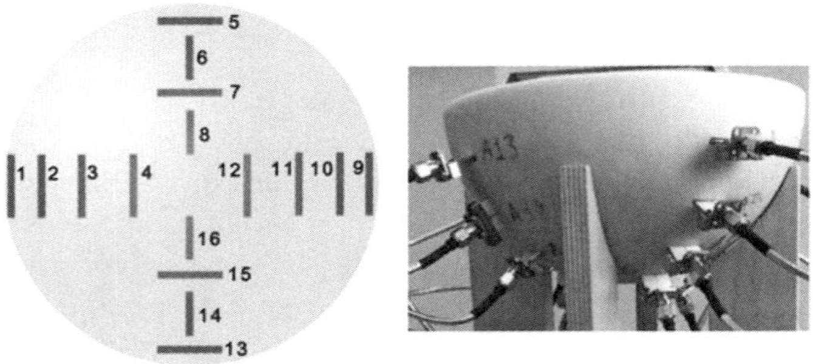

Figure 5.15 Example of hemispherical antenna setup from [139]. (©2016 IEEE.)

breast. Two types of UWB antennas have been considered for single and dual polarization. The single polarization antenna is a wideband monopole, whereas the dual-polarization antenna is a planar spiral single-arm antenna. The measured S_{11} parameter for such antennas is reported in Figure 5.16(b). An important issue for radiating structures of this kind concerns the materials to be adopted, which must be biocompatible, flexible, and soft as much as possible.

Another example of a multistatic time-domain system is the one developed in [143], which includes 16 antennas positioned on a hemispherical support. The system uses an interrogating field in the form of a UWB pulse (with frequency ranging from 3 GHz to 10 GHz). A single master clock is used for commuting all the antennas. The so-called equivalent time sampling technique [144] is employed for obtaining a very limited acquisition time (1.32 μs for a single acquisition from the 16 elements).

5.5 Brain Stroke Imaging

Brain stroke is clinically characterized by an abrupt onset of a focal neurological deficit, due to a vascular occlusion secondary to atherosclerotic thrombosis or embolism, or to a cerebral hemorrhage due to the rupture of a cerebral blood vessel. The essential clinical features are a loss of strength in one side of the body, speech disorders, imbalance or sensory disturbances. The most effective therapy in stroke due to vascular occlusion is intravenous or intra-arterial thrombolytic agents such as tissue plasminogen activators [145]. However, this therapy has to be administered within the first 3 hours from the onset of symptoms [146], or at least in the first 4.5 or 6 hours from the appearance of disturbances. Therefore, it is of main relevance that the patient is admitted in a short time period to a hospital where there is an active stroke unit, because it is mandatory to have a correct differential diagnosis between ischemic (almost 80-85% of all strokes) and hemorrhagic (almost 15-20%) strokes, which are clinically indistinguish-

(a)

(b)

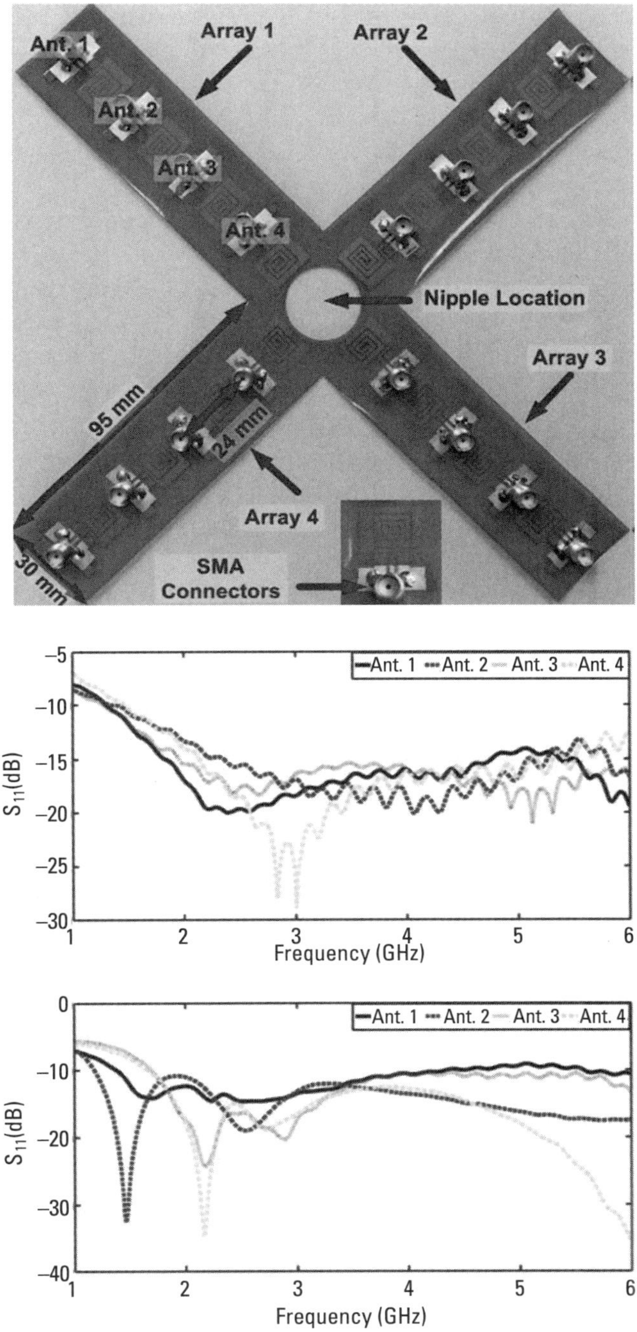

Figure 5.16 Example of a flexible measurement setup for breast cancer detection and the plots of the related S_{11} parameters [141]. (a) View of the antenna array and (b) measured S_{11} for different antenna positions. (©2015 IEEE.)

able but can easily be differentiated by CT examination. In ischemic strokes, the thrombolytic therapy—if carried out in the first hours—is very effective, while in hemorrhagic strokes it is contra-indicated. The time interval between the onset of symptoms and the start of the therapy is therefore crucial, but usually a lot of time is lost for the arrival of the patient in the hospital, the clinical and neurological evaluation, and the CT examination. If the period between the appearance of symptoms and the final diagnosis exceeds 3–6 hours, the thrombolytic therapy cannot be done anymore, due to its complication and loss of efficacy, considering that the cerebral tissue without oxygen does not survive for a long period and will become necrotic. The possibility to have a correct differential diagnosis between an ischemic and a hemorrhagic stroke in a short period of time is therefore of main relevance for the administration of an adequate therapy and for a positive clinical outcome.

The above considerations have led to a strong research activity focused on using alternative diagnostic methodologies. In this framework, microwave imaging has been proposed as a promising technology for the development of innovative, noninvasive, and portable diagnostic systems, which could be included in the medical apparatuses of the ambulances.

The specific goal of these systems, possibly to be realized in the form of helmets, is to point out in a very short time the possible presence of hemorrhagic spots, which clearly characterize (as mentioned) one of the two kinds of strokes. The detection of the ischemic stroke, however, is considered more difficult at microwaves, due to the smaller contrast with respect to the healthy tissues.

As far as we know, the first idea of using microwaves for detecting brain injury and hemorrhage has been reported in [147]. That paper (which was based on previous works devoted to breast cancer imaging) is basically founded on the consideration that a blood spot inside the brain represents a discontinuity in the dielectric properties of a brain, allowing its detection based on reflection concepts (i.e., by using a monostatic configuration; see Chapter 2). The authors reported results concerning a porcine brain located inside a human skull. The reflection coefficient (amplitude and phase), measured between 2 GHz and 6 GHz, seemed to indicate a significant difference when a blood mass was present. The authors also suggested the possibility of using a classification method to further process the received signals.

The idea of using microwaves for brain imaging has been reconsidered in [148], with reference to the design of triangular patch antennas to be included in a possible imaging system shaped as a helmet. In the same year, an approach has been proposed in [41], which extends to brain strokes a method previously developed for the imaging of biological tissues [149, 150]. The authors recognized that the microwave diagnosis of brain strokes is a difficult task, since the brain is covered by a dielectric shield (i.e., the human skull, whose dielectric properties constitute a high dielectric contrast with respect to the dielectric properties of the other tissues (see Section 5.2). Optimal imaging conditions have been defined in [41] by using a simplified 2-D model of the head (a multilayer circular cylinder) and 32×32 or 64×64 ideal antennas positioned on a circumference of 11 cm of diameter. The inversion process has been performed by using a Newton-type method and a frequency hopping technique in the range 0.5 GHz–2.0 GHz, without considering the tissue dispersion. On the basis of the obtained results, the authors suggested as operating range the frequency band 0.7 GHz–1.0 GHz, with about 40–60 dB of signal-to-noise ratio on the measured data. An interesting review paper on this matter has been reported in [151]. It is worth noting that the choice of the optimal frequency range is a fundamental task for this kind of application and several works have been published on this topic [152–154]. As an example, by using a phantom composed of liquids with dielectric properties similar to those of biological tissues, it has been experimentally shown in [152] that "the wave penetrates well up to 1.3 GHz as there is a strong attenuation around 2 GHz, and low levels of transmitted power above 3 GHz".

Succeeding these preliminary approaches, the detection and imaging of brain strokes at microwave frequencies have been pursued by the following different strategies. Some of the most relevant ones are mentioned in the next sections.

5.5.1 Approaches Based on Classification

One of the possible approaches for the detection of brain strokes by using microwave radiation is represented by classification methods [155–157]. Actually, these techniques do not aim at providing images of the brain. Therefore, they are just briefly mentioned in the present book. With these methods, accurate models of the electromagnetic scattering inside a head are not strictly required, and—in addition to a proper measurement system design—most efforts should be devoted to the search for a good training set. However, it is worth noting that these approaches have led to very interesting results, and some concepts can be really useful for the development of microwave systems for imaging purposes as well. On the one hand, such techniques are based on system designs that are very close to the ones adopted in imaging apparatuses (e.g., antennas, operating frequency bands, measurement devices) [148, 156, 158, 159]. On the other hand, due to these similarities and to their intrinsic speed, classification methods could be integrated with imaging algorithms in order to improve both the quality and the reliability of results.

For instance, in the classification-based approach described in [155], a set of antennas is used to produce microwave radiation inside the patient's head and the scattering parameters are collected by using a vector network analyzer.

Subsequently, the measured data are given as the input of a previously trained classifier, which determines the possible presence of a brain stroke. The reported results are promising, and after an initial phase that included numerical simulations and tests on human head phantoms [160], considerable clinical trials are actually in progress [161].

5.5.2 Qualitative Approaches

Among the large family of imaging methods designed for brain strokes and intracranial hemorrhage detection, many of the adopted imaging procedures are based on qualitative methods [162–164]. Radar and beamforming concepts are exploited in most cases, even though with reference to differential imaging applied to brain stroke follow-up, linearized approaches and the linear sampling method have been used as well [40, 154, 165, 166].

Radar-based methods, sometimes formulated in the frequency domain, are often combined with preprocessing algorithms, which include advanced clutter rejection techniques [167], as well as the exploitation of symmetries [168, 169]. In particular, confocal delay-and-sum (DAS) techniques [164, 168, 170–172] and multistatic microwave imaging via space-time beamforming (MIST) have been used in this scenario [173], [174]. The basics of these approaches have been presented in Section 5.4.1.

An example of the application of such techniques is provided in Figure 5.17, where DAS and MIST methods are used in conjunction with a preprocessing algorithm based on the principal component analysis (PCA) [169].

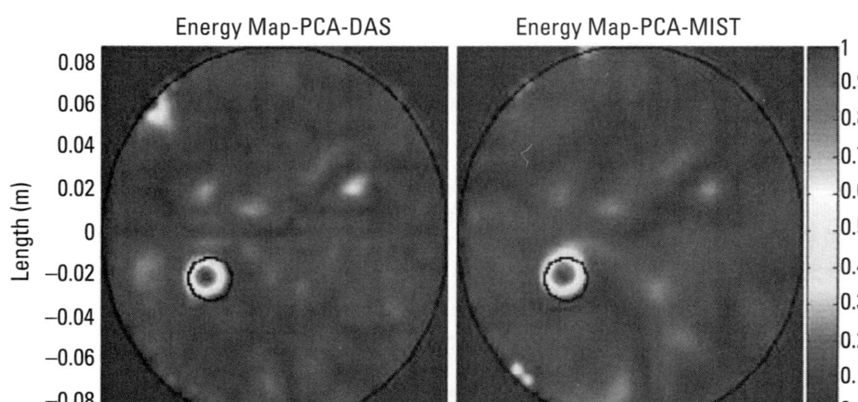

Figure 5.17 Brain stroke imaging. Energy maps (normalized qualitative images) obtained by using DAS and MIST beamforming methods [169]. A preprocessing algorithm is used. (©2017 Springer Nature.)

Compressive sensing (CS) approaches have also been combined with radar-based stroke imaging [175].

In the framework of qualitative techniques, both monostatic [163, 168] and multistatic measurement systems have been successfully adopted [38, 167]. Some of these radar-based approaches have been fruitfully integrated into experimental prototypes, and are under validation with realistic head phantoms and human volunteers [66, 167, 176, 177]. For instance, the experimental results on a human head phantom obtained by applying the method described in [65] are reported in Figure 5.18.

5.5.3 Quantitative Approaches Based on Inverse Scattering

Despite the difficulties that arise in applying microwave imaging techniques to brain stroke detection, the quantitative reconstruction of the dielectric properties of the patient's head is also attracting the attention of the research community [37, 39, 40, 178–181]. The interest comes from the possibility of charac-

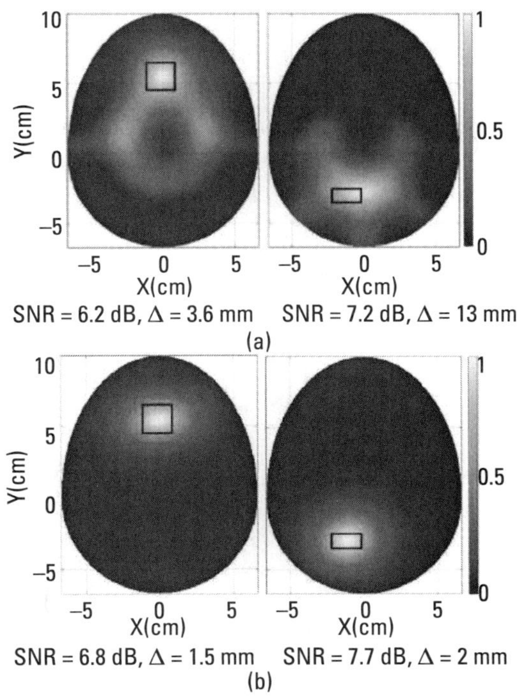

Figure 5.18 Brain stroke imaging. Reconstructed images from experimental data [65], using (a) real 8-element array, and (b) virtual 16-element array. Left images are related to the $2 \times 2 \times 2$ cm^3 and right images are related to the $2 \times 1 \times 1$ cm^3 bleeding target. The black rectangles show the exact location of target. (©2017 IEEE.)

terizing and locating brain strokes from a quantitative image of the dielectric properties of the head. Both 2-D and 3-D approaches have been investigated, following several different strategies, as outlined below.

With reference to the formulation reported in Section 3.2, several 2-D approaches have been developed. Although a human head is clearly a 3-D target very far from a cylindrical structure, under some assumptions about the radiation properties of the antennas it is still possible to adopt 2-D quantitative reconstruction strategies [182]. Most of the developed approaches use different versions of nonlinear Newton-based and Born/distorted Born iterative methods [37, 41, 92, 153, 179, 181, 183–185]. However, these techniques may need a proper initial guess for ensuring a good convergence of the procedure, as is also discussed in the previous Section 5.4.2. Therefore, the possible use of segmented MRI-derived images as a starting guess has been investigated in [186, 187]. Contrast source formulations have been also adopted [119]. It is worth noting that quantitative imaging techniques always use multistatic configurations [37, 39, 181, 188].

An example of a 2-D quantitative reconstruction is reported below. With reference to Figure 5.19, a tomographic configuration is considered, in which a slice of the VoxelMan head model (see Section 5.3.2) is used (in particular, the slice with index #40, which is located at 5.46 cm from the top of the head). The tissue parameters from [10] are adopted (see Section 5.2). The investigation domain D that surrounds the head is centered at the Cartesian axes origin and has an elliptic cross section characterized by an x-oriented minor axis of 16 cm and a y-oriented major axis of 20 cm. The values of the electric field inside both the investigation and measurement domains have been numerically computed by applying the method of moments. In the forward problem solution, the region

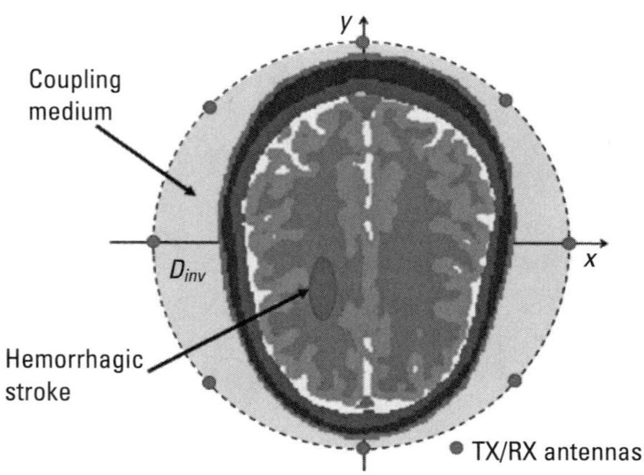

Figure 5.19 Schematic representation of the 2-D imaging setup for brain stroke detection [183]. (©2017 IEEE.)

D has been partitioned into 5,199 square subdomains of side length $l_f = 2.2$ mm. Conversely, the inversion process has been performed through a coarser discretization, that is, 1,300 cells of side $l_i = 4.4$ mm. The simple problem discretization reported in Section 3.2 has been adopted. The head model is surrounded by a lossless matching medium with relative dielectric permittivity $\varepsilon_{r,b}$.

A multistatic measurement system is used for collecting the total field data around the head. In particular, $V = 30$ ideal line current antennas have been simulated. They are equally (angularly) spaced on an ellipse of minor and major axes of 18 cm and 22 cm, respectively. The investigation domain D is sequentially illuminated by each antenna and the remaining $M = V - 1 = 29$ points are assumed to contain the electric field probes. The numerically computed scattered electric field (i.e., the differential field $E_{diff,z}$ with respect to the reference dielectric configuration) has been corrupted by an additive white Gaussian noise with zero mean and signal-to-noise ratio $SNR = 25$ dB. An elliptically shaped target with dielectric properties of blood has been inserted for simulating the presence of a hemorrhagic brain stroke. This inclusion is centered at $\mathbf{r}_s = (2,0)$ cm, has minor axis a_{hs}, and major axis $b_{hs} = 2a_{hs}$. The inverse scattering problem has been solved by using the inexact-Newton method in L_p Banach spaces described in Section 3.3. The outer Newton and the inner CG iterations are terminated when the normalized difference between the residuals in two subsequent steps is under the thresholds $d_{IN}^{th} = 1\%$ and $d_{CG}^{th} = 1\%$, respectively. Furthermore, maximum numbers $N_{IN} = 100$ and $N_{CG} = 100$ for outer and inner iterations have been fixed. The L^p Banach-space norm parameter p has been varied between the values 1.1 and 2 (the last value being the Hilbert space approach).

The considered test case is about the reconstruction of a brain stroke of axis $a_{hs} = 1$ cm (see Figure 5.20). A coupling medium with relative dielectric permittivity $\varepsilon_{r,b} = 20$ is employed. An operating frequency $f = 0.6$ GHz has been used and the healthy head has been considered as the reference model for the inversion procedure. The differences between the reconstructed and reference electric conductivity (i.e., $\Delta\sigma = \sigma - \sigma_{ref}$) are reported in Figure 5.20. Both Hilbert space ($p = 2$) and Banach space ($p = 1.3$) reconstructions are shown. As can be seen in Figure 5.20(b), significant ringing effects in the background affect Hilbert spaces solutions, and the reconstructed dielectric properties are underestimated. Instead, the corresponding inversion results operated in Banach spaces, presented in Figure 5.20(c), are clean from background artifacts and the estimation of the dielectric properties is more accurate.

Although most of the systems so far proposed are related to 2-D configurations, some preliminary 3-D reconstruction results have been obtained by using the scattering formulation reported in Section 3.1. It should be stressed that one of the main potential features of microwave imaging is related to the fact that the 2-D approach can be in principle straightforwardly extended to

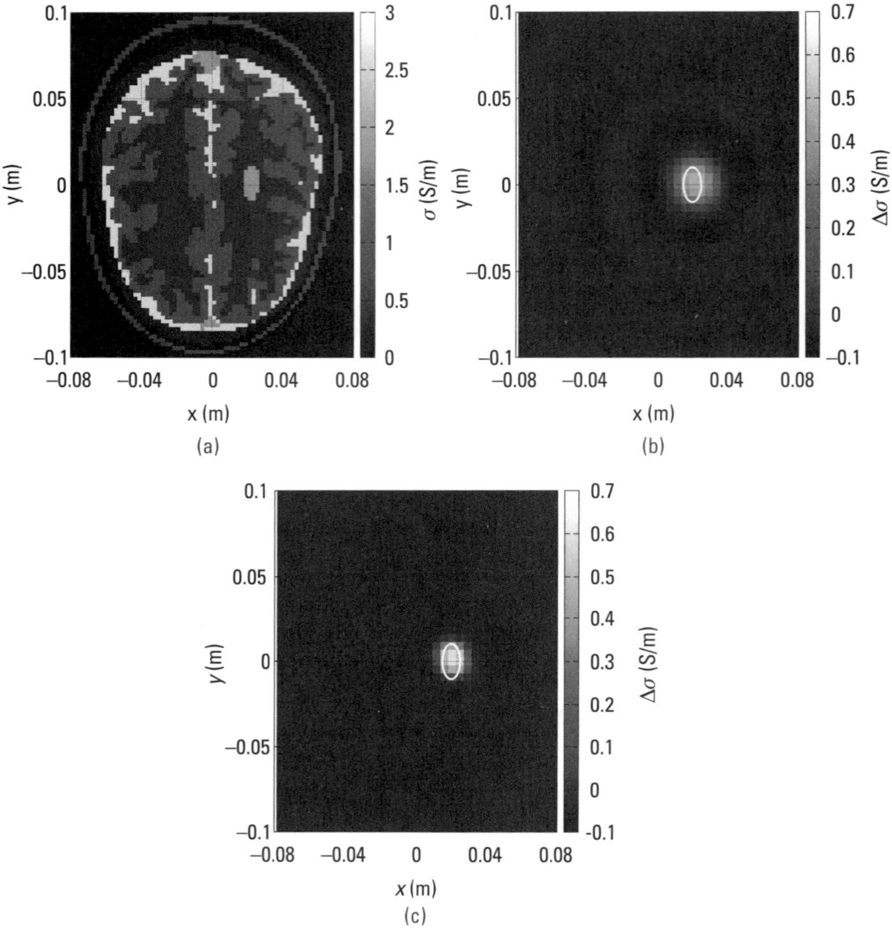

Figure 5.20 Head model with hemorrhagic stroke of minor axis $a_{hs} = 1$ cm. Distributions of the actual and the reconstructed differential dielectric properties in the investigation domain. (a) Actual relative dielectric permittivity, (b) reconstructed differential conductivity in Hilbert spaces, and (c) reconstructed differential conductivity in Banach spaces ($p = 1.3$).

3-D imaging. Actually, the main difficulty remains related to the computational resources needed to face a full vector 3-D inverse scattering problem [182].

A 3-D image of a brain, showing the reconstruction of the relative dielectric permittivity, is reported in Figure 5.21 [188]. A gradient-based method with a Tikhonov regularization is used [189]. It refers to the reconstruction of a hemorrhagic stroke in a human head phantom composed by a shell with a dielectric permittivity $4 - j0.4$, which has been filled with a brain-mimicking liquid with relative dielectric permittivity equal to $44.0 - j18.5$. The filling liquid is also used as coupling medium. The stroke-modeling material has a relative dielectric permittivity equal to $59.0 - j18.0$. The working frequency is

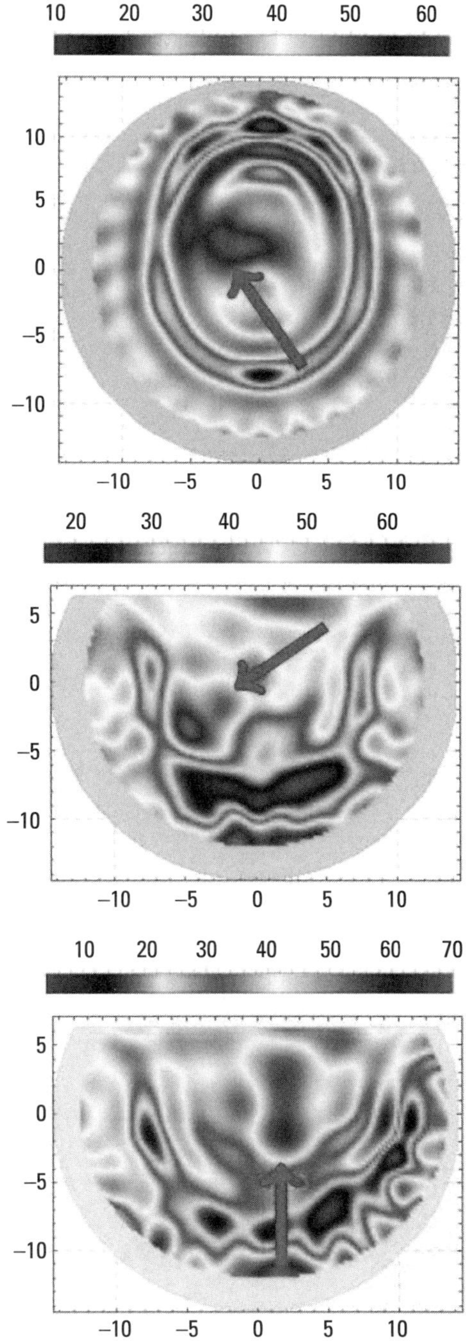

Figure 5.21 3-D reconstruction of a human head phantom with a hemorrhagic stroke (axial, coronal, and sagittal slices) [188]. (©2016 IEEE.)

1.0 GHz. The experimental setup described in Section 5.5.4 has been used. As can be seen, in this case, the presence of the blood spot is clearly visible in the reconstructed image.

5.5.4 Overview of Experimental Imaging Apparatuses

The company *EMTensor GmbH* (Vienna, Austria) has developed several microwave systems for head imaging [39, 182]. In particular, the apparatus described in [182, 188] is composed by 177 antennas arranged in 8 circular rings on a spherical metallic chamber with diameter 29 cm (see Figure 5.22[a]). The antennas are open-ended rectangular waveguides with dimensions of 21 mm × 7 mm × 53 mm and loaded with a material with a relative dielectric permittivity equal to 60. The fundamental cutoff frequency is 922 MHz. Two matrices

(a)

(b)

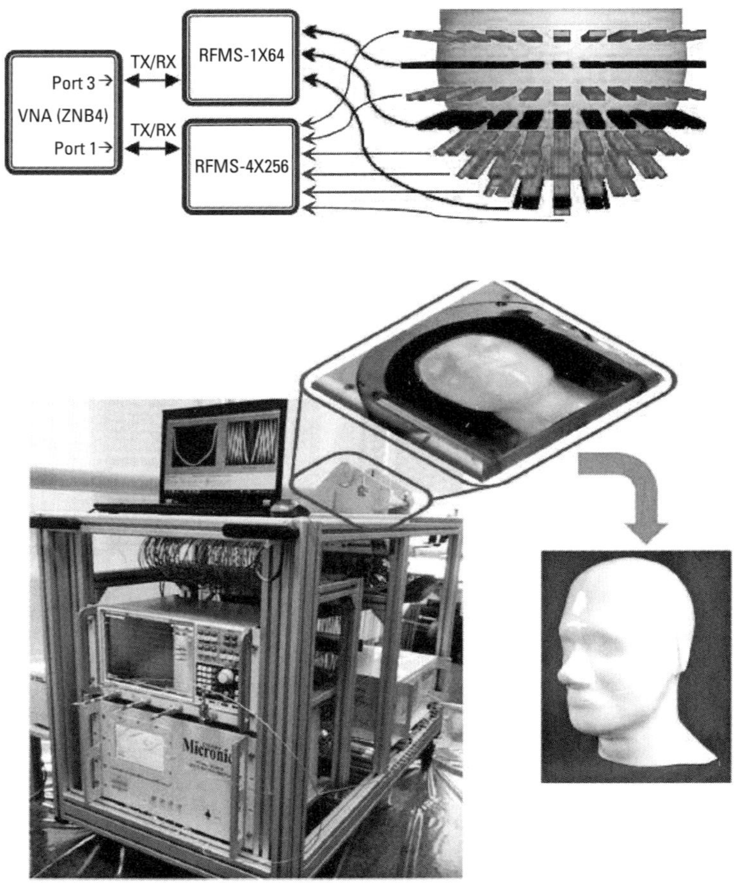

Figure 5.22 Microwave head imaging system [188]. (a) Antenna and (b) measurement setups. (©2016 IEEE.)

of microwave switches are used to connect the antennas to a vector network analyzer. The measurements are performed in two steps. First, the antennas on the rings colored in gray are sequentially used as transmitter and the black ones as receivers. Subsequently, the transmitting and receiving ports of a vector network analyzer are exchanged and the measurements are repeated for the new combinations. A photograph of the experimental setup is shown in Figure 5.22(b).

Another imaging system is shown in Figure 5.23. Such a system includes a vector network analyzer as microwave transceiver, a microwave switch matrix, and an adjustable measuring platform [168, 177]. The platform can accommodate up to 16 antennas whose radial positions can be adjusted by moving the holders along linear slits. In [65], slot-loaded folded UWB dipole antennas are used. The dimensions of the antennas are $7 \times 3 \times 1.5$ cm^3 and they operate in the band 1.1–3.23 GHz. A frequency-stepped strategy (with a frequency step of 6 MHz) is employed for acquiring the scattered field data. Moreover, a statistical technique based on the Kriging method is used for building a virtual antenna array with a larger number of antennas than the real ones available in the imaging system [65].

Concerning the antennas used for brain stroke imaging, they of course depend on both the frequency range and the required bandwidth. In air-coupled systems, ultrawideband antennas are often used, especially with qualitative reconstruction techniques [65, 177, 190–192]. Antennas matched on the human

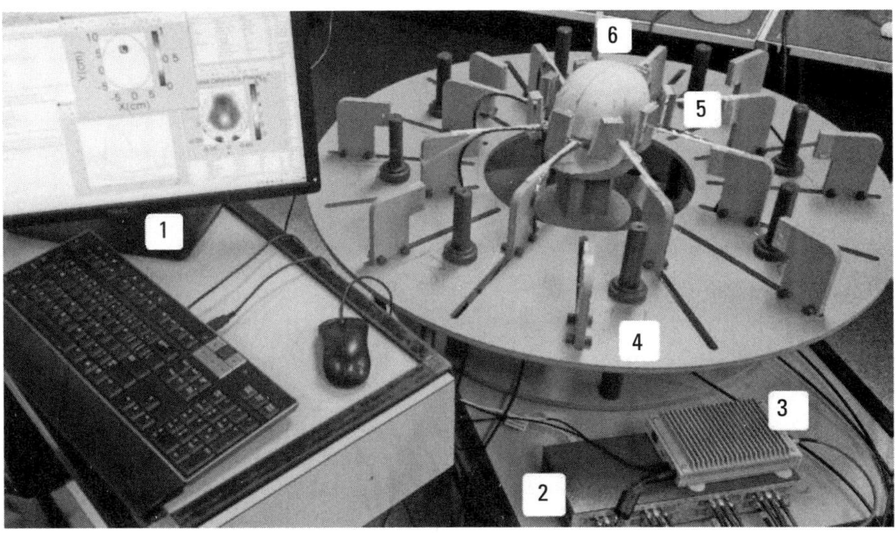

Figure 5.23 Microwave head imaging system [65]: (1) computer, (2) mini-circuit switches, (3) microwave transceiver, (4) imaging platform, (5) antenna array, and (6) human head phantom. (©2017 IEEE.)

head have been also proposed [193], whereas dielectric-filled open waveguides have been utilized in the quantitative imaging chambers described in [39, 182].

5.6 Other Medical Applications

Besides the specific application previously discussed (i.e., breast and brain diagnostics), the research activity on microwave biomedical imaging proceeds fast and covers theoretical, numerical, and experimental aspects. Key aspects are continuously reconsidered and evaluated more and more in depth. Just as an example, it has been pointed out several times in the need for an a priori starting configuration for a fast and accurate reconstruction process (as discussed in the previous section—mainly with reference to breast imaging—this problem has been considered in several studies, some of them cited through this book). In this framework, the Born and Rytov approximations (Chapter 3) have been recently reconsidered. For example, by using a phantom constituted by a chicken wing (composed of bone and muscle), which has been placed in the opening of an absorber sheet and surrounded with lard, very fast images based on the above approximations (3 seconds of CPU time) have been obtained [194].

Moreover, microwaves have been considered for imaging purposes in other medical applications, apart from breast and brain diagnostics. For example, in [195, 196], the imaging of human forearms has been addressed. A metallic cylindrical chamber is used to collect the scattered field data for frequencies between 0.8 GHz and 1 GHz, which are inverted by using a method based on the contrast source formulation (Chapter 3). An approach to bone imaging has also been presented [197]. In particular, a scanning apparatus has been tested with a multilayer complex limb-mimicking phantom and the image formation has been obtained by a procedure based on the noncoherent migration approach. Moreover, the application of Newton inversion schemes to the imaging of the calcaneus (heel bone) has also been evaluated [198].

Furthermore, imaging techniques can be applied for monitoring other medical treatments, such as microwave ablation [199–201], in which the electromagnetic energy is focused on tumors by using interstitial antennas in order to produce tissue-heating effects and, consequently, tissue necrosis. This therapy has been mainly proposed for solid tumors that are nonsurgically tractable. It is worth noting that more consolidated approaches such as NMR and ultrasound imaging techniques have also been proposed for monitoring this treatment. However, they have been found to exhibit some drawbacks related to the high cost (in the case of MRI), to the possible heating of the contrast agents, and to the distortion of the images induced by the temperature as well. For these reasons, monitoring techniques using microwaves are considered quite competitive solutions.

Recently, microwave imaging techniques have also been proposed as potential candidates to diagnosis the cervical myelopathy (or cervical spine damage) [202], which is a disorder of the spinal cord that can be due to the squeezing or compression of the spinal cord inside the neck (especially in elder patients) [202, 203].

References

[1] Larsen, L. E., and J. H. Jacobi (eds.), *Medical Applications of Microwave Imaging,* New York: IEEE Press, 1986.

[2] Bolomey, J. C., et al., "Microwave Diffraction Tomography for Biomedical Applications," *IEEE Trans. Microw. Theory Tech.*, Vol. 30, No. 11, November 1982, pp. 1998–2000.

[3] Paulsen, K. D., P. M. Meaney, and L. C. Gilman, *Alternative Breast Imaging: Four Model-Based Approaches,* New York: Springer Science & Business Media, 2005.

[4] Foster, K. R., and H. P. Schwan, "Dielectric Properties of Tissues and Biological Materials: A Critical Review," *Crit. Rev. Biomed. Eng.*, Vol. 17, No. 1, 1989, pp. 25–104.

[5] Stuchly, M. A., et al., "Dielectric Properties of Animal Tissues in Vivo at Frequencies 10 MHz–1 GHz," *Bioelectromagnetics,* Vol. 2, No. 2, 1981, pp. 93–103.

[6] Stuchly, M. A., and S. S. Stuchly, "Dielectric Properties of Biological Substances—Tabulated," *J. Microw. Power,* Vol. 15, No. 1, January 1980, pp. 19–25.

[7] Gabriel, C., S. Gabriel, and E. Corthout, "The Dielectric Properties of Biological Tissues: I. Literature Survey," *Phys. Med. Biol.*, Vol. 41, No. 11, November 1996, pp. 2231–2249.

[8] Gabriel, S., R. W. Lau, and C. Gabriel, "The Dielectric Properties of Biological Tissues: II. Measurements in the Frequency Range 10 Hz to 20 GHz," *Phys. Med. Biol.*, Vol. 41, No. 11, November 1996, pp. 2251–2269.

[9] Schmid, G., G. Neubauer, and P. R. Mazal, "Dielectric Properties of Human Brain Tissue Measured Less Than 10 h Postmortem at Frequencies from 800 to 2450 MHz," *Bioelectromagnetics,* Vol. 24, No. 6, September 2003, pp. 423–430.

[10] Gabriel, S., R. W. Lau, and C. Gabriel, "The Dielectric Properties of Biological Tissues: III Parametric Models for the Dielectric Spectrum of Tissues," *Phys. Med. Biol.*, Vol. 41, No. 11, November 1996, pp. 2271–2293.

[11] "https://www.fcc.gov/general/body-tissue-dielectric-parameters."

[12] Sasaki, K., K. Wake, and S. Watanabe, "Development of Best Fit Cole-Cole Parameters for Measurement Data from Biological Tissues and Organs Between 1 MHz and 20 GHz," *Radio Sci.*, Vol. 49, No. 7, July 2014, pp. 459–472.

[13] Ellison, W. J., "Permittivity of Pure Water, at Standard Atmospheric Pressure, over the Frequency Range 0–25 THz and the Temperature Range 0–100°C," *J. Phys. Chem. Ref. Data,* Vol. 36, No. 1, March 2007, pp. 1–18.

[14] Rekanos, I. T., and T. G. Papadopoulos, "FDTD Modeling of Wave Propagation in Cole-Cole Media with Multiple Relaxation Times," *IEEE Antennas Wirel. Propag. Lett.*, Vol. 9, 2010, pp. 67–69.

[15] Guo, B., J. Li, and H. Zmuda, "A New FDTD Formulation for Wave Propagation in Biological Media with Cole-Cole Model," *IEEE Microw. Wirel. Compon. Lett.*, Vol. 16, No. 12, December 2006, pp. 633–635.

[16] Rekanos, I. T., and T. V. Yioultsis, "Approximation of Grünwald–Letnikov Fractional Derivative for FDTD Modeling of Cole–Cole Media," *IEEE Trans. Magn.*, Vol. 50, No. 2, February 2014, pp. 181–184.

[17] Kelley, D. F., and R. J. Luebbers, "Piecewise Linear Recursive Convolution for Dispersive Media Using FDTD," *IEEE Trans. Antennas Propag.*, Vol. 44, No. 6, June 1996, pp. 792–797.

[18] Giannakis, I., and A. Giannopoulos, "A Novel Piecewise Linear Recursive Convolution Approach for Dispersive Media Using the Finite-Difference Time-Domain Method," *IEEE Trans. Antennas Propag.*, Vol. 62, No. 5, May 2014, pp. 2669–2678.

[19] Fujii, M., "Maximum Frequency Range Limit of Multi-Pole Debye Models of Human Body Tissues," *IEEE Microw. Wirel. Compon. Lett.*, Vol. 22, No. 2, February 2012, pp. 73–75.

[20] Mustafa, S., A. M. Abbosh, and P. T. Nguyen, "Modeling Human Head Tissues using Fourth-Order Debye Model in Convolution-Based Three-Dimensional Finite-Difference Time-Domain," *IEEE Trans. Antennas Propag.*, Vol. 62, No. 3, March 2014, pp. 1354–1361.

[21] Stauffer, P. R., et al., "Phantom and Animal Tissues for Modelling the Electrical Properties of Human Liver," *Int. J. Hyperth.*, Vol. 19, No. 1, February 2003, pp. 89–101.

[22] Lopresto, V., R. Pinto, G. A. Lovisolo, and M. Cavagnaro, "Changes in the Dielectric Properties of *Ex Vivo* Bovine Liver During Microwave Thermal Ablation at 2.45 GHz," *Phys. Med. Biol.*, Vol. 57, No. 8, April 2012, pp. 2309–2327.

[23] Lazebnik, M., M. C. Converse, J. H. Booske, and S. C. Hagness, "Ultrawideband Temperature-Dependent Dielectric Properties of animal Liver Tissue in the Microwave Frequency Range," *Phys. Med. Biol.*, Vol. 51, No. 7, April 2006, pp. 1941–1955.

[24] Rossmanna, C., and D. Haemmerich, "Review of Temperature Dependence of Thermal Properties, Dielectric Properties, and Perfusion of Biological Tissues at Hyperthermic and Ablation Temperatures," *Crit. Rev. Biomed. Eng.*, Vol. 42, No. 6, 2014, pp. 467–492.

[25] Martin, R. C. G., C. R. Scoggins, and K. M. McMasters, "Safety and Efficacy of Microwave Ablation of Hepatic Tumors: A Prospective Review of a 5-Year Experience," *Ann. Surg. Oncol.*, Vol. 17, No. 1, January 2010, pp. 171–178.

[26] Ji, Z., and C. L. Brace, "Expanded Modeling of Temperature-Dependent Dielectric Properties for Microwave Thermal Ablation," *Phys. Med. Biol.*, Vol. 56, No. 16, August 2011, pp. 5249–5264.

[27] Chaudhary, S. S., R. K. Mishra, A. Swarup, and J. M. Thomas, "Dielectric Properties of Normal & Malignant Human Breast Tissues at Radiowave & Microwave Frequencies," *Indian J. Biochem. Biophys.*, Vol. 21, No. 1, February 1984, pp. 76–79.

[28] Campbell, A. M., and D. V. Land, "Dielectric Properties of Female Human Breast Tissue Measured In Vitro at 3.2 GHz," *Phys. Med. Biol.*, Vol. 37, No. 1, January 1992, pp. 193–210.

[29] Lazebnik, M., et al., "A Large-Scale Study of the Ultrawideband Microwave Dielectric Properties of Normal Breast Tissue Obtained from Reduction Surgeries," *Phys. Med. Biol.*, Vol. 52, No. 10, May 2007, pp. 2637–2656.

[30] Lazebnik, M., et al., "A Large-Scale Study of the Ultrawideband Microwave Dielectric Properties of Normal, Benign and Malignant Breast Tissues Obtained from Cancer Surgeries," *Phys. Med. Biol.*, Vol. 52, No. 20, October 2007, pp. 6093–6115.

[31] Shea, J. D., P. Kosmas, B. D. V. Veen, and S. C. Hagness, "Contrast-Enhanced Microwave Imaging of Breast Tumors: A Computational Study Using 3D Realistic Numerical Phantoms," *Inverse Probl.*, Vol. 26, No. 7, July 2010, p. 074009.

[32] Chen, Y., I. J. Craddock, and P. Kosmas, "Feasibility Study of Lesion Classification via Contrast-Agent-Aided UWB Breast Imaging," *IEEE Trans. Biomed. Eng.*, Vol. 57, No. 5, May 2010, pp. 1003–1007.

[33] Bellizzi, G., O. M. Bucci, and I. Catapano, "Microwave Cancer Imaging Exploiting Magnetic Nanoparticles as Contrast Agent," *IEEE Trans. Biomed. Eng.*, Vol. 58, No. 9, September 2011, pp. 2528–2536.

[34] Scapaticci, R., et al., "An Effective Procedure for MNP-Enhanced Breast Cancer Microwave Imaging," *IEEE Trans. Biomed. Eng.*, Vol. 61, No. 4, April 2014, pp. 1071–1079.

[35] Martellosio, A., et al., "Dielectric Properties Characterization From 0.5 to 50 GHz of Breast Cancer Tissues," *IEEE Trans. Microw. Theory Tech.*, Vol. 65, No. 3, March 2017, pp. 998–1011.

[36] Zhao, L., et al., "A New High-Resolution Electromagnetic Human Head Model: A Useful Resource for a New Specific-Absorption-Rate Assessment Model," *IEEE Antennas Propag. Mag.*, Vol. 58, No. 5, October 2016, pp. 32–42.

[37] Ireland, D., K. Bialkowski, and A. Abbosh, "Microwave Imaging for Brain Stroke Detection Using Born Iterative Method," *IET Microw. Antennas Propag.*, Vol. 7, No. 11, August 2013, pp. 909–915.

[38] Zamani, A., A. M. Abbosh, and A. T. Mobashsher, "Fast Frequency-Based Multistatic Microwave Imaging Algorithm with Application to Brain Injury Detection," *IEEE Trans. Microw. Theory Tech.*, Vol. 64, No. 2, February 2016, pp. 653–662.

[39] Tournier, P.-H., et al., "Numerical Modeling and High-Speed Parallel Computing: New Perspectives on Tomographic Microwave Imaging for Brain Stroke Detection and Monitoring," *IEEE Antennas Propag. Mag.*, Vol. 59, No. 5, October 2017, pp. 98–110.

[40] Scapaticci, R., O. M. Bucci, I. Catapano, and L. Crocco, "Differential Microwave Imaging for Brain Stroke Followup," *Int. J. Antennas Propag.*, Vol. 2014, pp. 1–11.

[41] Semenov, S. Y., and D. R. Corfield, "Microwave Tomography for Brain Imaging: Feasibility Assessment for Stroke Detection," *Int. J. Antennas Propag.*, Vol. 2008, pp. 1–8.

[42] Gilmore, C., A. Zakaria, J. LoVetri, and S. Pistorius, "A Study of matching Fluid Loss in a Biomedical Microwave Tomography System: Matching Fluid Loss in MWT," *Med. Phys.*, Vol. 40, No. 2, January 2013, p. 023101.

[43] Bindu, G., A. Lonappan, V. Thomas, C. K. Aanandan, and K. T. Mathew, "Dielectric Studies of Corn Syrup for Applications in Microwave Breast Imaging," *Prog. Electromagn. Res.*, Vol. 59, 2006, pp. 175–186.

[44] Romeo, S., et al., "Dielectric Characterization Study of Liquid-Based Materials for Mimicking Breast Tissues," *Microw. Opt. Technol. Lett.*, Vol. 53, No. 6, June 2011, pp. 1276–1280.

[45] Meaney, P. M., C. J. Fox, S. D. Geimer, and K. D. Paulsen, "Electrical Characterization of Glycerin: Water Mixtures: Implications for Use as a Coupling Medium in Microwave Tomography," *IEEE Trans. Microw. Theory Tech.*, Vol. 65, No. 5, May 2017, pp. 1471–1478.

[46] Ackerman, M. J., "The Visible Human Project," *Proc. IEEE*, Vol. 86, No. 3, March 1998, pp. 504–511.

[47] Zastrow, E., et al., "Development of Anatomically Realistic Numerical Breast Phantoms with Accurate Dielectric Properties for Modeling Microwave Interactions With the Human Breast," *IEEE Trans. Biomed. Eng.*, Vol. 55, No. 12, December 2008, pp. 2792–2800.

[48] Neira, L. M., R. O. Mays, and S. C. Hagness, "Human Breast Phantoms: Test Beds for the Development of Microwave Diagnostic and Therapeutic Technologies," *IEEE Pulse*, Vol. 8, No. 4, July 2017, pp. 66–70.

[49] Lazebnik, M., M. Okoniewski, J. H. Booske, and S. C. Hagness, "Highly Accurate Debye Models for Normal and Malignant Breast Tissue Dielectric Properties at Microwave Frequencies," *IEEE Microw. Wirel. Compon. Lett.*, Vol. 17, No. 12, December 2007, pp. 822–824.

[50] Zubal, I. G., et al., "Computerized Three-Dimensional Segmented Human Anatomy," *Med. Phys.*, Vol. 21, No. 2, February 1994, pp. 299–302.

[51] Jones, D., "A Realistic Anthropomorphic Phantom for Calculating Organ Doses Arising from External Photon Irradiation," *Radiat. Prot. Dosimetry*, 1997, pp. 21–29.

[52] Petoussi-Hen, N., and M. Zankl, "Voxel Anthropomorphic Models as a Tool for Internal Dosimetry," *Radiat. Prot. Dosimetry*, Vol. 79, No. 1, October 1998, pp. 415–418.

[53] Xu, X. G., T. C. Chao, and A. Bozkurt, "VIP-MAN: an Image-Based Whole-Body Adult Male Model Constructed from Color Photographs of the Visible Human Project for Multi-Particle Monte Carlo Calculations," *Health Phys.*, Vol. 78, No. 5, May 2000, pp. 476–486.

[54] Christ, A., et al., "The Virtual Family—Development of Surface-Based Anatomical models of Two Adults and Two Children for Dosimetric Simulations," *Phys. Med. Biol.*, Vol. 55, No. 2, January 2010, pp. N23–N38.

[55] Massey, J. W., and A. E. Yilmaz, "AustinMan and AustinWoman: High-fidelity, Anatomical Voxel Models Developed from the VHP Color Images," in *Proc. 38th Annual International Conference of the IEEE Engineering in Medicine and Biology Society (IEEE EMBC)*, Orlando, FL, 2016, pp. 3346–3349.

[56] Oliveira, B. L., et al., "Development of Clinically Informed 3-D Tumor Models for Microwave Imaging Applications," *IEEE Antennas Wirel. Propag. Lett.*, Vol. 15, 2016, pp. 520–523.

[57] Burfeindt, M. J., et al., "MRI-Derived 3-D-Printed Breast Phantom for Microwave Breast Imaging Validation," *IEEE Antennas Wirel. Propag. Lett.*, Vol. 11, 2012, pp. 1610–1613.

[58] Garrett, J., and E. Fear, "A New Breast Phantom With a Durable Skin Layer for Microwave Breast Imaging," *IEEE Trans. Antennas Propag.*, Vol. 63, No. 4, April 2015, pp. 1693–1700.

[59] Joachimowicz, N., B. Duchene, C. Conessa, and O. Meyer, "Reference phantoms for microwave imaging," in *Proc. 2017 11th European Conference on Antennas and Propagation (EUCAP)*, Paris, France, 2017, pp. 2719–2722.

[60] Joachimowicz, N., C. Conessa, T. Henriksson, and B. Duchene, "Breast Phantoms for Microwave Imaging," *IEEE Antennas Wirel. Propag. Lett.*, Vol. 13, 2014, pp. 1333–1336.

[61] Govinda Raju, G. R., "Dielectric Constant of Binary Mixtures of Liquids," in *Proc. 1988 Conference on Electrical Insulation and Dielectric Phenomena*, Ottawa, Canada, 1988, pp. 357–363.

[62] Nguyen, P. T., A. M. Abbosh, and S. Crozier, "Thermo-Dielectric Breast Phantom for Experimental Studies of Microwave Hyperthermia," *IEEE Antennas Wirel. Propag. Lett.*, Vol. 15, 2016, pp. 476–479.

[63] Spitzer, V., M. J. Ackerman, A. L. Scherzinger, and D. Whitlock, "The Visible Human Male: A Technical Report," *J. Am. Med. Inform. Assoc.*, Vol. 3, No. 2, March 1996, pp. 118–130.

[64] "IEEE Recommended Practice for Determining the Peak Spatial-Average Specific Absorption Rate (SAR) in the Human Head from Wireless Communications Devices: Measurement Techniques," *IEEE Std 1528-2013*, September 2013, pp. 1–246.

[65] Zamani, A., A. M. Abbosh, and S. Crozier, "Multistatic Biomedical Microwave Imaging Using Spatial Interpolator for Extended Virtual Antenna Array," *IEEE Trans. Antennas Propag.*, Vol. 65, No. 3, March 2017, pp. 1121–1130.

[66] Mohammed, B. J., and A. M. Abbosh, "Realistic Head Phantom to Test Microwave Systems for Brain Imaging," *Microw. Opt. Technol. Lett.*, Vol. 56, No. 4, April 2014, pp. 979–982.

[67] Nikolova, N., "Microwave Imaging for Breast Cancer," *IEEE Microw. Mag.*, Vol. 12, No. 7, December 2011, pp. 78–94.

[68] Fear, E. C., P. M. Meaney, and M. A. Stuchly, "Microwaves for Breast Cancer Detection?," *IEEE Potentials*, Vol. 22, No. 1, February 2003, pp. 12–18.

[69] Hassan, A. M., and M. El-Shenawee, "Review of Electromagnetic Techniques for Breast Cancer Detection," *IEEE Rev. Biomed. Eng.*, Vol. 4, 2011, pp. 103–118.

[70] Fear, E. C., et al., "Enhancing Breast Tumor Detection with Near-Field Imaging," *IEEE Microw. Mag.*, Vol. 3, No. 1, March 2002, pp. 48–56.

[71] Li, X., et al., "An Overview of Ultra-Wideband Microwave Imaging via Space-Time Beamforming for Early-Stage Breast-Cancer Detection," *IEEE Antennas Propag. Mag.*, Vol. 47, No. 1, February 2005, pp. 19–34.

[72] Conceição, R. C., J. J. Mohr, and M. O'Halloran (eds.), *An Introduction to Microwave Imaging for Breast Cancer Detection*, Cham, Switzerland: Springer International Publishing, 2016.

[73] Di Meo, S., et al., "On the Feasibility of Breast Cancer Imaging Systems at Millimeter-Waves Frequencies," *IEEE Trans. Microw. Theory Tech.*, Vol. 65, No. 5, May 2017, pp. 1795–1806.

[74] Abbosh, A. M., B. Mohammed, and K. S. Bialkowski, "Differential Microwave Imaging of the Breast Pair," *IEEE Antennas Wirel. Propag. Lett.*, Vol. 15, 2016, pp. 1434–1437.

[75] Elahi, M. A., et al., "Comparison of Radar-Based Microwave Imaging Algorithms Applied to Experimental Breast Phantoms," in *Proc. 32nd General Assembly and Scientific Symposium of the International Union of Radio Science (URSI GASS)*, Montreal, QC, Canada, 2017, pp. 1–4.

[76] Byrne, D., M. O'Halloran, M. Glavin, and E. Jones, "Data Independent Radar Beamforming Algorithms for Breast Cancer Detection," *Prog. Electromagn. Res.*, Vol. 107, 2010, pp. 331–348.

[77] Curtis, C., B. R. Lavoie, and E. Fear, "An Analysis of the Assumptions Inherent to Near Field Beamforming for Biomedical Applications," *IEEE Trans. Comput. Imaging*, Vol. 3, No. 4, 2017, pp. 953–965.

[78] Fear, E. C., X. Li, S. C. Hagness, and M. A. Stuchly, "Confocal Microwave Imaging for Breast Cancer Detection: Localization of Tumors in Three Dimensions," *IEEE Trans. Biomed. Eng.*, Vol. 49, No. 8, 2002, pp. 812–822.

[79] Lim, H. B., N. T. T. Nhung, E.-P. Li, and N. D. Thang, "Confocal Microwave Imaging for Breast Cancer Detection: Delay-Multiply-and-Sum Image Reconstruction Algorithm," *IEEE Trans. Biomed. Eng.*, Vol. 55, No. 6, 2008, pp. 1697–1704.

[80] Klemm, M., et al., "Improved Delay-and-Sum Beamforming Algorithm for Breast Cancer Detection," *Int. J. Antennas Propag.*, Vol. 2008, pp. 1–9.

[81] Klemm, M., et al., "Microwave Radar-Based Breast Cancer Detection: Imaging in Inhomogeneous Breast Phantoms," *IEEE Antennas Wirel. Propag. Lett.*, Vol. 8, 2009, pp. 1349–1352.

[82] Bond, E. J., X. Li, S. C. Hagness, and B. D. Van Veen, "Microwave Imaging via Space-time Beamforming for Early Detection of Breast Cancer," *IEEE Trans. Antennas Propag.*, Vol. 51, No. 8, 2003, pp. 1690–1705.

[83] O'Halloran, M., E. Jones, and M. Glavin, "Quasi-Multistatic MIST Beamforming for the Early Detection of Breast Cancer," *IEEE Trans. Biomed. Eng.*, Vol. 57, No. 4, April 2010, pp. 830–840.

[84] O'Halloran, M., M. Glavin, and E. Jones, "Performance and Robustness of a Multistatic MIST Beamforming Algorithm for Breast Cancer Detection," *Prog. Electromagn. Res.*, Vol. 105, 2010, pp. 403–424.

[85] Xie, Y., et al., "Multistatic Adaptive Microwave Imaging for Early Breast Cancer Detection," *IEEE Trans. Biomed. Eng.*, Vol. 53, No. 8, 2006, pp. 1647–1657.

[86] Byrne, D., and I. J. Craddock, "Time-Domain Wideband Adaptive Beamforming for Radar Breast Imaging," *IEEE Trans. Antennas Propag.*, Vol. 63, No. 4, April 2015, pp. 1725–1735.

[87] O'Halloran, M., M. Glavin, and E. Jones, "Effects of Fibroglandular Tissue Distribution on Data-Independent Beamforming Algorithms," *Prog. Electromagn. Res.*, Vol. 97, 2009, pp. 141–158.

[88] Winters, D. W., et al., "Three-Dimensional Microwave Breast Imaging: Dispersive Dielectric Properties Estimation Using Patient-Specific Basis Functions," *IEEE Trans. Med. Imaging*, Vol. 28, No. 7, July 2009, pp. 969–981.

[89] Chew, W. C., and Y. M. Wang, "Reconstruction of Two-Dimensional Permittivity Distribution Using the Distorted Born Iterative Method," *IEEE Trans. Med. Imaging*, Vol. 9, No. 2, June 1990, pp. 218–225.

[90] Leone, G., and F. Soldovieri, "Analysis of the Distorted Born Approximation for Subsurface Reconstruction: Truncation and Uncertainties Effects," *IEEE Trans. Geosci. Remote Sens.*, Vol. 41, No. 1, January 2003, pp. 66–74.

[91] Remis, R. F., and P. M. Van Den Berg, "On the Equivalence of the Newton-Kantorovich and Distorted Born Methods," *Inverse Probl.*, Vol. 16, No. 1, February 2000, p. L1.

[92] Mojabi, P., and J. LoVetri, "Microwave Biomedical Imaging Using the Multiplicative Regularized Gauss–Newton Inversion," *IEEE Antennas Wirel. Propag. Lett.*, Vol. 8, 2009, pp. 645–648.

[93] Rubek, T., P. M. Meaney, P. Meincke, and K. D. Paulsen, "Nonlinear Microwave Imaging for Breast-Cancer Screening Using Gauss–Newton's Method and the CGLS Inversion Algorithm," *IEEE Trans. Antennas Propag.*, Vol. 55, No. 8, August 2007, pp. 2320–2331.

[94] Miao, Z., and P. Kosmas, "Multiple-frequency DBIM-TwIST Algorithm for Microwave Breast Imaging," *IEEE Trans. Antennas Propag.*, Vol. 65, No. 5, May 2017, pp. 2507–2516.

[95] Gilmore, C., P. Mojabi, and J. LoVetri, "Comparison of an Enhanced Distorted Born Iterative Method and the Multiplicative-Regularized Contrast Source Inversion Method," *IEEE Trans. Antennas Propag.*, Vol. 57, No. 8, August 2009, pp. 2341–2351.

[96] Bertero, M., and P. Boccacci, *Introduction to Inverse Problems in Imaging*, Bristol, UK: Institute of Physics Publishing, 1998.

[97] De Zaeytijd, J., A. Franchois, C. Eyraud, and J.-M. Geffrin, "Full-Wave Three-Dimensional Microwave Imaging with a Regularized Gauss–Newton Method—Theory and Experiment," *IEEE Trans. Antennas Propag.*, Vol. 55, No. 11, November 2007, pp. 3279–3292.

[98] Williams, T. C., J. M. Sill, and E. C. Fear, "Breast Surface Estimation for Radar-Based Breast Imaging Systems," *IEEE Trans. Biomed. Eng.*, Vol. 55, No. 6, June 2008, pp. 1678–1686.

[99] Sarafianou, M., et al., "Evaluation of Two Approaches for Breast Surface Measurement Applied to a Radar-Based Imaging System," *IEEE Trans. Antennas Propag.*, Vol. 64, No. 2, February 2016, pp. 609–617.

[100] Winters, D. W., et al., "Estimating the Breast Surface Using UWB Microwave Monostatic Backscatter Measurements," *IEEE Trans. Biomed. Eng.*, Vol. 55, No. 1, January 2008, pp. 247–256.

[101] Gao, F., B. D. Van Veen, and S. C. Hagness, "Sensitivity of the Distorted Born Iterative Method to the Initial Guess in Microwave Breast Imaging," *IEEE Trans. Antennas Propag.*, Vol. 63, No. 8, pp. 3540–3547, August 2015.

[102] P. M. Meaney, K. D. Paulsen, B. W. Pogue, and M. I. Miga, "Microwave Image Reconstruction Utilizing Log-Magnitude and Unwrapped Phase to Improve High-Contrast Object Recovery," *IEEE Trans. Med. Imaging,* Vol. 20, No. 2, pp. 104–116, February 2001.

[103] A. H. Golnabi, P. M. Meaney, and K. D. Paulsen, "Tomographic Microwave Imaging with Incorporated Prior Spatial Information," *IEEE Trans. Microw. Theory Tech.*, Vol. 61, No. 5, pp. 2129–2136, May 2013.

[104] L. M. Neira, B. D. Van Veen, and S. C. Hagness, "High-Resolution Microwave Breast Imaging Using a 3-D Inverse Scattering Algorithm with a Variable-Strength Spatial Prior Constraint," *IEEE Trans. Antennas Propag.*, Vol. 65, No. 11, November 2017, pp. 6002–6014.

[105] Scapaticci, R., P. Kosmas, and L. Crocco, "Wavelet-Based Regularization for Robust Microwave Imaging in Medical Applications," *IEEE Trans. Biomed. Eng.*, Vol. 62, No. 4, April 2015, pp. 1195–1202.

[106] Ferraye, R., J.-Y. Dauvignac, and C. Pichot, "An Inverse Scattering Method Based on Contour Deformations by Means of a Level Set Method Using Frequency Hopping Technique," *IEEE Trans. Antennas Propag.*, Vol. 51, No. 5, May 2003, pp. 1100–1113.

[107] Chew, W. C., and J. H. Lin, "A frequency-Hopping Approach for Microwave Imaging of Large Inhomogeneous Bodies," *IEEE Microw. Guid. Wave Lett.*, Vol. 5, No. 12, December 1995, pp. 439–441.

[108] Shea, J. D., P. Kosmas, S. C. Hagness, and B. D. Van Veen, "Three-Dimensional Microwave Imaging of Realistic Numerical Breast Phantoms via a Multiple-Frequency Inverse Scattering Technique: Microwave Imaging of Realistic Numerical Breast Phantoms," *Med. Phys.*, Vol. 37, No. 8, July 2010, pp. 4210–4226.

[109] Burfeindt, M. J., J. D. Shea, B. D. Van Veen, and S. C. Hagness, "Beamforming-Enhanced Inverse Scattering for Microwave Breast Imaging," *IEEE Trans. Antennas Propag.*, Vol. 62, No. 10, October 2014, pp. 5126–5132.

[110] Bevacqua, M. T., and R. Scapaticci, "A Compressive Sensing Approach for 3D Breast Cancer Microwave Imaging with Magnetic Nanoparticles as Contrast Agent," *IEEE Trans. Med. Imaging,* Vol. 35, No. 2, February 2016, pp. 665–673.

[111] Winters, D. W., B. D. Van Veen, and S. C. Hagness, "A Sparsity Regularization Approach to the Electromagnetic Inverse Scattering Problem," *IEEE Trans. Antennas Propag.*, Vol. 58, No. 1, January 2010, pp. 145–154.

[112] Azghani, M., P. Kosmas, and F. Marvasti, "Microwave Medical Imaging Based on Sparsity and an Iterative Method with Adaptive Thresholding," *IEEE Trans. Med. Imaging,* Vol. 34, No. 2, February 2015, pp. 357–365.

[113] Scapaticci, R., I. Catapano, and L. Crocco, "Wavelet-Based Adaptive Multiresolution Inversion for Quantitative Microwave Imaging of Breast Tissues," *IEEE Trans. Antennas Propag.*, Vol. 60, No. 8, August 2012, pp. 3717–3726.

[114] Azghani, M., P. Kosmas, and F. Marvasti, "Fast Microwave Medical Imaging Based on Iterative Smoothed Adaptive Thresholding," *IEEE Antennas Wirel. Propag. Lett.*, Vol. 14, 2015, pp. 438–441.

[115] Azghani, M., and F. Marvasti, "L2-Regularized Iterative Weighted Algorithm for Inverse Scattering," *IEEE Trans. Antennas Propag.*, Vol. 64, No. 6, June 2016, pp. 2293–2300.

[116] Caorsi, S., G. L. Gragnani, M. Pastorino, and M. Rebagliati, "A Model-Driven Approach to Microwave Diagnostics in Biomedical Applications," *IEEE Trans. Microw. Theory Tech.*, Vol. 44, No. 10, October 1996, pp. 1910–1920.

[117] Caorsi, S., G. L. Gragnani, and M. Pastorino, "Reconstruction of Dielectric Permittivity Distributions in Arbitrary 2-D Inhomogeneous Biological Bodies by a Multiview Microwave Numerical Method," *IEEE Trans. Med. Imaging*, Vol. 12, No. 2, June 1993, pp. 232–239.

[118] Caorsi, S., A. Massa, M. Pastorino, and A. Rosani, "Microwave Medical Imaging: Potentialities and Limitations of a Stochastic Optimization Technique," *IEEE Trans. Microw. Theory Tech.*, Vol. 52, No. 8, August 2004, pp. 1909–1916.

[119] Gilmore, C., et al., "Microwave Biomedical Data Inversion Using the Finite-Difference Contrast Source Inversion Method," *IEEE Trans. Antennas Propag.*, Vol. 57, No. 5, May 2009, pp. 1528–1538.

[120] Catapano, I., et al., "On Quantitative Microwave Tomography of Female Breast," *Prog. Electromagn. Res.*, Vol. 97, 2009, pp. 75–93.

[121] Ruvio, G., R. Solimene, A. Cuccaro, and M. J. Ammann, "Comparison of Noncoherent Linear Breast Cancer Detection Algorithms Applied to a 2-D Numerical Model," *IEEE Antennas Wirel. Propag. Lett.*, Vol. 12, 2013, pp. 853–856.

[122] Bozza, G., M. Brignone, and M. Pastorino, "Application of the No-Sampling Linear Sampling Method to Breast Cancer Detection," *IEEE Trans. Biomed. Eng.*, Vol. 57, No. 10, 2010, pp. 2525–2534.

[123] Colgan, T. J., S. C. Hagness, and B. D. Van Veen, "A 3-D Level Set Method for Microwave Breast Imaging," *IEEE Trans. Biomed. Eng.*, Vol. 62, No. 10, October 2015, pp. 2526–2534.

[124] Ali, M. A., and M. Moghaddam, "3D Nonlinear Super-Resolution Microwave Inversion Technique Using Time-Domain Data," *IEEE Trans. Antennas Propag.*, Vol. 58, No. 7, July 2010, pp. 2327–2336.

[125] Johnson, J. E., T. Takenaka, and T. Tanaka, "Two-Dimensional Time-Domain Inverse Scattering for Quantitative Analysis of Breast Composition," *IEEE Trans. Biomed. Eng.*, Vol. 55, No. 8, August 2008, pp. 1941–1945.

[126] Johnson, J. E., et al., "Advances in the 3-D Forward–Backward Time-Stepping (FBTS) Inverse Scattering Technique for Breast Cancer Detection," *IEEE Trans. Biomed. Eng.*, Vol. 56, No. 9, September 2009, pp. 2232–2243.

[127] Kosmas, P., and C. M. Rappaport, "FDTD-Based Time Reversal for Microwave Breast Cancer Detection-Localization in Three Dimensions," *IEEE Trans. Microw. Theory Tech.*, Vol. 54, No. 4, June 2006, pp. 1921–1927.

[128] Hossain, M. D., and A. S. Mohan, "Cancer Detection in Highly Dense Breasts Using Coherently Focused Time-Reversal Microwave Imaging," *IEEE Trans. Comput. Imaging*, Vol. 3, No. 4, December 2017, pp. 928–939.

[129] Hossain, M. D., and A. S. Mohan, "Coherent Time-Reversal Microwave Imaging for the Detection and Localization of Breast Tissue Malignancies," *Radio Sci.*, Vol. 50, No. 2, February 2015, pp. 87–98.

[130] Kosmas, P., and C. M. Rappaport, "Time Reversal with the FDTD Method for Microwave Breast Cancer Detection," *IEEE Trans. Microw. Theory Tech.*, Vol. 53, No. 7, July 2005, pp. 2317–2323.

[131] Epstein, N. R., P. M. Meaney, and K. D. Paulsen, "3D Parallel-Detection Microwave Tomography for Clinical Breast Imaging," *Rev. Sci. Instrum.*, Vol. 85, No. 12, December 2014.

[132] Byrne, D., M. Sarafianou, and I. J. Craddock, "Compound Radar Approach for Breast Imaging," *IEEE Trans. Biomed. Eng.*, Vol. 64, No. 1, January 2017, pp. 40–51.

[133] Fear, E. C., et al., "Microwave Breast Imaging with a Monostatic Radar-Based System: A Study of Application to Patients," *IEEE Trans. Microw. Theory Tech.*, Vol. 61, No. 5, May 2013, pp. 2119–2128.

[134] Porter, E., et al., "Time-Domain Multistatic Radar System for Microwave Breast Screening," *IEEE Antennas Wirel. Propag. Lett.*, Vol. 12, 2013, pp. 229–232.

[135] Bassi, M., et al., "An Integrated Microwave Imaging Radar with Planar Antennas for Breast Cancer Detection," *IEEE Trans. Microw. Theory Tech.*, Vol. 61, No. 5, May 2013, pp. 2108–2118.

[136] Casu, M. R., et al., "A COTS-Based Microwave Imaging System for Breast-Cancer Detection," *IEEE Trans. Biomed. Circuits Syst.*, Vol. 11, No. 4, August 2017, pp. 804–814.

[137] Asefi, M., and J. LoVetri, "Use of Field-Perturbing Elements to Increase Nonredundant Data for Microwave Imaging Systems," *IEEE Trans. Microw. Theory Tech.*, Vol. 65, No. 9, September 2017, pp. 3172–3179.

[138] Henriksson, T., N. Joachimowicz, C. Conessa, and J.-C. Bolomey, "Quantitative Microwave Imaging for Breast Cancer Detection Using a Planar 2.45 GHz System," *IEEE Trans. Instrum. Meas.*, Vol. 59, No. 10, October 2010, pp. 2691–2699.

[139] Porter, E., M. Coates, and M. Popovic, "An Early Clinical Study of Time-Domain Microwave Radar for Breast Health Monitoring," *IEEE Trans. Biomed. Eng.*, Vol. 63, No. 3, March 2016, pp. 530–539.

[140] Kanj, H., and M. Popovic, "A Novel Ultra-Compact Broadband Antenna for Microwave Breast Tumor Detection," *Prog. Electromagn. Res.*, Vol. 86, 2008, pp. 169–198.

[141] Bahramiabarghouei, H., et al., "Flexible 16 Antenna Array for Microwave Breast Cancer Detection," *IEEE Trans. Biomed. Eng.*, Vol. 62, No. 10, October 2015, pp. 2516–2525.

[142] Porter, E., et al., "A Wearable Microwave Antenna Array for Time-Domain Breast Tumor Screening," *IEEE Trans. Med. Imaging*, Vol. 35, No. 6, June 2016, pp. 1501–1509.

[143] Lee, S., and S. Kwon, "Instantaneous Microwave Imaging with Time-Domain Measurements for Breast Cancer Detection," *Electron. Lett.*, Vol. 49, No. 10, May 2013, pp. 639–641.

[144] Kikkawa, T., et al., "IR-UWB-CMOS Circuits for Breast Cancer Detection," in *Proc. 6th European Conference on Antennas and Propagation (EuCAP),* Prague, Czech Republic, pp. 1758–1760, 2012.

[145] Kwiatkowski, T. G., et al., "Effects of Tissue Plasminogen Activator for Acute Ischemic Stroke at One Year," *N. Engl. J. Med.,* Vol. 340, No. 23, June 1999, pp. 1781–1787.

[146] Hacke, W., et al., "Thrombolysis with Alteplase 3 to 4.5 Hours after Acute Ischemic Stroke," *N. Engl. J. Med.,* Vol. 359, No. 13, September 2008, pp. 1317–1329.

[147] Riechers, R. G., K. M. Pasala, and G. S. F. Ling, "Microwave Detection System for Locating Hemorrhage Sites within the Cranium and Other Regions," in *Proc. 15th Annual AESS/IEEE Dayton Section Symposium. Sensing the World: Analog Sensors and Systems Across the Spectrum,* Fairborn, OH, 1998, pp. 1–12.

[148] Trefna, H., and M. Persson, "Antenna Array Design for Brain Monitoring," in *Proc. IEEE Antennas and Propagation Society International Symposium,* San Diego, CA, 2008, pp. 1–4.

[149] Semenov, S. Y., et al., "Microwave Tomography for Detection/Imaging of Myocardial Infarction I: Excised Canine Hearts," *Ann. Biomed. Eng.,* Vol. 31, No. 3, March 2003, pp. 262–270.

[150] Semenov, S. Y., et al., "Microwave Tomography: Two-Dimensional System for Biological Imaging," *IEEE Trans. Biomed. Eng.,* Vol. 43, No. 9, September 1996, pp. 869–877.

[151] Semenov, S., "Microwave Tomography: Review of the Progress Towards Clinical Applications," *Philos. Transact. A Math. Phys. Eng. Sci.,* Vol. 367, No. 1900, August 2009, pp. 3021–3042.

[152] Bjelogrlic, M., et al., "Experimental Verification of Optimal Frequency Range for Microwave Head Imaging," in *Proc. 2017 International Conference on Electromagnetics in Advanced Applications (ICEAA),* Verona, Italy, pp. 1008–1011.

[153] Bisio, I., et al., "A Numerical Study Concerning Brain Stroke Detection by Microwave Imaging Systems," *Multimed. Tools Appl.,* June 2017, pp. 1–23.

[154] Scapaticci, R., L. Di Donato, I. Catapano, and L. Crocco, "A Feasibility Study on Microwave Imaging for Brain Stroke Monitoring," *Prog. Electromagn. Res. B,* Vol. 40, 2012, pp. 305–324.

[155] Persson, M., et al., "Microwave-Based Stroke Diagnosis Making Global Prehospital Thrombolytic Treatment Possible," *IEEE Trans. Biomed. Eng.,* Vol. 61, No. 11, November 2014, pp. 2806–2817.

[156] Fhager, A., Y. Yu, T. McKelvey, and M. Persson, "Stroke Diagnostics with a Microwave Helmet," in *Proc. of the 7th European Conference on Antennas and Propagation (EuCAP),* Gothenburg, Sweden, pp. 845–846, 2013.

[157] Khorshidi, M. A., T. McKelvey, M. Persson, and H. D. Trefna, "Classification of Microwave Scattering Data Based on a Subspace Distance with Application to Detection of Bleeding Stroke," in *Proc. 3rd IEEE International Workshop on Computational Advances in Multi-Sensor Adaptive Processing (CAMSAP),* Aruba, Netherlands, 2009, pp. 301–304.

[158] Fhager, A., and M. Persson, "A Microwave Measurement System for Stroke Detection," in *Proc. of the Antennas and Propagation Conference (LAPC),* pp. 1–2, Loughborough, UK, 2011.

[159] Fhager, A., and M. Persson, "Stroke Detection and Diagnosis with a Microwave Helmet," in *Proc. of the 6th European Conference on Antennas and Propagation (EUCAP)*, pp. 1796–1798, Prague, 2012.

[160] Candefjord, S., et al., "Microwave Technology for Detecting Traumatic Intracranial Bleedings: Tests on Phantom of Subdural Hematoma and Numerical Simulations," *Med. Biol. Eng. Comput.*, Vol. 55, No. 8, August 2017, pp. 1177–1188.

[161] Ljungqvist, J., et al., "Clinical Evaluation of a Microwave-Based Device for Detection of Traumatic Intracranial Hemorrhage," *J. Neurotrauma*, Vol. 34, No. 13, February 2017, pp. 2176–2182.

[162] Guardiola, M., L. Jofre, and J. Romeu, "3D UWB Tomography for Medical Imaging Applications," in *Proc. IEEE Antennas and Propagation Society International Symposium*, Toronto, Canada, 2010, pp. 1–4.

[163] Mustafa, S., B. Mohammed, and A. Abbosh, "Novel Preprocessing Techniques for Accurate Microwave Imaging of Human Brain," *IEEE Antennas Wirel. Propag. Lett.*, Vol. 12, 2013, pp. 460–463.

[164] Mohammed, B. J., K. S. Bialkowski, and A. M. Abbosh, "Radar-Based Time-Domain Head Imaging Using Database of Effective Dielectric Constant," *Electron. Lett.*, Vol. 51, No. 20, 2015, pp. 1574–1576.

[165] Bjelogrlic, M., et al., "Numerical Assessment of Brain Stroke Follow-Up via Differential Microwave Imaging," in *Proc. International Conference on Electromagnetics in Advanced Applications (ICEAA)*, Turin, Italy, 2015, pp. 484–485.

[166] Stevanovic, M. N., R. Scapaticci, and L. Crocco, "Brain Stroke Monitoring Using Compressive Sensing and Higher Order Basis Functions," in *Proc. 2017 11th European Conference on Antennas and Propagation (EUCAP)*, Paris, France, pp. 2742–2745.

[167] Zamani, A., and A. Abbosh, "Hybrid Clutter Rejection Technique for Improved Microwave Head Imaging," *IEEE Trans. Antennas Propag.*, Vol. 63, No. 11, November 2015, pp. 4921–4931.

[168] Mohammed, B. J., A. M. Abbosh, S. Mustafa, and D. Ireland, "Microwave System for Head Imaging," *IEEE Trans. Instrum. Meas.*, Vol. 63, No. 1, January 2014, pp. 117–123.

[169] Ricci, E., et al., "PCA-Based Artifact Removal Algorithm for Stroke Detection Using UWB Radar Imaging," *Med. Biol. Eng. Comput.*, Vol. 55, No. 6, June 2017, pp. 909–921.

[170] Ireland, D., and M. E. Bialkowski, "Microwave Head Imaging for Stroke Detection," *Prog. Electromagn. Res. M*, Vol. 21, 2011, pp. 163–175.

[171] Jalilvand, M., et al., "Hemorrhagic Stroke Detection via UWB Medical Imaging," in *Proc. 5th European Conference on Antennas and Propagation (EUCAP)*, Rome, Italy, 2011, pp. 2911–2914.

[172] Mohammed, B. J., K. Bertling, A. M. Abbosh, and A. D. Raki, "Microwave Head Imaging System Using Analogue Fibre-Optic Link for Improved Detection and Localisation," *Electron. Lett.*, Vol. 52, No. 16, 2016, pp. 1366–1368.

[173] Ricci, E., S. D. Domenico, E. Cianca, and T. Rossi, "Artifact Removal Algorithms for Stroke Detection Using a Multistatic MIST Beamforming Algorithm," in *Proc. 37th*

Annual International Conference of the IEEE Engineering in Medicine and Biology Society (EMBC), Milan, Italy, 2015, pp. 1930–1933.

[174] Ricci, E., et al., "Performance Evaluation of Novel Microwave Imaging Algorithms for Stroke Detection Using an Accurate 3D Head Model," *Wirel. Pers. Commun.*, Vol. 96, No. 3, October 2017, pp. 3317–3331.

[175] Guo, L., and A. M. Abbosh, "Compressive Sensing for Stroke Detection in Microwave-Based Head Imaging," in *Proc. IEEE Antennas and Propagation Society International Symposium*, Memphis, 2014, pp. 1911–1912.

[176] Mobashsher, A. T., A. Mahmoud, and A. M. Abbosh, "Portable Wideband Microwave Imaging System for Intracranial Hemorrhage Detection Using Improved Back-projection Algorithm with Model of Effective Head Permittivity," *Sci. Rep.*, Vol. 6, February 2016, p. 20459.

[177] Mobashsher, A. T., A. M. Abbosh, and Y. Wang, "Microwave System to Detect Traumatic Brain Injuries Using Compact Unidirectional Antenna and Wideband Transceiver With Verification on Realistic Head Phantom," *IEEE Trans. Microw. Theory Tech.*, Vol. 62, No. 9, September 2014, pp. 1826–1836.

[178] Semenov, S., "Electromagnetic Sensing and Imaging for Medical Applications," in *Proc. IEEE Conference on Antenna Measurements Applications (CAMA)*, Antibes Juan-les-Pins, France, 2014, pp. 1–4.

[179] Guo, L., and A. M. Abbosh, "Microwave Imaging of Nonsparse Domains Using Born Iterative Method with Wavelet Transform and Block Sparse Bayesian Learning," *IEEE Trans. Antennas Propag.*, Vol. 63, No. 11, November 2015, pp. 4877–4888.

[180] Kanfoud, I. E., et al., "Whole-Microwave System Modeling for Brain Imaging," in *Proc. IEEE Conference on Antenna Measurements Applications (CAMA)*, Chiang Mai, Thailand, 2015, pp. 1–4.

[181] Jalilvand, M., C. Wu, J. Schmid, and T. Zwick, "Quantitative Imaging of Numerically Realistic Human Head Model Using Microwave Tomography," *Electron. Lett.*, Vol. 50, No. 4, February 2014, pp. 255–256.

[182] Hopfer, M., et al., "Electromagnetic Tomography for Detection, Differentiation, and Monitoring of Brain Stroke: A Virtual Data and Human Head Phantom Study," *IEEE Antennas Propag. Mag.*, Vol. 59, No. 5, October 2017, pp. 86–97.

[183] Estatico, C., A. Fedeli, M. Pastorino, and A. Randazzo, "Feasibility Assessment of a Banach-Space Inversion Procedure for Biomedical Applications," in *Proc. 11th European Conference on Antennas and Propagation (EUCAP)*, Paris, France, 2017, pp. 677–681.

[184] Guo, L., and A. Abbosh, "Microwave Head Imaging Using Multi-Frequency Tomography," in *Proc. International Conference on Electromagnetics in Advanced Applications (ICEAA)*, Cairnes, Australia, 2016, pp. 666–669.

[185] Fedeli, A., M. Pastorino, and A. Randazzo, "Electromagnetic Biomedical Imaging in Banach Spaces: A Numerical Case Study," in *Proc. 32nd General Assembly and Scientific Symposium of the International Union of Radio Science (URSI GASS)*, Montreal, Canada, 2017, pp. 1–3.

[186] Irishina, N., and A. Torrente, "Brain Stroke Detection by Microwaves Using Prior Information from Clinical Databases," *Abstr. Appl. Anal.*, Vol. 2013, May 2013, p. e412638.

[187] Mahmood, Q., et al., "A Comparative Study of Automated Segmentation Methods for Use in a Microwave Tomography System for Imaging Intracerebral Hemorrhage in Stroke Patients," *J. Electromagn. Anal. Appl.*, Vol. 07, No. 05, May 2015, p. 152.

[188] Semenov, S., et al., "Electromagnetic Tomography for Brain Imaging: 3-D Reconstruction of Stroke in a Human Head Phantom," in *Proc. IEEE Conference on Antenna Measurements Applications (CAMA)*, Syracuse, NY, 2016, pp. 1–4.

[189] Bulyshev, A. E., et al., "Three-Dimensional Vector Microwave Tomography: Theory and Computational Experiments," *Inverse Probl.*, Vol. 20, No. 4, 2004, p. 1239.

[190] Abbosh, A., and A. T. Mobashsher, "CPW-Fed Low-Profile Directional Antenna Operating in Low Microwave Band for Wideband Medical Diagnostic Systems," *Electron. Lett.*, Vol. 50, No. 4, February 2014, pp. 246–248.

[191] Abbosh, A., and A. T. Mobashsher, "Three-Dimensional Folded Antenna with Ultra-Wideband Performance, Directional Radiation and Compact Size," *IET Microw. Antennas Propag.*, Vol. 8, No. 3, November 2013, pp. 171–179.

[192] Mobashsher, A. T., and A. Abbosh, "Slot-Loaded Folded Dipole Antenna with Wideband and Unidirectional Performance for L-Band Applications," *IEEE Antennas Wirel. Propag. Lett.*, Vol. 13, 2014, pp. 798–801.

[193] Li, X., M. Jalilvand, Y. L. Sit, and T. Zwick, "A Compact Double-Layer On-Body Matched Bowtie Antenna for Medical Diagnosis," *IEEE Trans. Antennas Propag.*, Vol. 62, No. 4, April 2014, pp. 1808–1816.

[194] Tajik, D., D. S. Shumakov, and N. K. Nikolova, "An Experimental Comparison Between the Born and Rytov Approximations in Microwave Tissue Imaging," in *Proc. IEEE MTT-S International Microwave Symposium (IMS)*, Honololu, HI, pp. 1391–1393, 2017.

[195] Ostadrahimi, M., et al., "Enhancement of Gauss-Newton Inversion Method for Biological Tissue Imaging," *IEEE Trans. Microw. Theory Tech.*, Vol. 61, No. 9, September 2013, pp. 3424–3434.

[196] Gilmore, C., A. Zakaria, S. Pistorius, and J. LoVetri, "Microwave Imaging of Human Forearms: Pilot Study and Image Enhancement," *Int. J. Biomed. Imaging*, Vol. 2013, 2013, pp. 1–17.

[197] Ruvio, G., et al., "Microwave Bone Imaging: A Preliminary Scanning System for Proof-Of-Concept," *Healthc. Technol. Lett.*, Vol. 3, No. 3, September 2016, pp. 218–221.

[198] Meaney, P. M., et al., "Clinical Microwave Tomographic Imaging of the Calcaneus: A First-in-Human Case Study of Two Subjects," *IEEE Trans. Biomed. Eng.*, Vol. 59, No. 12, December 2012, pp. 3304–3313.

[199] Bucci, O. M., et al., "Microwave Ablation Monitoring via Microwave Tomography: A Numerical Feasibility Assessment," in *Proc. 10th European Conference on Antennas and Propagation (EuCAP)*, Davos, Switzerland, 2016, pp. 1–5.

[200] Bellizzi, G. G., et al., "A Full-Wave Numerical Assessment of Microwave Tomography for Monitoring Cancer Ablation," in *Proc. 11th European Conference on Antennas and Propagation (EUCAP)*, Paris, France, 2017, pp. 3722–3725.

[201] Chen, G., et al., "Real-Time 3D Microwave Monitoring of Interstitial Thermal Therapy," *IEEE Trans. Biomed. Eng.*, Vol. 65, No. 3, March 2018, pp. 528–538.

[202] Muqatash, S. A., M. Khamechi, and A. Sabouni, "Detection of the hnique," in *Proc. IEEE International Symposium on Antennas and Propagation & USNC/URSI National Radio Science Meeting*, San Diego, CA, pp. 783–784, 2017.

[203] Cramer, G. D., S. A. Darby, and G. D. Cramer, *Clinical Anatomy of the Spine, Spinal Cord, and ANS, Third Edition*, St. Louis, Missouri: Elsevier, 2014.

6

Microwave Imaging for Subsurface Prospection

This chapter discusses microwave techniques for subsurface imaging, with particular reference to the characterization of the subsoil and to the detection of buried targets. In this context, systems based on ground penetrating radars (GPRs) are traditionally applied. Consequently, the basic concepts of GPR systems are briefly recalled and some of the main standard data processing techniques used in this field are introduced. Subsequently, newly proposed methods based on inverse scattering are discussed.

6.1 Potentialities and Limitations of Microwave Imaging for Subsurface Prospection

The imaging of targets buried in a (eventually inhomogeneous) dielectric half space is a problem of great practical importance in a wide spectrum of applications. In this framework, GPR is a widely-recognized tool for performing subsurface prospection by using electromagnetic radiation in the microwave frequency band [1–5]. GPR systems illuminate the inspected region by a pulsed incident field, and the scattered signal due to the interaction between such electromagnetic radiation and the buried targets is collected in a proper set of points (the main measurement configurations for GPR applications are discussed in Section 6.4). It is worth noting that in most cases the subsurface region presents a significant dielectric contrast, which causes strong reflections of the incident field at the interface. Moreover, since the materials composing the inspected region are usually lossy, a high attenuation of the electromagnetic field can be present. The penetration of the electromagnetic waves into the subsurface region can be thus limited by these two factors. Moreover, the penetration depth also depends on frequency (see Chapter 2). Consequently, when dealing with GPR systems, it is important to properly choose the working frequency for the specific application at hand. In particular, lower frequencies (around 50–500 MHz) allow a higher penetration and are typically used for geophysical applications, such as mapping of the subsoil and detection of deeply buried targets. Intermediate values (around 800–900 MHz) are usually considered for shallow-depth investigations. Higher frequencies (from 1–3 GHz) are mainly considered for nondestructive testing applications. It is worth noting that recently approaches for enhancing the detectability of buried targets (e.g., by reducing the reflection from the air/subsurface interface) have been studied. For example, in [6] the use of passive superstrates deposited on the air-ground interface are proposed for increasing the measurable scattered field.

As introduced at the beginning of the Section, GPR is nowadays successfully applied in a wide range of different applicative fields. Firstly, we can mention the characterization of the subsoil for geophysical purposes. For example, GPR has been used for locating water tables [7–9] and for detecting the existence and level of water in the subsoil [10–13]. Moreover, it has been used to analyze the soil structure, for example, for determining the clay content, salinity, bulk density, and texture. This latter application is also important in the field of agriculture, in order to identify the optimal growing conditions and to monitor irrigation [14, 15]. Another application of GPR is related to the detection of pollutants in the soil (e.g., hydrocarbons and other organic wastes) [2, 16, 17]. GPR is also widely applied to study glaciers and frozen regions. In most of these applications, as previously reported, low working frequency are typically used. In this framework, GPR has also been proposed for subsurface mapping and water search in planetary explorations, for example for the rover of the ExoMars planetary mission [18].

Another important applicative context in which GPR is acquiring an ever growing importance is the characterization of subsoil for civil applications [4]. For example, it is widely adopted for detecting buried utilities, such as pipelines and cables [19–22], for estimating road and pavement layer thicknesses [23–27], and to detect voids and cavities. GPR has also been applied for the assessment of railways, for example for the characterization of the ballast and the detection of changes in the tracks [28–30]. In these cases, middle-high frequencies are usually considered (ranging from about 500 MHz to 1 GHz).

Another applicative field in which GPR has found a significant use is landmine and unexploded devices detection [31–37], since it can be used to detect both metallic and plastic devices. In this case, high frequencies (e.g., around

1–2 GHz) are preferred for searching shallow targets, whereas for deeper objects lower frequencies (e.g., between 500 and 800 MHz) can be used.

Imaging of archeological and cultural-heritage diagnostics is also successfully performed by using GPR. In particular, it is used to obtain images of buried sites and to monitor mosaics, frescoes, statues and other artifacts [38–41]. In this case, the working frequency is usually in the range 1–2 GHz.

Other recent applications are related to forensic surveys, for example, in the detection of buried bodies and graves [42], and the searching/rescuing of peoples trapped under debris in disaster areas or avalanches. Moreover, GPR has also been proposed in forestry applications for mapping of the tree roots [43] and for the analysis of the health status of trees [44] (see also Chapter 4). Finally, it is worth remarking that GPR is also applied in structural monitoring applications, for example, in assessing the conditions of bridge decks and retaining walls [45–48].

Concerning the imaging modalities, common GPR systems often provide two-dimensional (2-D) or three-dimensional (3-D) images of the raw measured data (the so-called *B*-scan and *C*-scan, respectively), as detailed in Section 6.3. Such images, although allowing qualitative evaluations by experienced users about the presence of targets, are however usually quite difficult to interpret, especially for nontrained personnel. Consequently, a fundamental point in subsurface imaging systems is the development of proper processing approaches for extracting the needed information and for creating simple-to-interpret images. Qualitative imaging methods, also referred as migration approaches, are often used in the framework of GPR. However, as in other applicative fields, such approaches do not usually allow identifying the type of targets, but they only provide information about their presence and some geometrical features. Inverse scattering techniques thus represent valuable alternatives, since they permit in principle to obtain maps of the dielectric properties of the subsurface region. However, despite a large number of efficient techniques that have been developed, there are still some open issues that must be faced in order to enhance the performance of GPR systems in real applications. For example, in inverse scattering methods simple antenna models are usually assumed (e.g., line-current sources or short dipoles), which however do not take into account the real radiation patterns and couplings with the subsurface medium. Some attempts to study and overcome such limitation have been proposed, for example in [49–52]. Moreover, inverse scattering approaches, when developed in the frequency domain, require one to specify the Green's function for the considered subsurface scenario. If simplified configurations are used (e.g., homogeneous half-spaces or planarly-layered media), closed form expressions are available (which however present some numerical complexity, as discussed in the following). In more realistic cases, the subsoil has however a greater variability (e.g., oblique separation planes among layers). Moreover, usually the air/ground interface is a rough surface, which requires the use of proper formulations of the electromagnetic problem and of the related Green's functions in order to take into account the scattering from nonplanar structures [53–55].

Finally, it is important to remark that forward simulation of buried scenarios plays an important role in the characterization and testing of subsurface imaging systems. To this end, solvers based on the finite-difference time-domain (FDTD) method are often used [56, 57], since they are able to accurately model the time-domain behavior of the electromagnetic field in the subsurface medium. Finite-element time-domain methods have also been proposed [58]. Moreover, fast analytical/semi-analytical solutions have been developed with the aim of providing fast solvers for canonical targets [59, 60]. Frequency domain methods can be used, too [61–64]; in this case however, the time-domain response needs to be obtained via an inverse Fourier transform.

6.2 Dielectric Properties of the Soil

The knowledge of the dielectric properties of the soil is of great importance in both understanding and designing microwave imaging systems for buried objects detection. Soil is a complex material, since it is usually composed by several different parts (e.g., rocks, sand, clay, silt, and water). The dielectric properties thus depend upon several factors, such as bulk and particle densities, mass fractions of the various components, volumetric water content, and temperature. In the literature, the problem of characterizing the dielectric behavior of such a complex material has been extensively studied (see for example [1], [65–70]). In particular, dielectric mixing models [71], combining the dielectric properties of the various materials composing the soil, have been adopted to model the actual behavior of the dielectric properties of terrains.

In this section, the semi-empirical model proposed in [66, 69] for soils composed by sand, silt, and clay is reported as an example of characterization of these composite materials. Such model has been initially developed for frequencies ranging from 1.4 to 18 GHz and subsequently extended to the band 0.3–1.3 GHz. The real and imaginary parts of the relative dielectric permittivity of the soil are approximated by means of the following semi-empirical mixing formula [66, 69]

$$\varepsilon_r' = \begin{cases} \left(\dfrac{1 + \dfrac{\rho_b}{\rho_s}}{\left(\varepsilon_s^{\alpha} - 1\right) + m_v^{\beta'} \varepsilon_{fw}'^{\alpha} - m_v} \right)^{\frac{1}{\alpha}} & f \in [1.4 - 18] \text{ GHz} \\ 1.15 \left(\dfrac{1 + \dfrac{\rho_b}{\rho_s}}{\left(\varepsilon_s^{\alpha} - 1\right) + m_v^{\beta'} \varepsilon_{fw}'^{\alpha} - m_v} \right)^{\frac{1}{\alpha}} - 0.68 & f \in [0.3 - 1.3] \text{ GHz} \end{cases} \tag{6.1}$$

$$\varepsilon_r'' = \left(m_v^{\beta''} \varepsilon_{fw}''^{\alpha} \right)^{\frac{1}{\alpha}} \tag{6.2}$$

where ρ_b and ρ_s are the bulk and specific densities of the soil, $\alpha = 0.65$ is a constant value (determined empirically from measured data), m_v is the volumetric moisture content, and ε_s is given by [66, 69]

$$\varepsilon_s = (1.01 + 0.44\rho_s)^2 - 0.062 \tag{6.3}$$

In (6.2) and (6.3), $\varepsilon_{fw} = \varepsilon_{fw}' - j\varepsilon_{fw}''$ is the dielectric permittivity of free water, which is modeled with a single-pole Debye model similar to the one described in Chapter 2, modified in order to correctly take into account the soil mixture conductivity. In particular, the static conductivity is expressed as

$$\sigma_s = \sigma_{eff} \frac{(\rho_s - \rho_b)}{\rho_s m_v} \tag{6.4}$$

where

$$\sigma_{eff} = \begin{cases} -1.645 + 1.939\rho_b - 2.25622S + 1.594C & f \in [1.4 - 18] GHz \\ 0.0467 + 0.2204\rho_b - 0.4111S + 0.6614C & f \in [0.3 - 1.3] GHz \end{cases}$$

S and C being the mass fractions of sand and clay. Finally, the parameters β' and β'' have been empirically found to be equal to [66, 69]

$$\beta' = 1.2748 - 0.519S - 0.152C \tag{6.5}$$

$$\beta'' = 1.33797 - 0.603S - 0.166C \tag{6.6}$$

An example of the behavior of the dielectric properties of sandy soil in the frequency range of 0.3–1.3 GHz is shown in Figure 6.1. The considered soil (at a temperature of 20°C) is characterized by $S = 50\%$, $C = 15\%$, $\rho_s = 2.66$ g/cm^3, and $\rho_b = 1.5$ g/cm^3. Two values of the volumetric moisture content m_v have been considered (5% and 25%). As can be seen, the real part of the permittivity exhibits little variations in this frequency range, but, as expected, there is a strong dependence upon the volumetric water content. On the contrary, the imaginary part shows a reduction when the frequency increases.

The dielectric characterization of soil materials at microwave frequencies has been the subject of various measurement campaigns, also because of the interest that this topic attracts not only for GPR applications, but also for remote sensing purposes. For further details, the reader can refer to the specific scientific literature about this topic (for instance, [72–77]). Just as an example, Figures 6.2 and 6.3 report the values of the measured dielectric properties of some kinds of rocks that can be found in the subsoil [73, 77]. In some cases, these materials exhibit an anisotropic behavior, already encountered in Chapter 4.

6.3 Ground Penetrating Radar Basics

The theory of GPR is discussed in details in different books specifically dedicated to this topic (see for example [1–3]). Here we just want to recall the main basic concepts, with the aim of introducing the use of microwave imaging

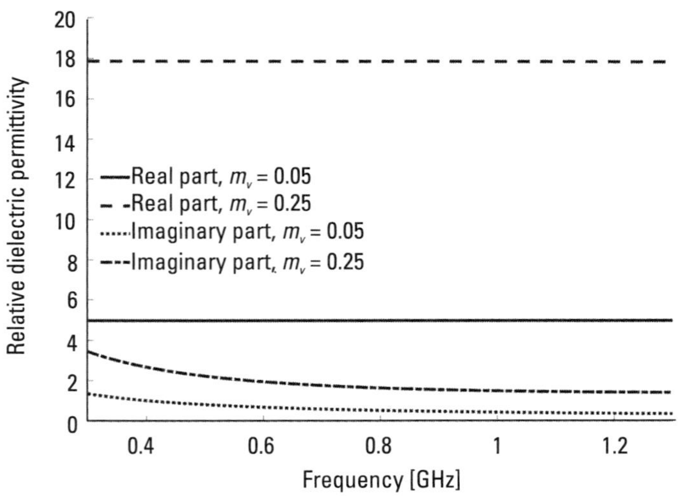

Figure 6.1 Example of dielectric properties (real and imaginary parts) of two sandy soil samples [66] characterized by mass fractions of sand and clay equal to S = 50% and C = 15%, , bulk and specific densities equal to $\rho_s = 2.66$ g/cm^3 and $\rho_b = 1.5$ g/cm^3, and volumetric moisture contents m_v equal to 5% and 25%.

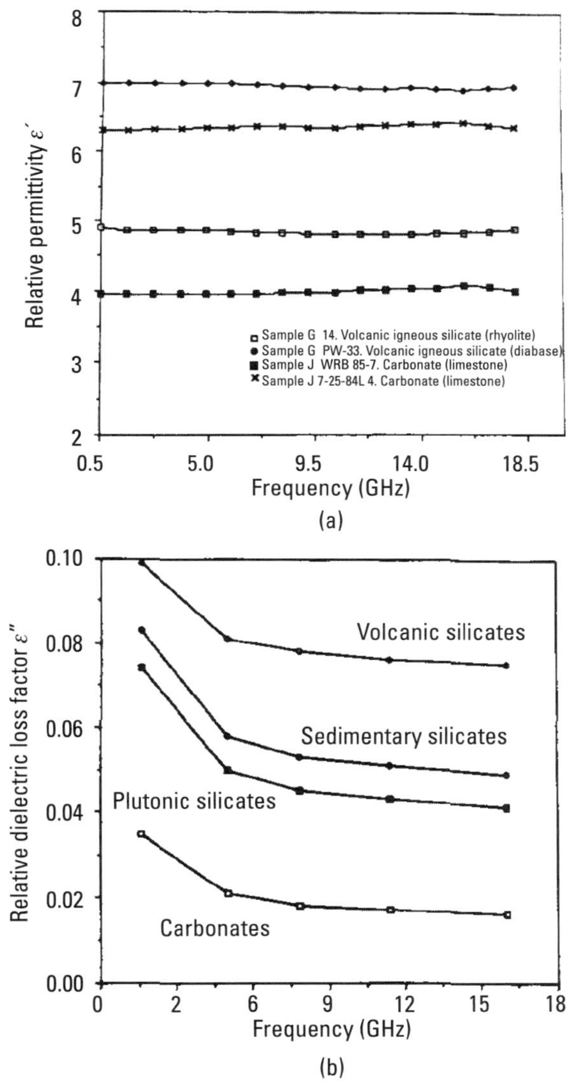

Figure 6.2 Example of measured (a) relative dielectric permittivity and (b) loss factor at microwave frequencies for several different kinds of dry rocks [73]. (©1990 IEEE.)

techniques, especially those based on inverse scattering, in this specific context. In this chapter (when not otherwise specified), a cylindrical configuration (see Figure 6.4) under transverse-magnetic illumination conditions is considered (i.e., the dielectric properties both of the media composing the two regions and of the target possibly buried in the lower half space are assumed independent of the z coordinate and the electric field vector is supposed to be directed along

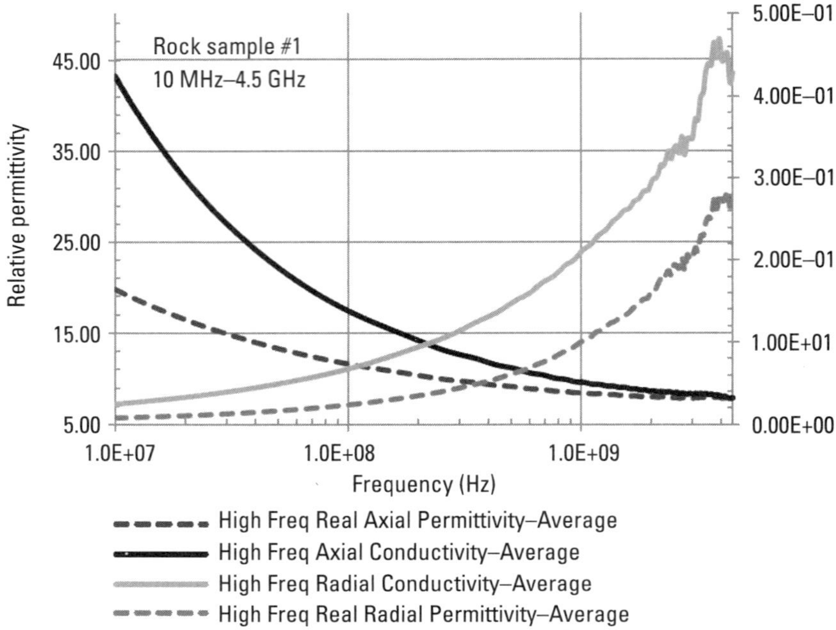

Figure 6.3 Measured axial and radial relative permittivity and conductivity tensor components of a shale source rock sample [77]. (©2017 IEEE.)

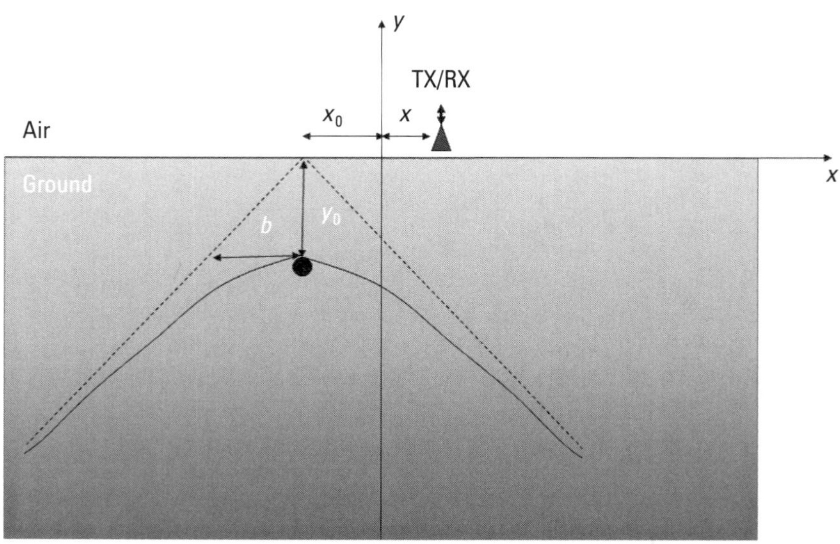

Figure 6.4 Schematic representation of a monostatic GPR system.

the same axis). This hypothesis has been discussed in Chapter 3. In several practical conditions, such an assumption is quite reasonable, since the buried targets (e.g., pipes or other utilities) often exhibit elongated shapes. As discussed in Chapter 3, under such hypotheses, the electromagnetic scattering problem turns out to be a scalar and 2-D problem (both in free space and with reference to half space or stratified geometries). The air-ground interface is planar, perpendicular to the y axis, and passing through the origin of the considered reference system (as shown in Figure 6.4). A monostatic configuration (see Section 6.4), with an ideal antenna located over the air-ground interface at position x, is considered for explicative purposes. It is worth noting that, in real applications, the monostatic configuration is usually approximated by using two antennas separated by a small offset d_{TR} (quasi-monostatic configuration). The antenna (in transmitting mode) illuminates the scenario with an incident electric field vector, whose z-component is denoted by $e_{inc,z}(\mathbf{r}_t, t)$, and the z-component of the total electric field vector, $e_{tot,z}(\mathbf{r}_t, t)$, resulting from the interaction with the buried target is collected almost at the same point.

An example of the behavior of the z-component of the total electric field vector simulated by using the FDTD method [56, 57] is shown in Figure 6.5. Such representation is usually referred as A-scan. The considered target is a perfect electric conducting (PEC) circular cylinder whose cross section is centered, in the cross plane, at position $\mathbf{r}_{t,c} = (-0.1, -0.25)$m and it has diameter of $d_c = 0.15$m. A Ricker pulse [1] with a central frequency of 1.5 GHz has been used

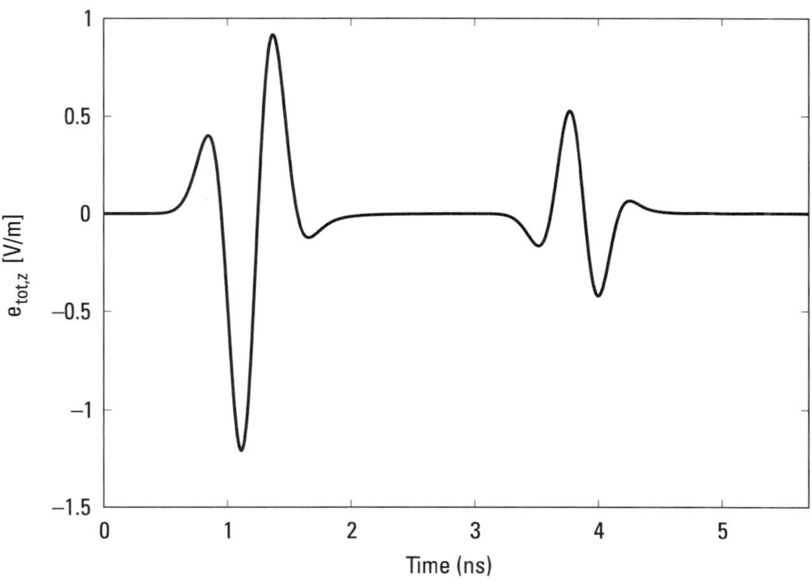

Figure 6.5 Example of a simulated A-scan in the presence of a PEC cylinder.

to feed the antenna (which is modeled as an ideal line-current source). The total electric field is collected above the cylinder. An ideal probe, separated by a small offset of $d_{TR} = 4$ cm from the antennas to avoid numerical problems, is used to collect the field. A homogeneous lossless soil with relative dielectric permittivity equal to $\varepsilon_{r,b} = 6$ has been assumed. In such a figure, it is possible to identify different contributions. The first one is due to the reflections from the air-ground interface and to the direct coupling between the transmitting and receiving antennas. A second pulse, due to the interaction with the target, is also present after a time t_r, whose value depends on the distance between the object surface and the antennas and on the velocity of propagation of electromagnetic waves in the considered media.

In the common usage of GPR systems, the measurements are not collected at a single position, but the antennas scan the subsurface region by moving along a measurement path over the air-ground interface. As a result, several *A*-scans, corresponding to the different positions of the antennas, are collected. Such *A*-scans are usually combined to obtain a *B*-scan representation, that is, a 2-D map showing the amplitude of the received signal versus the position of the receiving antenna (on the abscissa) and time (on the ordinate). In some cases, the data can also be acquired on a planar surface over the ground. In this case, a 3-D representation, usually referred as a *C*-scan, is obtained by stacking the *B*-scans obtained on parallel acquisition lines.

An example of a simulated *B*-scan, for the previously considered example, is reported in Figure 6.6. In this case, it is supposed that the antennas are moved on a probing line of length $L_w = 0.8$m (from –0.4m to 0.4m) with a step of $l_w = 0.01$m. As can be seen, in the upper part of the image, a band is present, which is due to the transmitting/receiving coupling and to the reflection form the air-ground interface. Moreover, the cylindrical scatterer produces a hyperbola in the *B*-scan image.

The presence of the hyperbolic structure in the *B*-scan can be qualitatively explained as follows. Let us consider a buried cylindrical target with a small (ideally negligible) size, whose cross section is located at position $\mathbf{r}_{t,0} = (x_0, y_0)$. The distance between the transmitting/receiving antennas and the target (see Figure 6.4) is given by

$$d = \sqrt{\left(x - x_0\right)^2 + y_o^2} \tag{6.7}$$

Accordingly, the time needed by the electromagnetic wave to travel from the GPR to the target (forth and back) is given by $t_r = 2d/v$, where v is the velocity of propagation in the soil (which is equal to $v = v_0 / \sqrt{\varepsilon'_{r,b}}$, v_0 being the speed of electromagnetic waves in vacuum, if a lossless homogeneous medium with a real relative dielectric permittivity $\varepsilon'_{r,b}$ is considered). By taking the square of

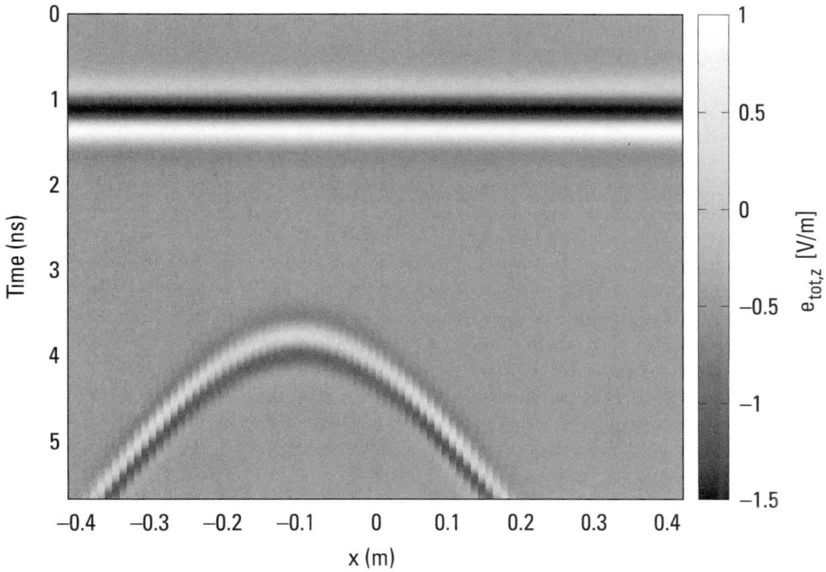

Figure 6.6 Example of a *B*-scans obtained with the common offset setup. Single PEC cylinder.

both sides of (6.7), after some simple mathematical steps the following equation is obtained [1]

$$\frac{t_r^2}{a^2} - \frac{(x - x_0)^2}{b^2} = 1 \tag{6.8}$$

which represents a hyperbola with respect to the variables (x, t_r), with $a = t_0 = 2y_0/v$, $b = t_0 v/2$, and $t_0 = 2y_0/v$.

The previous relationship assumes an ideal target of small (ideally negligible) size. In the case of target with a circular cross section of radius R, it is possible to derive a similar relationship, that is, [3]

$$\frac{\left(t_r + \frac{2R}{v}\right)^2}{a^2} - \frac{(x - x_0)^2}{b^2} = 1 \tag{6.9}$$

which again represents a hyperbola with $a = t_0 + 2R/v$ and $b = t_0 v/2 + R$. In this case, the asymptotes intersect at a height $2R/v$ over the air-ground interface. Anyway, the location of the apex of the hyperbola and the slope of the asymptotes do not change.

Clearly, in real applications, the underground configuration is more complex, due to different nonidealities. First of all, the soil is rarely homogeneous. Moreover, the air-ground interface is not strictly planar. However, in real *B*-scans, hyperbolas similar to the previously discussed ones are usually present, too, as can be seen in Figure 6.7, where different hyperbolas due to several buried pipes can be clearly identified. This figure shows an example of experimental data measured at the Institut Français des Sciences et Technologies des Transports, de l'Aménagement et des Réseaux (IFSTTAR) geophysical test site in Nantes, France [78]. A GPR scan was performed by using a GSSI SIR-3000 GPR with a central frequency of 900 MHz. The soil is composed by gneiss 14/20 gravel, whose estimated dielectric properties at the considered frequency are $\varepsilon_{r,b} \cong 4.5$ and $\sigma_b \cong 0.01\text{S/m}$. The transmitting and receiving antennas are separated by a fixed distance of $d_{TR} = 0.14\text{m}$. The data are collected along a line above the air-ground interface with spacing between the measurement positions equal to $l_w = 0.03\text{m}$.

6.4 Measurement Configurations for Subsurface Sensing

In Section 6.3, the particular case of a monostatic measurement setup has been considered for explicative purposes. Such a configuration is often assumed in subsurface imaging. From a practical point of view, as mentioned, measure-

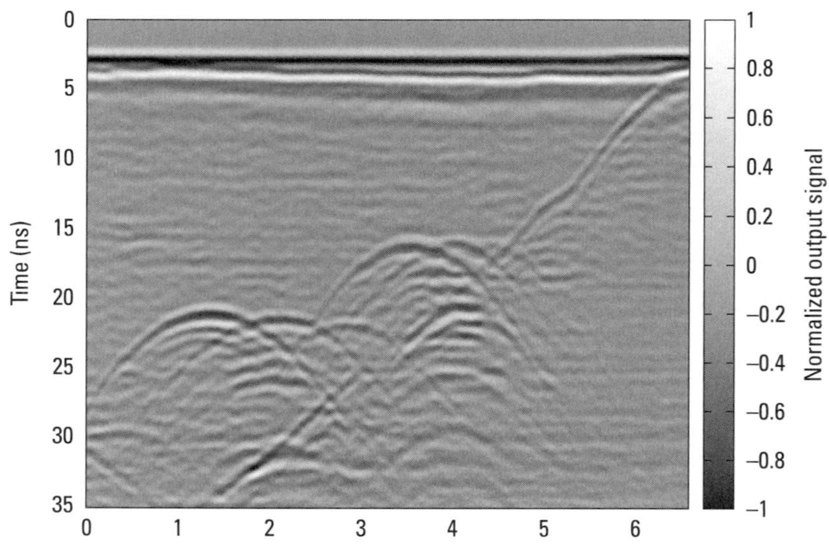

Figure 6.7 Example of *B*-scan obtained from the measurements available in the IFSTTAR geophysical test-site data set [78].

ments are obtained by using a pair of antennas with a (small) fixed separation d_{TR} (i.e., a quasi-monostatic setup is used to approximate the ideal monostatic case), as schematically shown in Figure 6.8(a). The pair of antennas is moved along a probing line of length L_w, parallel to the air-soil interface at a given height h_w, and the measurements are collected at fixed distances l_w. Often, in the GPR community, such kind of configuration is referred as common offset

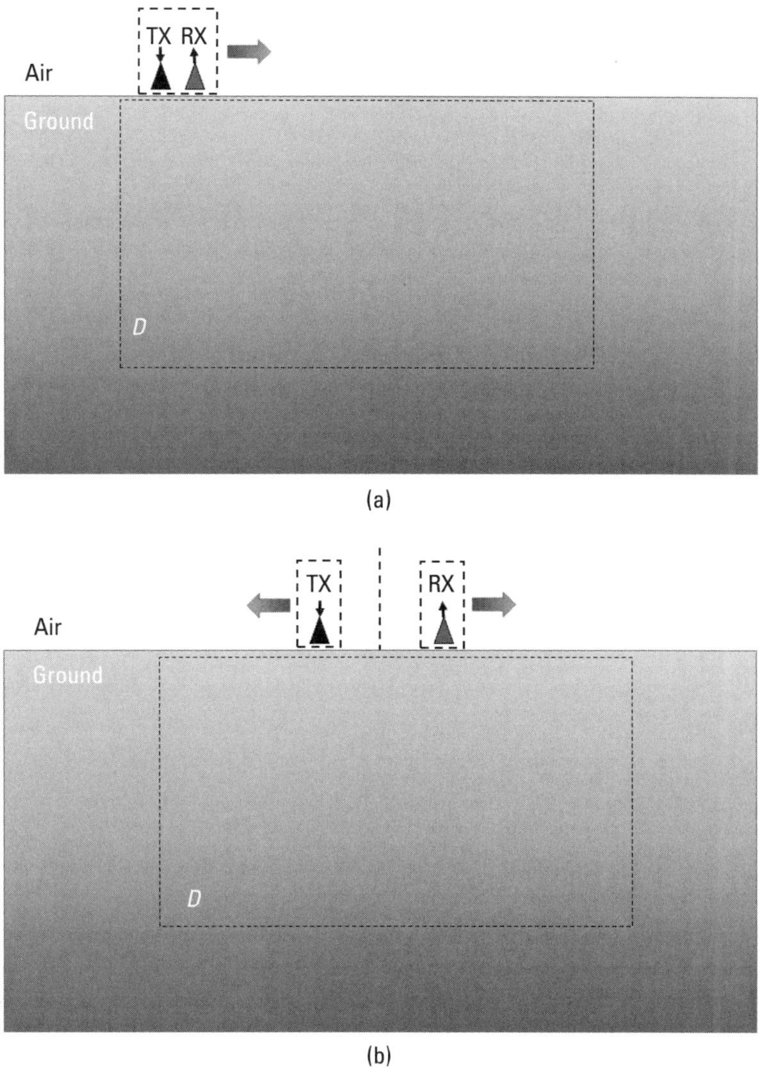

Figure 6.8 Conventional measurement configurations adopted for subsurface imaging. (a) Common offset setup; (b) Common midpoint setup; (c) Multistatic multiview (MIMO) setup; (d) Cross-borehole configuration.

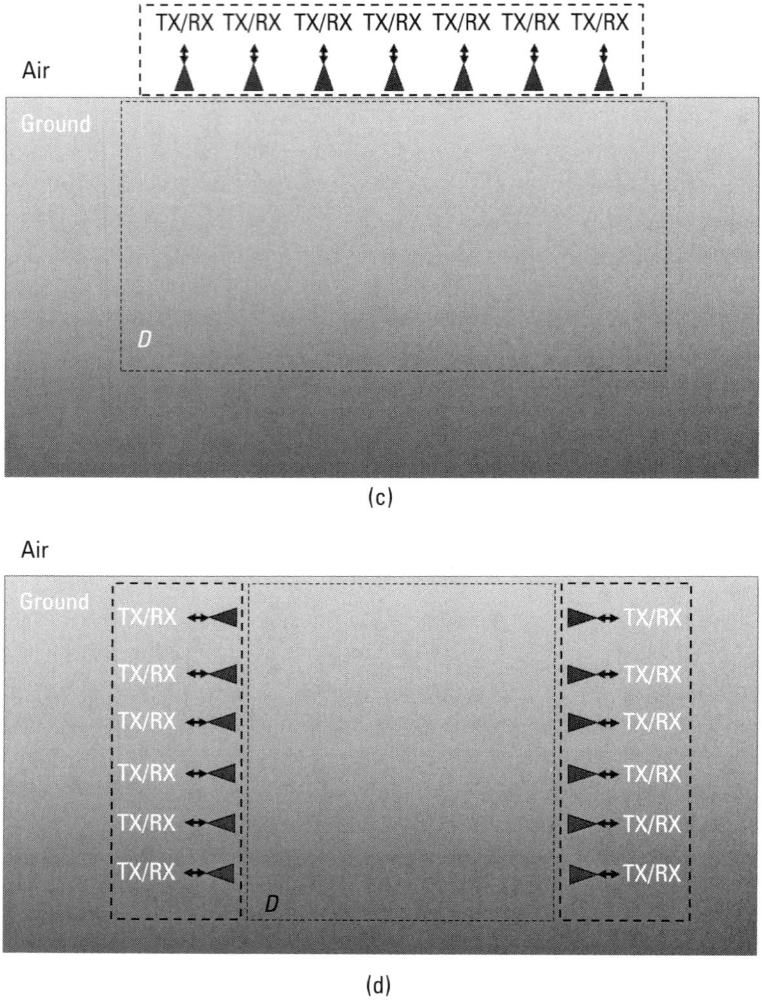

Figure 6.8 (continued)

(CO), since the separation between the two antennas used to scan the inspected region is kept constant.

The common offset setup is not the only one used in subsurface sensing. In the common midpoint (CM) approach, the separation between the two antennas d_{TR} in subsequent measurements is increased, but keeping the midpoint fixed, as sketched in Figure 6.8(b). It is worth noting that the CO setup allows collecting only monostatic (or quasi-monostatic) measurements, that is, the main scattering contribution is due to the specular reflections. Conversely, with the CM configuration the acquisition of bistatic data is possible. Other different measurement configurations are also adopted. For example, in the common source and common receiver setups, one of the antennas (the transmitting or the receiving one, respectively) is kept at a fixed location, whereas the other one is moved in order to scan the inspected area. In all these situations, a single antenna pair is used to scan the overall inspected region. Consequently, the hardware complexity of the measurement setup is quite reduced, since the system does not require complex switching and control hardware subsystems. Therefore, the realization of low-cost measurement setups is possible, but a limited amount of information can be retrieved.

In order to increase the amount of attainable information, it is possible to use multistatic measurement setups (Figure 6.8[c]). Here, a set of M antennas is adopted. Each one acts, in general, both as transmitter and receiver. In particular, when an antenna transmits, the others collect the electric field, resulting from the interactions between the incident field and the buried targets. Hence, multiview data are obtained, since the targets are seen from different directions. In recent years, such a setup has also been indicated as a multiple-input multiple-output (MIMO) configuration, denoting the overall system as a MIMO GPR. Obviously, in this configuration as well, only limited aspect data can be collected, since all the antennas are located on one side of the investigation domain. However, the possibility of increasing the information content of data results in an increased system complexity: In order to select the pair of transmitting/receiving antennas for each measurement, it is necessary to use a microwave switch and a more complicated control system.

A further extension is represented by borehole and cross-borehole measurement setups. In this case, the antennas are located inside one or two boreholes, which are created on the lateral sides of the inspected region (the cross-borehole case is shown in Figure 6.8[d]). Even here, multistatic data can be obtained by using all (or a part of) the antennas as both transmitters and receivers.

Forward-looking measurement configurations have also been proposed [79–82]. In this case, the transmitting/receiving antennas are usually mounted on a vehicle (the transmitting and receiving antennas may have different heights, $h_{w,TX}$ and $h_{w,RX}$, respectively). The systems inspect the scenario in the forward direction, as schematically shown in Figure 6.9. The investigation domain is located at a given distance along the horizontal direction, instead of being below the antennas. Clearly, such configuration poses additional problems and limitations due to the angle of incidence and the available received power. Nevertheless, as an example, the above configuration can be a good choice in demining applications, since it allows introducing a certain standoff between the personnel operating the GPR apparatus and the inspected region.

When it is necessary to scan very large areas, or when a direct access to the region to be inspected is not possible (e.g., in hostile or desert regions), airborne GPR systems can be adopted. In this case, the antenna pair is mounted on a helicopter or an airplane [83–85]. The above imaging modality introduces further

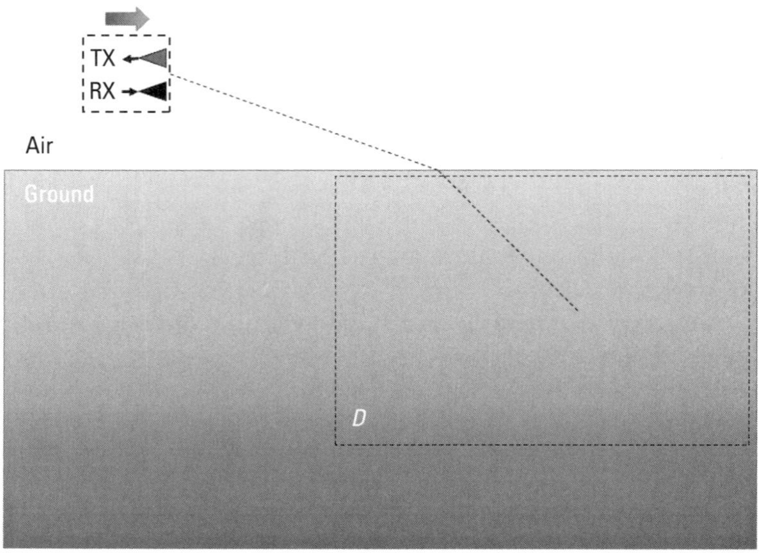

Figure 6.9 Forward-looking measurement configuration for subsurface imaging.

difficulties, concerning both the construction of the various hardware components (e.g., the need for designing and realizing antennas with proper radiation characteristics and appropriate for transportation on airplanes/helicopters) and the development of suitable imaging procedures, which must take into account additional parameters related to the high altitude with respect to the ground, the speed of movement, the surface topography, and so forth.

6.5 Migration Processing Techniques

The most commonly used processing techniques in the GPR field belong to the class of migration processing approaches. Such techniques are usually able to provide information only about the presence and—in some cases—the dimensions of the target to be retrieved. Different approaches of this kind can be used (see, for example, [1, 2, 86–89] and the references therein). Some of the main ones are discussed in this section, with reference to cylindrical configurations, such as the ones adopted for the examples reported in Section 6.3.

Let us consider again the typical *B*-scan image of Section 6.3 (Figure 6.6), where hyperbolas are generated in the quasi-monostatic case by a small cylindrical target. The basic idea of migration is to collapse the hyperbola in a single point, by back-propagating the collected measurements. Such a basic idea was initially developed in the framework of seismic imaging [90]. The methods implementing this original concept are usually referred as wave interference migration or *A*-scan driven approaches. Basically, once the position of the receiving antenna is fixed, the target producing the pulse seen in the *A*-scan must be located on a circle (in the transverse plane) centered on the receiving antenna and with a radius equal to $vt_r/2$, where t_r is the time at which the pulse echo is received. The circles related to the all the different positions of the receiving antenna intersect in the point where the target is located.

A second way of performing such kind of migration is the so-called pixel-driven approach or diffraction summation technique. In this case, for each pixel of the *B*-scan image, a synthetic hyperbola is built starting from the knowledge of the position of the receiving points and of the propagation velocity. Then, the pixels belonging to such hyperbola are identified and the corresponding values of the scattered field in the *B*-scan image are summed together. The previous qualitative ideas can be expressed from a mathematical point of view by the following formula [91]

$$I(\mathbf{r}_t) = \int_{D_{obs}} \int_{T} e_{scatt,z}(x',t)\,\delta\left(t - \frac{2}{v}\sqrt{(x-x')^2 + y^2}\right) dt\, dx', \mathbf{r}_t \in D \quad (6.10)$$

where $I(\mathbf{r}_t)$ is an indicator function, which is used to construct the migrated image, δ is the Dirac's delta function, and T is the time duration of the *A*-scans. The above approach is also known as synthetic aperture focusing technique (SAFT) and, although initially developed starting from quite intuitive concepts, it can be formally derived from the scattering equations [91].

An example of results obtained by applying the migration approach formulated through (6.10) is reported in Figure 6.10. In particular, with reference to the configuration considered in Section 6.3, in which a buried PEC circular cylinder has to be retrieved, the normalized indicator function $I_n(\mathbf{r}_t) = |I(\mathbf{r}_t)| / \max_{\mathbf{r}_t \in D} I(\mathbf{r}_t)$ inside the investigation domain is shown in Figure 6.10. The scattered electric field is obtained by using the time gating procedure described in Section 6.6 (in particular, the first 2 ns of the received signal are muted). As can be seen, the target is quite correctly detected.

The migration equation in (6.10) can be written in terms of the frequency-domain data by exploiting the Fourier transform $E_{scatt,z}(x,\omega)$ of $e_{scatt,z}(x,t)$ with respect to t, that is

$$E_{scatt,z}(x,\omega) = \int_0^T e_{scatt,z}(x,t)\,e^{-j\omega t}\,dt \quad (6.11)$$

where the field is assumed to be equal to zero before and after the time duration of the *A*-scan. In this case, neglecting some unnecessary constants (which are eliminated in the normalization phase of the indicator function), one obtains

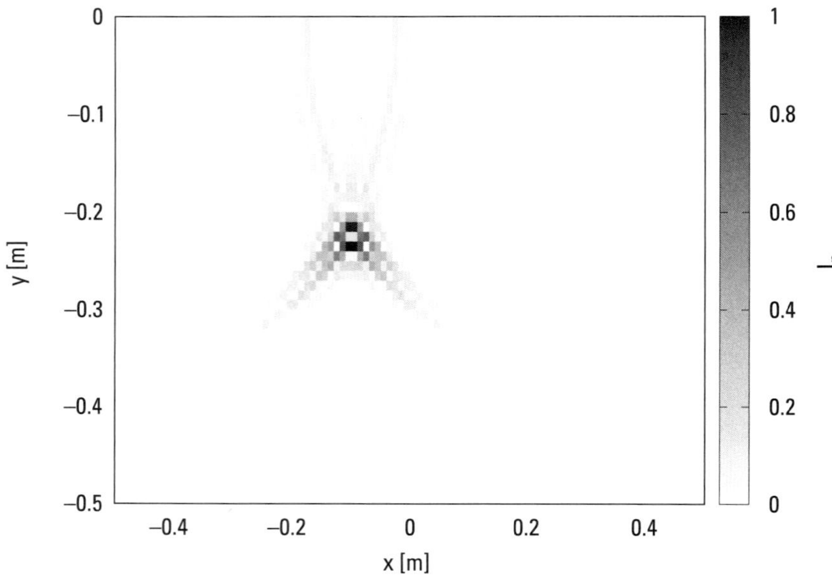

Figure 6.10 Example of an application of diffraction summation migration to the B-scan shown in Figure 6.6.

$$I(\mathbf{r}_t) = \int_{D_{obs}} \int_{\Omega} E_{scatt,z}(x',\omega) e^{j\frac{2\omega}{v}\sqrt{(x-x')^2+y^2}} d\omega dx', \mathbf{r}_t \in D \quad (6.12)$$

where Ω is the considered angular frequency band (defined by the spectrum of the incident pulse).

Another adopted imaging scheme is the so-called frequency-wavenumber (F-K) migration, in which the indicator function is given by [92],

$$I(\mathbf{r}_t) = \int_{K_x} \int_{\Omega} \tilde{E}_{scatt,z}(k_x,\omega) e^{j(k_x x + k_y y)} d\omega \, dk_x, \mathbf{r}_t \in D \quad (6.13)$$

where $\tilde{E}_{scatt,z}(k_x,\omega)$ is the Fourier transform of the z-component of the electric field vector with respect to the x coordinate and K_x is the related spatial bandwidth. The F-K migration formula can be derived from a scalar wave equation [92, 87]. Equation (6.13) can be efficiently computed by using the fast Fourier transform (FFT), since, after a change in the integration variable from ω to k_y, one obtains

$$I(\mathbf{r}_t) = \int_{K_x} \int_{K_y} \frac{v^2 k_y}{\omega} \hat{E}_{scatt,z}(k_x,k_y) e^{j(k_x x + k_y y)} dk_x dk_y, \mathbf{r}_t \in D \quad (6.14)$$

where $k_y = \sqrt{4k^2 - k_x^2}$ and $\hat{E}_{scatt,z}\left(k_x, k_y\right) = \tilde{E}_{scatt,z}\left(k_x, (v/2)\sqrt{k_x^2 + k_y^2}\right)$. It is worth noting that, in applying (6.14), the field $\hat{E}_{scatt,z}$ must be resampled onto a regular grid in order to use standard FFT algorithms. As it is well known, the FFT computation can introduce approximation errors in the imaging process and proper interpolation procedures are needed. Therefore, nonuniform FFT (NUFFT) formulations have been also considered [93, 94]. Finally, it should be mentioned that if the amplitude factor in (6.14) is neglected, this migration scheme becomes equivalent to the SAR algorithm [87, 95] (see also Chapter 4).

An example of results obtained by using the F-K migration algorithm is provided in Figure 6.11 [87]. Experimental data in a monostatic configuration have been acquired by using a VNA and a double-ridge horn antenna working in the frequency range 1–12.4 GHz. Two targets with dimensions comparable to landmines have been buried in a dry silica sand with a relative dielectric permittivity $\varepsilon'_{r,b} = 2.4$, at a depth of 10 cm from the surface. The measurements have been performed on a line of length $L_w = 11$m with steps of $l_w = 1$ cm.

The above methods require the knowledge of the speed of electromagnetic waves in the subsurface medium. In practical applications, this information

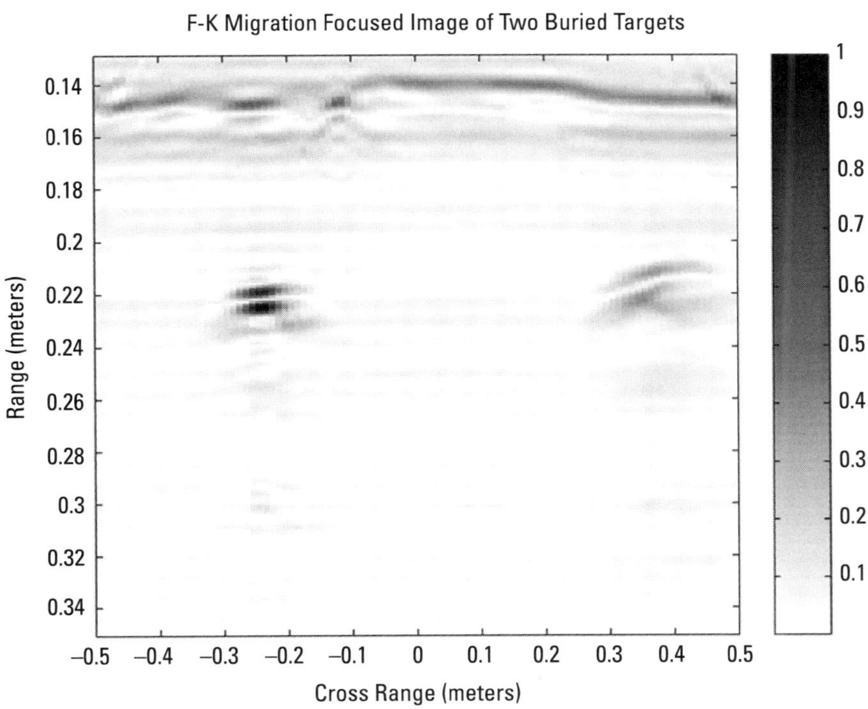

Figure 6.11 Example of a reconstruction of two buried targets obtained by using the F-K migration algorithm [87]. (© 2006 IEEE.)

may not be available, or may be difficult to be estimated. To overcome such limitations, the time domain reflectometry (TDR) [96, 97] can be applied in order to retrieve the dielectric properties of the soil. However, approaches aimed at extracting this information directly from the GPR signals have also been proposed [98–100].

6.6 Inverse Scattering-Based Approaches for Subsurface Prospecting

Inverse scattering methods can be used to directly provide maps of the physical and geometrical parameters of the subsurface scenario under test. In most cases inverse scattering techniques work in the frequency domain, whereas most GPR apparatuses use pulsed excitations. Therefore, it is necessary to extract the frequency-domain scattered field data from the corresponding time-domain data.

Moreover, inverse scattering procedures require the knowledge of the scattered field, which represents the known term of the inverse scattering equation (see Chapter 3). Such a quantity can be estimated directly from the available measurements in time domain. To this end, different approaches can be used. The simplest idea is to remove the first part of the time domain signal, which, as previously mentioned, contains the reflection contribution from the air-ground interface. Such an operation, which is commonly referred as time gating, is schematically shown in Figure 6.12, where the simulated *A*-scan of Figure 6.5 is considered again. The part of the signal included in the gray region in the left figure is simply removed by setting the corresponding values to zero. The estimated *z*-component of the scattered field vector is then obtained, as shown in the right side of Figure 6.12.

In general, the gating operation can be viewed as the application of an operator Γ acting to the measured time domain total field in order to obtain the scattered field $\tilde{e}_{scatt,z}$, that is,

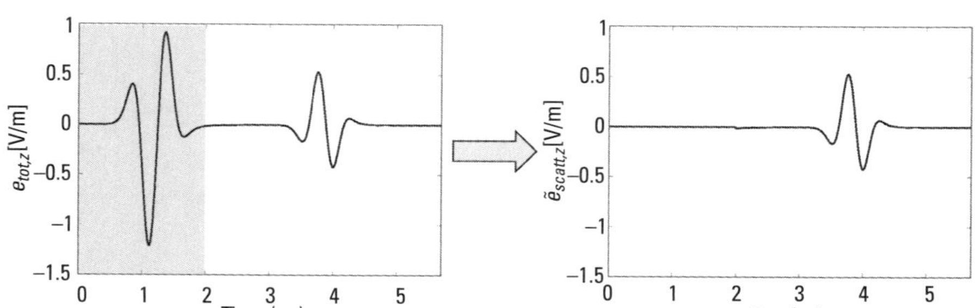

Figure 6.12 Example of time gating of the *A*-scan shown in Figure 6.5.

$$\tilde{e}_{scatt,z}(\mathbf{r}_t,t) = \Gamma(e_{tot,z})(\mathbf{r}_t,t) \tag{6.15}$$

In its simplest form (Figure 6.12), the gating operator yields accurate estimations only if the pulses backscattered by the targets are not overlapped to the reflection signal due to the interface (e.g., in the case in which the targets are not very near to the ground surface). Another popular strategy for estimating the scattered field is the average subtraction, in which (for each *A*-scan) the average value of the measured total field in all the other *A*-scans at the same time t is subtracted from the data (this strategy will be further discussed in Chapter 7). Other more advanced gating and filtering approaches, for example those based on entropy concepts, have also been proposed in the literature (for an example, see [101]).

In any case, after the scattered field is estimated, the data in the frequency domain are obtained by means of the Fourier transform in (6.11). Moreover, the presence of the background is taken into account by using the proper scalar or dyadic Green's function for the considered configuration. Usually, as already mentioned, simplified geometries are considered, for example, those constituted by homogeneous half spaces and planarly-layered media [63, 102, 103]. Rough surfaces are also sometimes taken into account [53–55]. The simplest case is represented by the homogeneous half-space configuration, for which the 2-D Green's function is given by [104, 105]

$$g_{hs}(x,y,x',y') = \begin{cases} \dfrac{j}{4\pi}\displaystyle\int_{-\infty}^{+\infty} \dfrac{e^{jk_x(x-x')}}{\gamma_0}\left(e^{-j\gamma_0|y-y'|} + Re^{-j\gamma_0(y+y')}\right)dk_x & y \geq 0 \\ \dfrac{j}{2\pi}\displaystyle\int_{-\infty}^{+\infty} \dfrac{e^{jk_x(x-x')}}{\gamma_0+\gamma_b} e^{j(\gamma_b y - \gamma_0 y')}dk_x & y < 0 \end{cases} \tag{6.16}$$

where $\gamma_o = \sqrt{k_0^2 - k_x^2}$ and $\gamma_b = \sqrt{k_b^2 - k_x^2}$, k_0 and $k_b = \omega\sqrt{\varepsilon_b \mu_0}$ being the wavenumber in vacuum and in the subsurface medium, respectively, and R is given by

$$R = \frac{\gamma_0 - \gamma_b}{\gamma_0 + \gamma_b} \tag{6.17}$$

The above expression is valid for $y' \geq 0$. The corresponding equation for $y' < 0$ can be directly obtained by exploiting the reciprocity of the Green's function. An analogous expression can be written for the dyadic Green's function needed in the 3-D cases [106]. The Sommerfeld's integrals in (6.16) present singularities that introduce numerical difficulties in their evaluations [105]. To overcome such numerical problems, deformations of the integration contour, singularity subtraction techniques, or ad hoc quadrature rules must be applied as described, for example, in [63, 107–112].

6.6.1 Linearized Approaches

When applicable, linear approximations can be very useful to simplify the inverse scattering problem, as discussed in Chapter 3. In this framework, the first-order Born approximation is effective for retrieving the dielectric profiles of weakly scattering objects in subsurface imaging [113, 114]. When metallic bodies have to be detected, the Kirchhoff approximation can be used in methods aimed at retrieving their external shapes [115, 116].

The solution of the inverse scattering problem, even in the linearized case, is ill-posed (see Chapter 3). A widely used approach for solving, in a regularized way, the linearized scattering equations arising in the subsurface imaging problem is the singular value decomposition (SVD) [117]. Considering for example the 2-D case, the linearized scattering is governed by (3.34) with the proper Green's function. By using the SVD, the contrast function inside the investigation domain is expressed as

$$c(\mathbf{r}_t)=\sum_{i=1}^{P}\frac{W_i}{\sigma_i}\langle E_{scatt,z},u_i\rangle v_i,\ \mathbf{r}_t\in D \tag{6.18}$$

where $<\bullet,\bullet>$ denotes the inner product, σ_i are the singular values of the integral operator (in descending order), u_i and v_i indicate the corresponding singular functions, and W_i is a filtering term [117]. The number of singular values, P, depends on the dimensionality of the problem. Moreover, if $W_i=\sigma_i^2\left(\sigma_i^2+\alpha\right)^{-1}$, a Tikhonov regularization is obtained, where α is the regularization parameter. If $W_i=1$ for $\sigma_i^2>\alpha$ and $W_i=0$ otherwise, the so-called truncated SVD method is obtained. In the latter case, the smallest index i for which $W_i\neq 0$ acts as regularization parameter. The choice of the regularization parameter is not an easy task, since it depends on both the operator and the amount of noise on data.

Figure 6.13 provides a reconstruction result obtained by using a linearized reconstruction method. It concerns a homogeneous half-space configuration (Figure 6.4), where the subsurface medium has dielectric properties given by $\varepsilon_{r,b}=3$ and $\sigma_b=0.001$ S/m (dry sandy soil). The investigation domain is a rectangular area of sides $L_x=2$m and $L_y=1$m, which has been discretized into $N=800$ square cells. Two separate lossless buried targets are present in the inspected region. The first one is a void circular cylinder with diameter $d_c=0.2$m and center at $\mathbf{r}_{t,c}=(0.45,-0.2)$m. The second target is a rectangular cylinder with center at $\mathbf{r}_{t,s}=(-0.2,-0.2)$m, sides of length $l_{x,s}=0.3$m and $l_{y,s}=0.05$m, relative

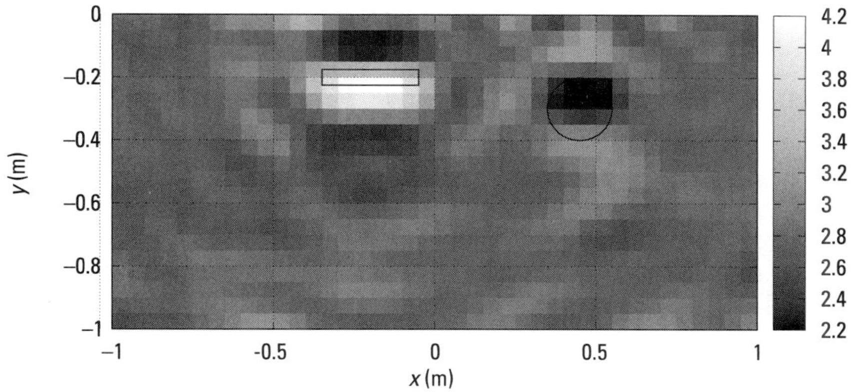

Figure 6.13 Example of the reconstruction of two buried dielectric targets obtained by using a linearized approach based on the Born approximation and the truncated SVD inversion method.

dielectric permittivity $\varepsilon_{r,s} = 7$, and electric conductivity $\sigma_s = 0.01$ S/m. A CO measurement configuration is assumed, with the transmitting and receiving antennas positioned along a line of length $L_w = 2$m at a height $h_w = 0.05$m. The distance between the two antennas is $d_{TR} = 0.3$m. The input data (numerically simulated [118]) are collected in $M = 30$ measurement points equally distributed on the probing line. The 2-D inverse problem has been solved under the Born approximation and using multifrequency data ($f_i = 150 + i50$ MHz, $i = 0,\ldots,4$). A Gaussian noise with zero mean value and SNR equal to 15 dB has been added to the computed input data. The reconstructed dielectric profile has been obtained by using the truncated SVD (only considering normalized singular values larger than 0.1). As can be seen from Figure 6.13, the two targets are correctly detected and localized. However, as expected, the values of the dielectric permittivity are only qualitatively reconstructed.

The reconstruction of two PEC cylinders is shown in Figure 6.14 [115]. The configuration is the same as in the previous example. The investigation area has sides $L_x = 2$m and $L_y = 1$m. The first cylinder has a circular cross section centered at $\mathbf{r}_{t,c} = (-0.5, -0.75)$m and diameter $d_c = 0.5$. The second one has a rectangular cross section centered in $\mathbf{r}_{t,s} = (0.5, -0.75)$m, with sides $l_{x,s} = 0.5$m and $l_{y,s} = 0.1$m. The soil is characterized by a relative dielectric permittivity $\varepsilon'_{r,b} = 4$ and an electric conductivity $\sigma_b = 0.01$ S/m. The z-component of total electric field vector has been collected at 51 points equally spaced on a probing line of length $L_w = 2$m located on the air/ground interface (i.e., $h_w = 0$). The reconstruction process is based on a linearized approach in which the Kirchhoff approximation and the truncated SVD are applied [115]. Multifrequency data have been used in this simulation, too. In particular, $F = 21$ frequencies equal-

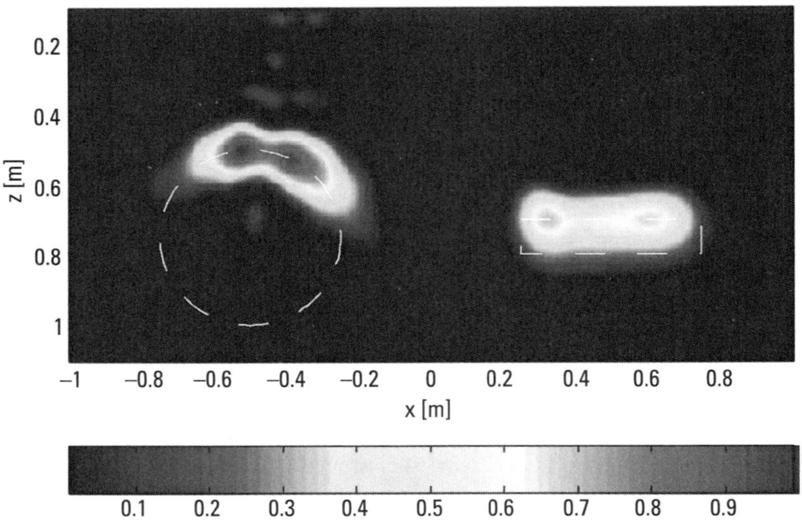

Figure 6.14 Example of the reconstruction of two buried metallic targets obtained by using a linearized approach based on the Kirchhoff approximation and on the truncated SVD [115]. (©2008 IEEE.)

ly spaced in the range 100–700 MHz have been considered and the forward scattering problem has been numerically solved by using the FDTD method.

Linearized schemes can also be used in association with the forward-looking GPR modality (Figure 6.9). An example of simulation results is reported in Figure 6.15 [80]. Two lossless circular cylinders with finite length are buried in a soil characterized by $\varepsilon'_{r,b} = 9$ and $\sigma_b = 1$mS/m. The first one is centered in $\mathbf{r}_{c,1}$ = (4,0.25, –0.12) with a diameter $d_{c,1} = 0.07$m and a relative dielectric permittivity $\varepsilon'_{r,1} = 2.3$. The second one is centered in $\mathbf{r}_{c,2}$ = (4.5, 0.25, 0.17)m with a diameter $d_{c,2} = 0.12$m and a relative dielectric permittivity $\varepsilon_{r,1} = 2.7$. Both cylinders are long 0.04m. Synthetic data are computed by using a 3-D forward solver based on the FDTD method. The transmitting and receiving antennas are located at heights $h_{w,TX}$= 1m and $h_{w,RX}$= 2m over the soil interface, respectively, and they are used to collect data over a probing line of length L_w= 2.75m with a spacing between consecutive measurements of l_w= 0.03m. The transmitting antenna is fed with a Ricker pulse [57] with central frequency of 600 MHz. A noise characterized by SNR= 45 dB has been added to the computed data. In the inversion procedure, the Born approximation is applied, considering a square investigation domain of sides L_x = 3m and L_y = 0.5m (corresponding to a 2-D slice of the original domain used for the forward computation). The electric field for F= 46 values of the frequency between 400 MHz and 1.3 GHz is extracted from the time-domain data and used in the inversion. In this case, the first-order Born approximation is applied together with a truncated SVD (only the singular values with normalized value higher than –10 dB are kept).

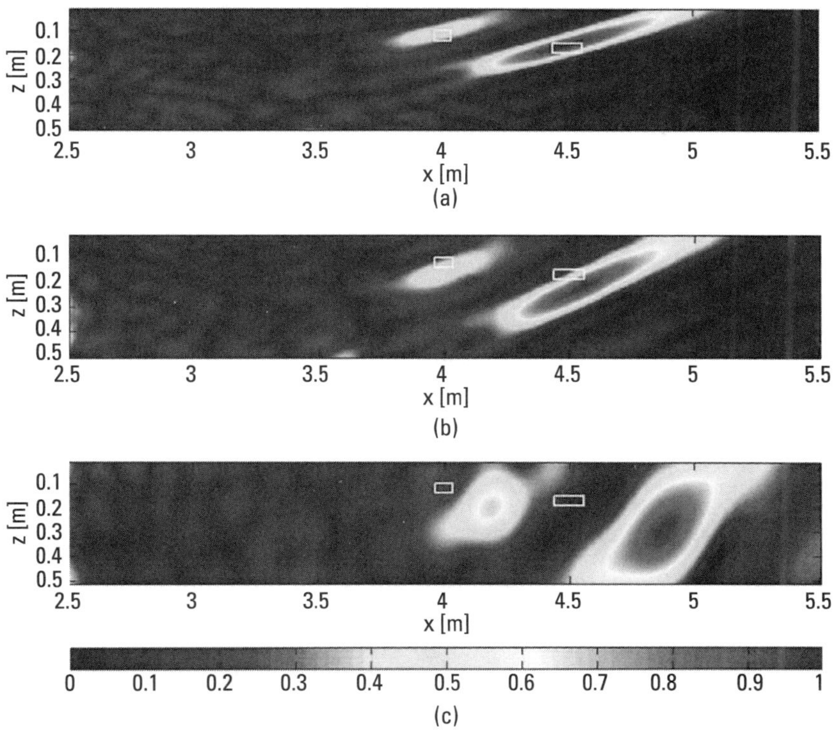

Figure 6.15 Example of tomographic reconstruction in forward-looking GPR applications [80]. (a) Exactly known soil; (b) wrong soil permittivity; (c) free-space. (© 2015 IEEE.)

Figure 6.15(a) reports the reconstruction (normalized values are plotted) obtained by using the correct value of the relative dielectric permittivity of the soil. The effects of an incorrect choice of this parameter are exemplified in Figure 6.15(b), in which $\varepsilon'_{r,b} = 4$ has been used, as well as in Figure 6.15(c), where the air/soil interface is neglected (a free-space configuration is assumed during the inversion process).

Compressive sensing techniques can also be used for reconstructing the distribution of the dielectric properties in the case of sparse configurations. As discussed in Chapter 3, common compressive sensing approaches require that the operator describing the mapping between the field and the unknown is linear. Consequently, they have been mainly applied with reference to the first-order Born approximation [119], or for retrieving the contrast source current density [120].

The previous inversion schemes rely on the use of the correct Green's function for the characterization of the subsurface region. Although this allows a straightforward derivation of the scattering equation, it could be a limit when dealing with complex geometries for which Green's function is not known. In

[121], an alternative technique based on the finite-difference frequency-domain (FDFD) method has been used in conjunction with the Born approximation. The obtained linear problem is solved by using a Tikhonov regularization. Another approach based on a linearized contrast source inversion and the FDFD method has been proposed in [122]. In [123], a Born-type linearization, which make use of Cauchy data on the boundary of a homogeneous neighborhood of the target, has been applied in order to avoid the computation of the Green's function.

Other schemes involving linearization can be used as well. For example, in [124, 125] an approach based on the quasi-linear approximation [126] has been applied. In this case, the inversion is performed by solving a set of linear problems. The use of the extended Born approximation [127–129] and its higher order extension [130] have also been tried for buried target detection [131].

6.6.2 Nonlinear Inversion Techniques

Inversion approaches based on linear approximations of the scattering problem are quite effective. However, as previously recalled, in many situations they only allow one to reconstruct qualitative information about the targets (e.g., some geometrical features). In order to obtain quantitative reconstructions, for example, the full distributions of the dielectric properties, nonlinear schemes should be adopted (as described in Chapter 3).

A first step in this direction is represented by the use of higher-order approximations (e.g., the *n*th order Born approximation discussed in Chapter 3). For example, the second-order Born approximation has been employed for subsurface imaging in [132–134]. An example of the results obtained by using this approximation is reported in Figure 6.16 [135], with reference to a mixed configuration in which the antennas are arranged both in boreholes and on the air/ground interface [133]. In particular, $M = 42$ measurement points uniformly distributed along three probing lines of length $L_w = 2\lambda_0$ and used to collect the field (the boreholes are located at $\pm\lambda_0$ on the x axis). A subset of $S = 7$ antennas (one every six elements) is used to illuminate the inspected scenario. The considered soil is characterized by $\varepsilon'_{r,b} = 4$ and $\sigma_b = 0.01$ S/m. The investigation domain is a square area of side $L_x = L_y = 1.5\lambda_0$, which has been discretized into $N = 400$ square subdomains. Two lossy dielectric circular cylinders are located in the inspected scenario. The diameters of the cross sections of the two targets are $d_{c,1} = 0.5\lambda_0$ and $d_{c,2} = 0.3\lambda_0$, and they are centered in $\mathrm{r}_{t,c,1} = (-0.375\lambda_0, -0.75\lambda_0)$ and $\mathrm{r}_{t,c,2} = (0.375\lambda_0, -0.75\lambda_0)$, respectively. Both cylinders have a value of the contrast function c equal to $0.37 + j0.04$. The scattered field data, at a single frequency equal to 300 MHz, have been computed by using a forward solver based on the method of moments and corrupted with

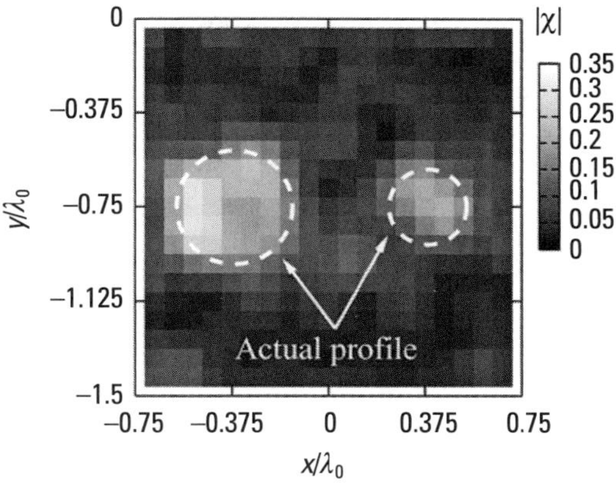

Figure 6.16 Example of tomographic reconstruction in a mixed configuration by using the second-order Born approximation [133]. (© 2007 IEEE.)

a Gaussian noise with zero mean value and $SNR = 25$ dB. The data are inverted by an inexact-Newton method in which the inner linear problem is solved by a truncated Landweber algorithm [135]. The reconstructed distribution of the amplitude of the contrast function is shown in Figure 6.16. As can be seen, both the targets are correctly identified and also the value of the contrast function is estimated with quite good accuracy. In [133] it has also been found that, at least for the considered configuration, the use of the second-order Born approximation was able to provide lower reconstruction errors with respect to the linearized inversion scheme.

However, to inspect strong scatterers, it is, in general, necessary to consider the full nonlinear scattering formulation (quantitative imaging). As already pointed out in other Chapters, particular care must be exercised in order to mitigate local minima problems due to the nonlinearity. To this end, different inversion approaches can be used, such as algorithms based on the contrast-source inversion [136, 55, 137, 138], in which the following residual functional is minimized by using a conjugate gradient method

$$\mathcal{R}(c) = \frac{\left\| E_{scatt,z} - G_{data}\left(J_{cs,z}\right)\right\|^2_{L^2(D_{obs})}}{\left\| E_{scatt,z}\right\|^2_{L^2(D_{obs})}} + \frac{\left\| cE_{inc,z} - J_{cs,z} + cG_{state}\left(J_{cs,z}\right)\right\|^2_{L^2(D_{inv})}}{\left\| cE_{inc,z}\right\|^2_{L^2(D_{inv})}} \tag{6.19}$$

where all the involved symbols are defined in Chapter 3. Born and distorted Born iterative methods can also be used [139–142] (these techniques have also been considered in the framework of biomedical applications in Chapter 5). Newton schemes and conjugate gradient approaches have also been successfully applied [143–151].

Figure 6.17 shows an example of the reconstruction of a subsurface scenario performed by using a nonlinear inversion algorithm based on a Newton scheme. In particular, a multi-frequency version of the L^p Banach space inversion procedure described in Section 3.3 is applied [145]. It is worth noting that, in order to exploit the information at multiple frequencies, some modifications should be introduced in the basic algorithm reported there. Since the contrast function depends upon the frequency, it is necessary to apply a transformation to some frequency-independent unknowns in order to apply the inversion procedure (similarly to what discussed in Chapter 5 with reference to the microwave imaging of the breast). In particular, here a simplified model of the dielectric properties is assumed, where the real parts of the dielectric permittivity and the electric conductivity are assumed to be independent from the frequency in the considered frequency range. Consequently, the following mapping is used for the contrast function

$$c(\mathbf{r}_t,\omega)=T_\omega(\mathbf{x})(\mathbf{r}_t,\omega)=\frac{1}{\varepsilon_{r,b}}\left(\underbrace{\varepsilon_r'(\mathbf{r}_t)-\varepsilon_{r,b}'}_{x_\varepsilon}-j\frac{\omega_1}{\omega}\underbrace{\frac{\sigma(\mathbf{r}_t)-\sigma_b}{\omega_1\varepsilon_0}}_{x_\sigma}\right) \tag{6.20}$$

where $\mathbf{x}(\mathbf{r}_t)=\left[x_\varepsilon(\mathbf{r}_t)x_\sigma(\mathbf{r}_t)\right]^t$ is the array of unknown functions to be retrieved. The operator equation (3.53), at frequency ω, can be then rewritten in terms of the new unknown array $\mathbf{x}$, that is, $E_{scatt,z}(\mathbf{r}_t,\omega) = F(T_\omega(\mathbf{x}))(\mathbf{r}_t,\omega) = F_\omega(\mathbf{x})(\mathbf{r}_t,\omega)$, $\mathbf{r}_t \in D_{obs}$. Combining the data collected for a discrete number of frequencies ω_f, $f= 1, \ldots, F$, the following set of equations is obtained

$$\begin{bmatrix} E_{scatt,z}(\mathbf{r}_t,\omega_1) \\ E_{scatt,z}(\mathbf{r}_t,\omega_2) \\ \vdots \\ E_{scatt,z}(\mathbf{r}_t,\omega_F) \end{bmatrix} = \begin{bmatrix} F_{\omega_1}(\mathbf{x})(\mathbf{r}_t) \\ F_{\omega_2}(\mathbf{x})(\mathbf{r}_t) \\ \vdots \\ F_{\omega_F}(\mathbf{x})(\mathbf{r}_t) \end{bmatrix} = F_{MF}(\mathbf{x})(\mathbf{r}_t),\quad \mathbf{r}_t \in D_{obs} \tag{6.21}$$

Equation (6.21) represents the new functional relationship to be inverted with respect to $\mathbf{x}$. To this end, the inversion procedure in L^p Banach spaces detailed in Chapter 3 can be used. Concerning the Fréchet derivative $F'_{MF,n}$ of

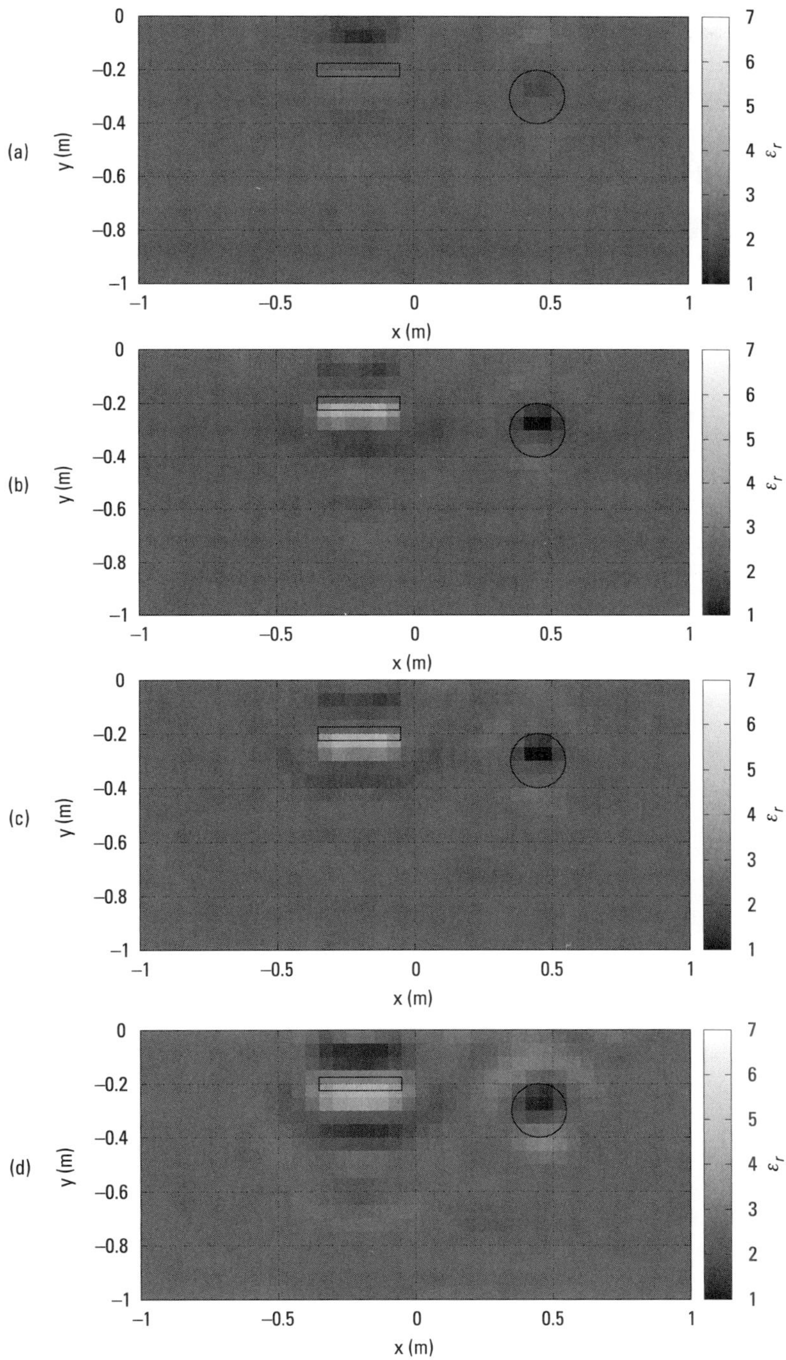

Figure 6.17 Example of a reconstruction of two buried targets by using a multifrequency Newton scheme in L^p Banach spaces. (a) F=1, p=1.4; (b) F=3, p=1.4; (b) F=5, p=1.4; (d) F=5, p=2.

the operator F_{MF} around the current solution $\mathbf{x}_n$ at the nth Newton linearization step, since the mapping between $\mathbf{x}$ and c is linear, it can be directly written in terms of the Fréchet derivative F_n' in (3.55), computed around $c_n = T_\omega(\mathbf{x}_n)$, that is,

$$F'_{MF,n} = \begin{bmatrix} F'_{\omega_1,n} \\ F'_{\omega_2,n} \\ \vdots \\ F'_{\omega_F,n} \end{bmatrix}, \; F'_{\omega,n} = \left[F'_n\big|_{c_n=T_\omega(\mathbf{x}_n)} - j\frac{\omega_1}{\omega} F'_n\big|_{c_n=T_\omega(\mathbf{x}_n)} \right] \tag{6.22}$$

For the example reported in Figure 6.17, the same configuration considered in the linearized reconstruction of Figure 6.13 is assumed. In this case, a variable number F of distinct frequencies have been considered. They are again equally spaced in a band of 200 MHz, which is centered at f = 300 MHz. A Gaussian noise with zero mean value and SNR = 15 dB has been added to the synthetic data computed by using the moment method. The parameters of the L^p Banach-space inversion procedure are: N_{IN} = 100 maximum Newton outer steps and N_{LW} = 100 maximum Landweber inner iterations.

The reconstruction results obtained by using different numbers of frequencies (F = 1, 3, 5) are reported in Figure 6.17. They have been obtained by using the optimal value for the parameter p, which in this case is equal to 1.4. For comparison purposes, the reconstruction obtained by using the same approach, but developed in a more standard Hilbert-space (p = 2), is also reported. As can be seen, when a single frequency processing has been used, the targets are barely visible and high errors in the dielectric properties are present. The simultaneous processing of multiple frequencies allows one to obtain, as expected, much better reconstructions. Finally, from Figure 6.17, it can be seen that the use of a different norm in the inversion procedure yields a more accurate reconstruction with respect to the regularization approach developed in Hilbert spaces, for which the retrieved targets are more smoothed and ringing artifacts appeared in the background.

The application of stochastic imaging methods [152], such as genetic algorithms, differential evolution method, and swarm approaches, has also been evaluated for the detection of buried objects [153–158]. As mentioned in Section 3.3, they are (in principle) able to avoid that the solution be trapped in local minima corresponding to false solutions. However, their main limitation is related to the high computational burden. Therefore, they can be applied in practice only if few parameters have to be retrieved. Accordingly, they have been considered in order to obtain only some features of the targets (e.g., the relevant parameters of circular and elliptic cylinders, as well as the parameters related to parametric representations of the external contours). Another possibility is to apply multiscale strategies, in which multiple low-resolution inversion processes are performed by focusing the reconstruction, at any step, only to the subdomain containing the target (the region-of-interest [RoI]) [134, 146, 151, 159] (these approaches will be also mentioned in Chapter 8).

Figure 6.18 reports an experimental reconstruction result obtained by using a stochastic method [160]. In particular, the particle swarm optimization (PSO) algorithm [161] has been combined with a multiscaling imaging approach in order to experimentally detect the presence of buried empty tanks and sandstone rocks. The data set measured at the Near Surface Geophysical Group test site [162], with the Mala X3M apparatus, is used. Figure 6.18 reports the retrieved distributions of the dielectric properties of the investigation domain (a square area of side $L_x = L_y = 0.8$ m starting 0.05m below the air/soil interface) in the case in which two empty tanks are present. These targets have dimensions 0.32m × 0.25m × 015m and they are buried at a depth of 0.15m (Figure 6.18[a]). The dielectric properties of the soil have been set equal to $\varepsilon'_{r,b} = 5$ and $\sigma_b = 38$ mS/m. The time-domain field data are collected by moving a pair of bow-tie antennas (in a CO mode, where the distance between transmitting and receiving antennas is $d_{TR} = 0.2$ m) on a probing line in which 21 measurement points with $l_w = 0.02$m are considered. In the inversion process, three frequencies are used, in the range 100–300 MHz. The reconstruction obtained is illustrated in Figure 6.18(b) and Figure 6.18(c). For comparison purposes, the results obtained by using a similar multiscale approach combined with the conjugate gradient inversion algorithm are also reported. In particular, both direct multifrequency processing (exploiting all the frequenices together) and frequency hopping techniques are considered.

The previously discussed approaches were all based on a frequency domain formulation. Methods working directly in the time domain, also referred as full-waveform inversions, have also been proposed [163–169]. For example, in [163] an algorithm based on the use of the FDTD is evaluated. Basically, the developed procedure is based on the iteratively minimization (by means of a gradient-based method) of the misfit between the time-domain measured data and the values computed by using the FDTD. An experimental validation of an approach belonging to this class of techniques can be found in [170].

6.6.3 Other Approaches

In the previous sections, linear and nonlinear inversion procedures based on the inversion of the scattering equations have been discussed. Other qualitative inversion schemes can be applied for subsurface imaging. In particular, with reference to the qualitative methods already mentioned in Section 3.3, time reversal techniques [171, 172], such as those based on the decomposition of the

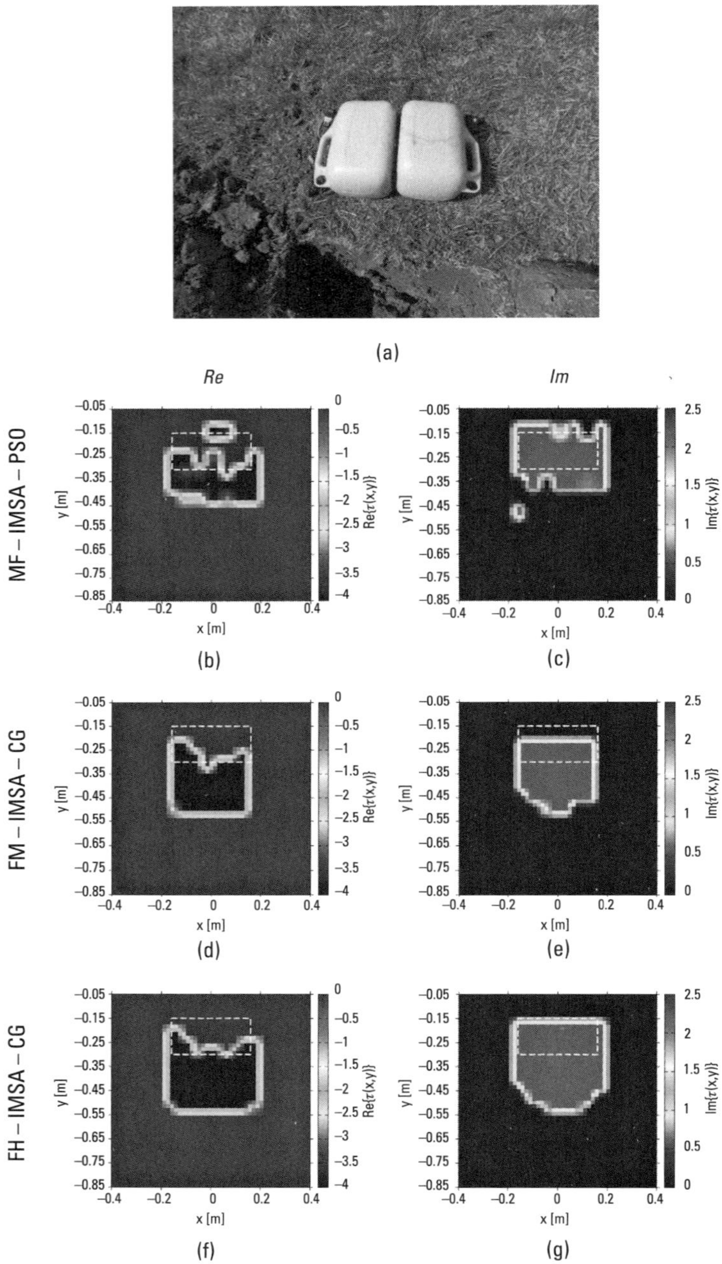

Figure 6.18 Experimental reconstruction of two empty tanks [160]. (a) Picture of the buried target. (b), (d), and (f) Real and (c), (e), and (g) imaginary parts of the retrieved dielectric profiles (difference with respect to the dielectric properties of the background) with (b) and (c) MF-IMSA-PSO [160], (d) and (e) MF-IMSA-CG, and (f) and (g) FH-IMSA-CG [151]. (© 2017 IEEE.)

time reversal operator (DORT) method [173–175] and on the multiple signal classification (MUSIC) method [172, 176, 177] have been exploited for this purpose. The linear sampling method (LSM) has also been used in [178–180] for the reconstruction of the external supports of targets eventually buried in a subsurface region. It is worth remarking that in its basic formulation, the LSM requires full-view measurements, which are not feasible in subsurface imaging. However, when limited-view configurations are adopted, it has been found that the approach is still effective (although with reduced performances, especially when small measurement apertures are considered) [179].

An example of the reconstruction of two dielectric targets obtained by using the LSM is reported below. A configuration similar to those used in Figure 6.13 and Figure 6.17 is considered. In this case, however, a multistatic measurement configuration (e.g., similar to those of MIMO GPR) is employed. In particular, $M = 30$ antennas (acting in both transmitting and receiving mode) are assumed to be uniformly spaced on a probing line of length $L_w = 2$m parallel to the soil at height $h_w = 0.05$m. The scattered field data have been numerically simulated for $F = 5$ frequencies uniformly distributed in a frequency band of 200 MHz and centered at 300 MHz. A Gaussian noise with zero mean value and $SNR = 15$ dB has been added to the computed data. Two targets are assumed to be present. The first one is a lossless circular cylinder with relative dielectric permittivity $\varepsilon_{r,c} = 8$, diameter $d_c = 0.2$m, and center at $\mathbf{r}_{t,c} = (0.45, -0.3)$m. The second target is a rectangular cylinder with center at $\mathbf{r}_{t,s} = (-0.2, -0.2)$m, sides of length $l_x = 0.3$m and $l_y = 0.2$m, relative dielectric permittivity $\varepsilon'_{r,s} = 7$, and electric conductivity $\sigma_s = 0.01$ S/m. The retrieved indicator function is shown in Figure 6.19. As can be seen, both objects are detected and their positions are correctly retrieved. On the contrary, due to the considered limited-view configuration, their shapes are quite distorted.

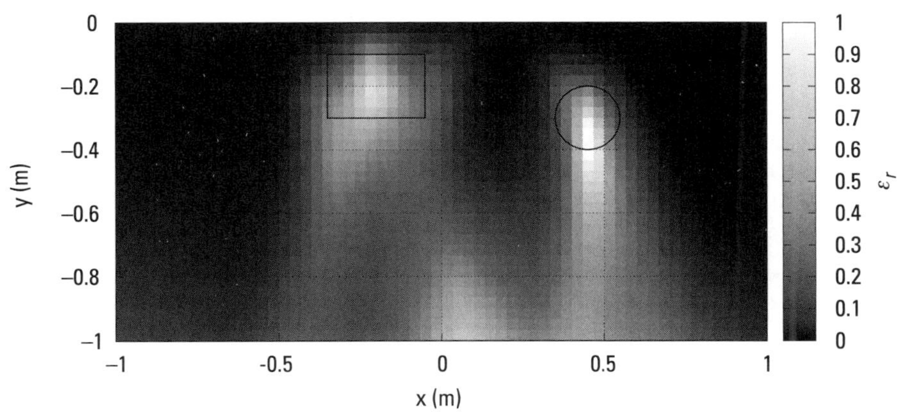

Figure 6.19 Example of a reconstruction of two buried dielectric targets obtained by using the LSM.

Learning-based approaches have also been proposed, mainly for extracting information about the presence, position, and size of a buried target. In these cases, it is (in principle) possible to directly train the network starting from measurements. Consequently, a complex model of the scattering phenomena is not needed. Furthermore, once the network is trained, the inversion process can be performed in a very short time. Neural networks have been initially considered for GPR applications in [181, 182]. More recently, support vector machines (SVMs) have also been successfully employed [183]. Subarray processing combined with triangularization techniques and machine learning approaches have also been found to be effective [184–186].

Finally, it is worth noting that in the field of GPR, when the objective is just to detect the presence and size of buried targets (e.g., for utility mapping), simplified approaches can be used. Hyperbola recognition techniques are widely employed approaches for addressing this problem. Actually, as discussed in Section 6.3, when a target (with approximately circular cross section) is present inside the investigation domain, a hyperbola appears in the raw *B*-scan, whose geometrical properties are related to the target parameters. Consequently, by extrapolating the hyperbola parameters (e.g., apex and slope of the asymptotes) it is possible to extract information about position, size and dielectric properties of the target. Such operation can be performed by means of image processing and pattern recognition algorithms [187–190].

6.7 Overview of Practical Implementations of Subsurface Imaging Systems

As previously highlighted, subsurface imaging systems are often based on the use of the GPR. Different commercial products are nowadays available. Some examples (not covering all the possible manufacturers/products) are reported in Table 6.1.

As discussed in Section 6.3, most common GPR systems are based on pulsed radars, that is, the incident field is a (usually very short) time-domain pulse [1], which is generated by an impulse generator. The receivers are based on high-frequency time-domain sampling oscilloscopes (Figure 6.20[a]). A typical pulse employed in GPR systems is composed by a monocycle signal (as in Figure 6.5), whose duration is in the range of nanoseconds. In the last years, several works have been devoted to the development of ever more efficient hardware design for GPR systems, able to generate and receive ultrawideband signals (for example, see [191–194]).

Continuous wave radars are also sometimes used [1]. In particular, frequency modulated continuous wave (FMCW) [195–197] and frequency stepped continuous wave (SFCW) [198–200] architectures can be used. In the

Table 6.1
Examples of Commercial GPR Systems

Manufacturer	Product	Central frequencies
GSSI	UtilityScan	400 MHz, 300/800 MHz, 350 MHz
	StructureScan	1.6 GHz, 2.6 GHz
	RoadScan	1 GHz, 2 GHz
	SIR	200 MHz, 400 MHz, 900 MHz
IDS	Opera Duo/Detector Duo	250 MHz, 700 MHz
	Stream EM, Stream C, Stream X (array of antennas)	200 MHz, 600 MHz
Mala	Ground Explorer	80 MHz, 160 MHz, 450 MHz, 750MHz
	Concrete Scanner	1.2 GHz, 1.6 GHz, 2.3 GHz
	X3M	100 MHz, 250 MHz, 500 MHz, 800 MHz
Sensors & Software	pulseEKKO	12.5 MHz, 25 MHz, 50 MHz, 100 MHz, 200 MHz, 250 MHz, 500 MHz, 1 GHz
	Noggin	100 MHz, 250 MHz, 500 MHz, 1 GHz
	IceMap	500 MHz
UTSI Electronics	Groundvue 3	250MHz, 400 MHz, 1 GHz, 1.5 GHz, 4 GHz
	Crack Detection Head	1.5 GHz

former case, the frequency of the sinusoidal signal is linearly varied between two fixed values (defining the frequency bandwidth). Such configurations are sometimes preferred since they can offer high dynamic range, low noise figure, and high radiated powers [4]. However, when adopted in standard GPR applications, for which the users expect a *B*-scan (or a *C*-scan) representation, there is the need of a quadrature receiver and a conversion block that performs an inverse Fourier transform of the measured data in order to synthesize a pulsed signal (Figure 6.20[b]). Such requirements complicate the design of the hardware part with respect to standard FMCW apparatuses. In SFCW systems, the frequency is changed in discrete steps, instead of being continuously swept in the bandwidth of operation. SFCW systems provide high accuracy, although they require more complex control logic [4].

An important element of any GPR system is represented by the antennas. They should be properly designed in order to satisfy critical requirements. First of all, they must ensure a wide frequency band of operation, since, as recalled before, very short pulses must be transmitted and received (leading to a fractional bandwidth usually greater than 100%). Moreover, linear phase characteristics, constants gain, and fixed polarizations are needed in the operating band. The design should also minimize the direct coupling between the transmitting and receiving antennas, which are usually located very near one to each other. Consequently, in the last years, a lot of work has been devoted to the development of antennas able to tackle such requirements. The main considered

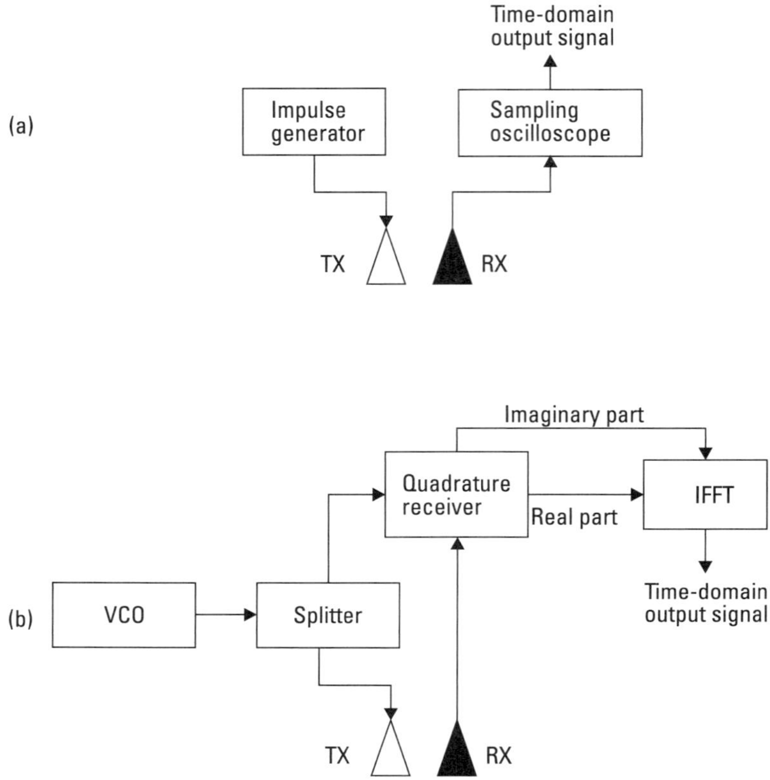

Figure 6.20 Block schemes of (a) pulsed and (b) FMCW GPR systems.

structures are bow-tie [201–206], Vivaldi [49, 207–210], and horn [211–215] antennas. Spiral antennas have also been used [216, 217].

It is worth noting that in many GPR applications the antennas are placed near the surface of the inspected medium. Consequently, their behavior can be affected by the interaction with the subsurface material. Such coupling must be properly taken into account during the design phase. Imaging procedures can be affected by the real antenna behavior, too. Although in many cases such effects are neglected by assuming ideal transmitting (such as dipoles or line-current sources) and receiving antennas, some research has been devoted to analyze the effects of the antennas [49, 218, 219]. In [50, 220], models of the whole GPR system, including the antenna and soil effects, have also been developed.

In this framework, an important issue is represented by the accurate simulation of the behavior of the antenna in presence of the ground. To this end, several numerical approaches can be applied, for example, based on frequency-domain solvers (used, for example, in [208, 221–223]) or FDTD methods (used, for example, in [224–230]). The latter approach has also been adopted for simulating realistic antennas used in commercial apparatuses. For example, in [231], realistic FDTD models of two commercial antennas (from MALA, working at 1.2 GHz, and from GSSI, working at 1.5 GHz) have been used. Moreover, a realistic model of the soil (including a rough terrain surface, vegetation, and water) has also been included. The 3-D models of the considered antennas, whose parameters have been devised by fitting real measurements in free space using the Taguchi's optimization method [232], are shown in Figure 6.21. As can be seen, both the actual antennas (composed by pairs of bow-tie antennas) and the casing/shielding structures are inserted into the model. Such models have been used in conjunction with the open-source FDTD software

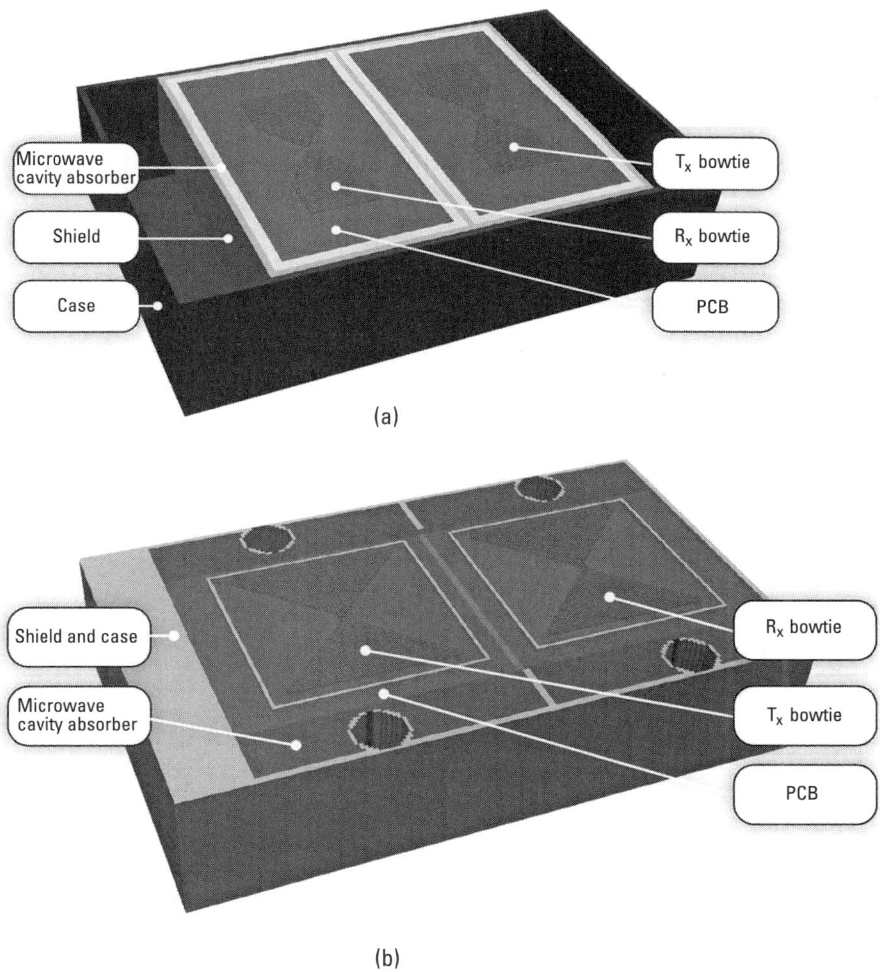

Figure 6.21 Examples of FDTD models of commercial GPR antennas. (a) GSSI 1.5 GHz and (b) MALA 1.2 GHz antennas [231]. (© 2016 IEEE.)

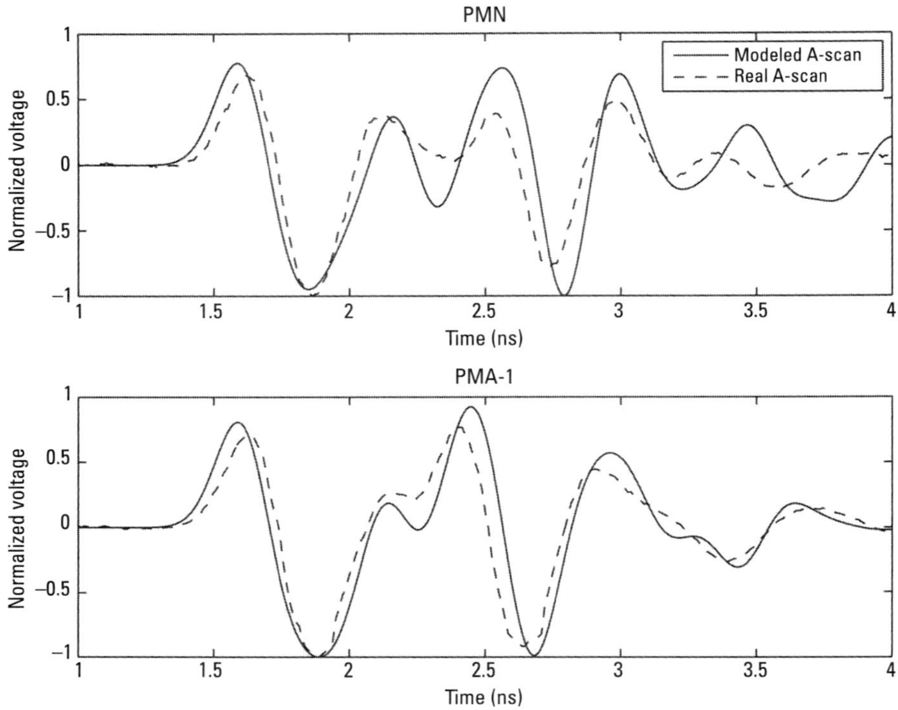

Figure 6.22 Examples of A-scan simulated with the FDTD models in Figure 6.21 in the presence of landmines and comparison with measured data [231]. (© 2016 IEEE.)

gprMax [57] to simulate the response of two types of landmines: A PMA-I mine (which has a parallelepiped shape with dimensions 30 mm × 140 mm × 65 mm and is composed by PEC, plastic [$\varepsilon_r = 2.5$], and rubber [$\varepsilon_r = 6$] parts) and a PMN mine (which has a cylindrical shape with diameter of 115 mm and height of 50 mm and is composed by PEC, plastic [$\varepsilon_r = 3$], bakelite [$\varepsilon_r = 3.5$], and rubber [$\varepsilon_r = 6$] parts). In both cases, the structures usually have minimum metallic content. The simulated *A*-scan obtained by using the 1.5 GHz antenna (a Gaussian incident pulse is used) is shown in Figure 6.22, together with the signal measured with the real apparatus. As can be seen, the agreement is quite good, confirming the possibility of properly modeling real antennas.

References

[1] Daniels, D. J., *Ground Penetrating Radar*, 2nd ed., London: Institution of Electrical Engineers, 2004.

[2] Jol, H. M., *Ground Penetrating Radar Theory and Applications*, Oxford, UK: Elsevier Science, 2009.

[3] Persico, R., *Introduction to Ground Penetrating Radar: Inverse Scattering and Data Processing*, Hoboken, NJ: Wiley, 2014.

[4] Benedetto, A., and L. Pajewski, *Civil Engineering Applications of Ground Penetrating Radar*, Cham, Switzerland: Springer International Publishing, 2015.

[5] Turk, A. S., A. K. Hocaolu, and A. A. Vertiy, *Subsurface Sensing*, Hoboken, NJ: Wiley, 2010.

[6] Valagiannopoulos, C. A., N. L. Tsitsas, and A. H. Sihvola, "Unlocking the Ground: Increasing the Detectability of Buried Objects by Depositing Passive Superstrates," *IEEE Trans. Geosci. Remote Sens.*, Vol. 54, No. 6, June 2016, pp. 3697–3709.

[7] Mahmoudzadeh, M. R., A. P. Francés, M. Lubczynski, and S. Lambot, "Using Ground Penetrating Radar to Investigate the Water Table Depth in Weathered Granites—Sardon Case Study, Spain," *J. Appl. Geophys.*, Vol. 79, April 2012, pp. 17–26.

[8] Pyke, K., S. Eyuboglu, J. J. Daniels, and M. Vendl, "A Controlled Experiment to Determine the Water Table Response Using Ground Penetrating Radar," *J. Environ. Eng. Geophys.*, Vol. 13, No. 4, 2008, pp. 335–342.

[9] Saintenoy, A., and J. W. Hopmans, "Ground Penetrating Radar: Water Table Detection Sensitivity to Soil Water Retention Properties," *IEEE J. Sel. Top. Appl. Earth Obs. Remote Sens.*, Vol. 4, No. 4, 2011, pp. 748–753.

[10] Huisman, J. A., S. S. Hubbard, J. D. Redman, and A. P. Annan, "Measuring Soil Water Content with Ground Penetrating Radar: A Review," *Vadose Zone J.*, Vol. 2, No. 4, November 2003, pp. 476–491.

[11] Guan, B., et al., "Near-Field Full-Waveform Inversion of Ground-Penetrating Radar Data to Monitor the Water Front in Limestone," *IEEE J. Sel. Top. Appl. Earth Obs. Remote Sens.*, Vol. 10, No. 10, 2017, pp. 4328–4336.

[12] Tosti, F., et al., "GPR Analysis of Clayey Soil Behaviour in Unsaturated Conditions for Pavement Engineering and Geoscience Applications," *Surf. Geophys.*, Vol. 14, No. 2, 2016, pp. 127–144.

[13] Busch, S., et al., "Coupled Hydrogeophysical Inversion of Time-Lapse Surface GPR Data to estimate Hydraulic Properties of a Layered Subsurface," *Water Resour. Res.*, Vol. 49, No. 12, 2013, pp. 8480–8494.

[14] Allred, B. J., J. J. Daniels, and M. R. Ehsani (eds.), *Handbook of Agricultural Geophysics*, Boca Raton: CRC Press, 2008.

[15] Algeo, J., R. L. Van Dam, and L. Slater, "Early-Time GPR: A Method to Monitor Spatial Variations in Soil Water Content During Irrigation in Clay Soils," *Vadose Zone J.*, Vol. 15, No. 11, 2016.

[16] Baek, S.-H., S.-S. Kim, J.-S. Kwon, and E. S. Um, "Ground Penetrating Radar for Fracture Mapping in Underground Hazardous Waste Disposal Sites: A Case Study from an Underground Research Tunnel, South Korea," *J. Appl. Geophys.*, Vol. 141, 2017, pp. 24–33.

[17] Wijewardana, Y. N. S., et al., "Ground-Penetrating Radar (GPR) Responses for Sub-Surface Salt Contamination and Solid Waste: Modeling and Controlled Lysimeter Studies," *Environ. Monit. Assess.*, Vol. 189, No. 2, 2017.

[18] Ciarletti, V., et al., "WISDOM GPR Designed for Shallow and High-Resolution Sounding of the Martian Subsurface," *Proc. IEEE,* Vol. 99, No. 5, May 2011, pp. 824–836.

[19] Millington, T. M., N. J. Cassidy, L. Crocco, and F. Soldovieri, "Tomographic GPR Imaging Using a Linear Inversion Algorithm Informed by FDTD Modelling: A Numerical Case Study of Buried Utility Pipes Monitoring," *Surf. Geophys.*, Vol. 11, No. 2, 2013, pp. 221–230.

[20] Ayala-Cabrera, D., M. Herrera, J. Izquierdo, and R. Pérez-García, "Location of Buried Plastic Pipes Using Multi-Agent Support Based on GPR Images," *J. Appl. Geophys.*, Vol. 75, No. 4, 2011, pp. 679–686.

[21] Allred, B. J., et al., "Detection of Buried Agricultural Drainage Pipe with Geophysical Methods," *Appl. Eng. Agric.*, Vol. 20, No. 3, 2004, pp. 307–318.

[22] Prego, F. J., M. Solla, I. Puente, and P. Arias, "Efficient GPR Data Acquisition to Detect Underground Pipes," *NDT&E Int.*, Vol. 91, 2017, pp. 22–31.

[23] Bastard, C. L., V. Baltazart, Y. Wang, and J. Saillard, "Thin-Pavement Thickness Estimation Using GPR with High-Resolution and Superresolution Methods," *IEEE Trans. Geosci. Remote Sens.*, Vol. 45, No. 8, August 2007, pp. 2511–2519.

[24] Sun, M., et al., "Time Delay and Interface Roughness Estimations by GPR for Pavement Survey," *Surf. Geophys.*, Vol. 13, No. 3, May 2015, pp. 279–287.

[25] Marecos, V., S. Fontul, M. de Lurdes Antunes, and M. Solla, "Evaluation of a Highway Pavement Using Non-Destructive Tests: Falling Weight Deflectometer and Ground Penetrating Radar," *Constr. Build. Mater.*, Vol. 154, November 2017, pp. 1164–1172.

[26] Stryk, J., et al., "Comparative Measurements of Ground Penetrating Radars Used for Road and Bridge Diagnostics in the Czech Republic and France," *Constr. Build. Mater.*, Vol. 154, November 2017, pp. 1199–1206.

[27] Benedetto, A., F. Tosti, L. Bianchini Ciampoli, and F. D'Amico, "An Overview of Ground-Penetrating Radar Signal Processing Techniques for Road Inspections," *Signal Process.*, Vol. 132, March 2017, pp. 201–209.

[28] Tosti, F., et al., "An Investigation into the railway Ballast Dielectric Properties Using Different GPR Antennas And Frequency systems," *NDT&E Int.*, Vol. 93, 2018, pp. 131–140.

[29] Anbazhagan, P., P. S. N. Dixit, and T. P. Bharatha, "Identification of Type and Degree of Railway Ballast Fouling Using Ground Coupled GPR Antennas," *J. Appl. Geophys.*, Vol. 126, 2016, pp. 183–190.

[30] Leng, Z., and I. L. Al-Qadi, "Railroad Ballast Evaluation Using Ground-Penetrating Radar," *Transp. Res. Rec.*, No. 2159, 2010, pp. 110–117.

[31] Daniels, D. J., "Ground Penetrating Radar for Buried Landmine and IED Detection," in *Unexploded Ordnance Detection and Mitigation,* J. Byrnes (ed.), Dordrecht: Springer, 2009, pp. 89–111.

[32] Yinon, J., *Forensic and Environmental Detection of Explosives,* Hoboken, NJ: Wiley, 1999.

[33] Lakshmi, C., and E. Logashanmugam, "Automatic Multiple Landmine Detection Using GPR," *ARPN J. Eng. Appl. Sci.*, Vol. 11, No. 11, 2016, pp. 6905–6911.

[34] Karlina, R., and M. Sato, "Model-Based Compressive Sensing Applied to Landmine Detection by GPR," *IEICE Trans. Electron.*, Vol. E99C, No. 1, 2016, pp. 44–51.

[35] González-Huici, M. A., I. Catapano, and F. Soldovieri, "A Comparative Study of GPR Reconstruction Approaches for Landmine Detection," *IEEE J. Sel. Top. Appl. Earth Obs. Remote Sens.*, Vol. 7, No. 12, 2014, pp. 4869–4878.

[36] Giannakis, I., A. Giannopoulos, and A. Yarovoy, "Model-Based Evaluation of Signal-to-Clutter Ratio for Landmine Detection Using Ground-Penetrating Radar," *IEEE Trans. Geosci. Remote Sens.*, Vol. 54, No. 6, 2016, pp. 3564–3573.

[37] Núñez-Nieto, X., M. Solla, P. Gómez-Pérez, and H. Lorenzo, "GPR Signal Characterization for Automated Landmine and UXO Detection Based on Machine Learning Techniques," *Remote Sens.*, Vol. 6, No. 10, 2014, pp. 9729–9748.

[38] Masini, N., and F. Soldovieri, *Sensing the Past: From Artifact to Historical Site*, Cham, Switzerland: Springer, 2017.

[39] Castaldo, R., et al., "GPR Microwave Tomography for Diagnostic Analysis of Archaeological Sites: The Case of a Highway Construction in Pontecagnano (Southern Italy)," *Archaeol. Prospect.*, Vol. 16, No. 3, 2009, pp. 203–217.

[40] Catapano, I., et al., "Full Three-Dimensional Imaging via Ground Penetrating Radar: Assessment in Controlled Conditions and on Field for Archaeological Prospecting," *Appl. Phys. A*, Vol. 115, No. 4, November 2013, pp. 1415–1422.

[41] Zhao, W., et al., "Advances in GPR Data Acquisition and Analysis for Archaeology," *Geophys. J. Int.*, Vol. 202, No. 1, 2015, pp. 62–71.

[42] Almeida, E. R., et al., "Microwave Tomography-Enhanced GPR in Forensic Surveys: The Case Study of a Tropical Environment," *IEEE J. Sel. Top. Appl. Earth Obs. Remote Sens.*, Vol. 9, No. 1, January 2016, pp. 115–124.

[43] Li, W., et al., "Tree Root Automatic Recognition in Ground Penetrating Radar Profiles Based on Randomized Hough Transform," *Remote Sens.*, Vol. 8, No. 5, May 2016, p. 430.

[44] Jezova, J., S. Lambot, A. Fedeli, and A. Randazzo, "Ground-Penetrating Radar for Tree Trunk Investigation," in *Proc. 2017 9th International Workshop on Advanced Ground Penetrating Radar (IWAGPR)*, Edinburgh, UK, 2017, pp. 1–6.

[45] Solla, M., H. Lorenzo, F. I. Rial, and A. Novo, "Ground-Penetrating Radar for the Structural Evaluation of Masonry Bridges: Results and Interpretational Tools," *Constr. Build. Mater.*, Vol. 29, 2012, pp. 458–465.

[46] Alani, A. M., M. Aboutalebi, and G. Kilic, "Applications of Ground Penetrating Radar (GPR) in Bridge Deck Monitoring and Assessment," *J. Appl. Geophys.*, Vol. 97, 2013, pp. 45–54.

[47] Diamanti, N., A. P. Annan, and J. D. Redman, "Concrete Bridge Deck Deterioration Assessment Using Ground Penetrating Radar (GPR)," *J. Environ. Eng. Geophys.*, Vol. 22, No. 2, 2017, pp. 121–132.

[48] Hugenschmidt, J., and A. Kalogeropoulos, "The Inspection of Retaining Walls Using GPR," *J. Appl. Geophys.*, Vol. 67, No. 4, April 2009, pp. 335–344.

[49] Comite, D., A. Galli, I. Catapano, and F. Soldovieri, "The Role of the Antenna Radiation Pattern in the Performance of a Microwave Tomographic Approach for GPR Imaging,"

IEEE J. Sel. Top. Appl. Earth Obs. Remote Sens., Vol. 10, No. 10, October 2017, pp. 4337–4347.

[50] Lambot, S., et al., "Modeling of Ground-Penetrating Radar for Accurate Characterization of subsurface Electric Properties," *IEEE Trans. Geosci. Remote Sens.*, Vol. 42, No. 11, 2004, pp. 2555–2568.

[51] Soldovieri, F., O. Lopera, and S. Lambot, "Combination of Advanced Inversion Techniques for an Accurate Target Localization via GPR for Demining Applications," *IEEE Trans. Geosci. Remote Sens.*, Vol. 49, No. 1, January 2011, pp. 451–461.

[52] Soldovieri, F., R. Persico, and G. Leone, "Effect of Source and Receiver Radiation Characteristics in Subsurface Prospecting within the Distorted Born Approximation," *Radio Sci.*, Vol. 40, No. 3, June 2005, p. RS3006.

[53] Kuo, C.-H., and M. Moghaddam, "Electromagnetic Scattering from a Buried Cylinder in Layered Media with Rough Interfaces," *IEEE Trans. Antennas Propag.*, Vol. 54, No. 8, August 2006, pp. 2392–2401.

[54] Altuncu, Y., "A Numerical Method for Electromagnetic Scattering by 3-D Dielectric Objects Buried under 2-D Locally Rough Surfaces," *IEEE Trans. Antennas Propag.*, Vol. 63, No. 8, August 2015, pp. 3634–3643.

[55] Gurbuz, T. U., B. Aslanyurek, E. P. Karabulut, and I. Akduman, "An Efficient Nonlinear Imaging Approach for Dielectric Objects Buried under a Rough Surface," *IEEE Trans. Geosci. Remote Sens.*, Vol. 52, No. 5, May 2014, pp. 3013–3022.

[56] Taflove, A., and S. C. Hagness, *Computational Eelectrodynamics: the Finite-Difference Time-Domain Method*, Boston: Artech House, 2005.

[57] Warren, C., A. Giannopoulos, and I. Giannakis, "gprMax: Open Source Software to Simulate Electromagnetic Wave Propagation for Ground Penetrating Radar," *Comput. Phys. Commun.*, Vol. 209, December 2016, pp. 163–170.

[58] Lou, Z., L. E. R. Petersson, J.-M. Jin, and D. J. Riley, "Total- and Scattered Field Decomposition Technique for the Finite-Element Time-Domain Modeling of Buried Scatterers," *IEEE Antennas Wireless Propag. Lett.*, Vol. 4, 2005, pp. 133–137.

[59] Frezza, F., et al., "Cylindrical-Wave Approach for Electromagnetic Scattering by Subsurface Metallic Targets in a Lossy Medium," *J. Appl. Geophys.*, Vol. 97, October 2013, pp. 55–59.

[60] Ponti, C., M. Santarsiero, and G. Schettini, "Electromagnetic Scattering of a Pulsed Signal by Conducting Cylindrical Targets Embedded in a Half-Space Medium," *IEEE Trans. Antennas Propag.*, Vol. 65, No. 6, June 2017, pp. 3073–3083.

[61] Chew, W. C., *Waves and Fields in Inhomogeneous Media*, Piscataway, NY: IEEE Press, 1995.

[62] Meng, M., et al., "Fast Analysis of Electromagnetic Scattering from Conducting Objects Buried under a Lossy Ground," *IEEE Geosci. Remote Sens. Lett.*, Vol. 13, No. 4, April 2016, pp. 555–559.

[63] Michalski, K. A., and D. Zheng, "Electromagnetic Scattering and Radiation by Surfaces of arbitrary Shape in Layered Media II: Implementation and Results for Contiguous Half-Spaces," *IEEE Trans. Antennas Propag.*, Vol. 38, No. 3, March 1990, pp. 345–352.

[64] Cui, T. J., and W. Wiesbeck, "Electromagnetic Scattering by Multiple Three-Dimensional Scatterers Buried under One-Dimensional Multi-Layered Media," in *Proc. International Geoscience and Remote Sensing Symposium, (IGARSS '96),* Lincoln, NE, 1996, Vol. 2, pp. 1340–1342.

[65] Behari, J., *Microwave Dielectric Behavior of Wet Soils,* New York: Springer, 2005.

[66] Peplinski, N. R., F. T. Ulaby, and M. C. Dobson, "Dielectric Properties of Soils in the 0.3-1.3-GHz Range," *IEEE Trans. Geosci. Remote Sens.,* Vol. 33, No. 3, May 1995, pp. 803–807.

[67] Wang, J. R., and T. J. Schmugge, "An Empirical Model for the Complex Dielectric Permittivity of Soils as a Function of Water Content," *IEEE Trans. Geosci. Remote Sens.,* Vol. GE-18, No. 4, October 1980, pp. 288–295.

[68] Hallikainen, M. T., et al., "Microwave Dielectric Behavior of Wet Soil-Part I: Empirical Models and Experimental Observations," *IEEE Trans. Geosci. Remote Sens.,* Vol. 23, No. 1, January 1985, pp. 25–34.

[69] Dobson, M. C., F. T. Ulaby, M. T. Hallikainen, and M. A. El-rayes, "Microwave Dielectric Behavior of Wet Soil-Part II: Dielectric Mixing Models," *IEEE Trans. Geosci. Remote Sens.,* Vol. GE-23, No. 1, January 1985, pp. 35–46.

[70] Matzler, C., "Microwave Permittivity of Dry Sand," *IEEE Trans. Geosci. Remote Sens.,* Vol. 36, No. 1, January 1998, pp. 317–319.

[71] Sihvola, A. H., *Electromagnetic Mixing Formulas and Applications,* London: Institution of Engineering and Technology, 1999.

[72] Hippel, A. R. V., *Dielectric Materials and Applications,* Boston: Artech House, 1995.

[73] Ulaby, F. T., et al., "Microwave Dielectric Properties of Dry Rocks," *IEEE Trans. Geosci. Remote Sens.,* Vol. 28, No. 3, May 1990, pp. 325–336.

[74] Wang, J. R., "The Dielectric Properties of Soil-Water Mixtures at Microwave Frequencies," *Radio Sci.,* Vol. 15, No. 05, September 1980, pp. 977–985.

[75] Jackson, T. J., and P. E. O'Neill, "Microwave Dielectric Model for Aggregated Soils," *IEEE Trans. Geosci. Remote Sens.,* Vol. GE-24, No. 6, November 1986, pp. 920–929.

[76] Sabburg, J., J. A. R. Ball, and N. H. Hancock, "Dielectric Behavior of Moist Swelling Clay Soils at Microwave Frequencies," *IEEE Trans. Geosci. Remote Sens.,* Vol. 35, No. 3, May 1997, pp. 784–787.

[77] Shehab, F., et al., "Uniaxial Complex Relative Permittivity Tensor Measurement of Rocks From 40 Hz to 4.5 GHz," *IEEE Trans. Geosci. Remote Sens.,* Vol. 55, No. 2, February 2017, pp. 1125–1139.

[78] Pajewski, L., R. Persico, and X. Derobert (eds.), *Catalog of GPR Test Sites,* 1st ed., Rome: TU1208 GPR Association, 2017.

[79] Wang, T., J. M. Keller, P. D. Gader, and O. Sjahputera, "Frequency Subband Processing and Feature Analysis of Forward-Looking Ground-Penetrating Radar Signals for Land-Mine Detection," *IEEE Trans. Geosci. Remote Sens.,* Vol. 45, No. 3, March 2007, pp. 718–729.

[80] Catapano, I., et al., "Forward-Looking Ground-Penetrating Radar via a Linear Inverse Scattering Approach," *IEEE Trans. Geosci. Remote Sens.*, Vol. 53, No. 10, October 2015, pp. 5624–5633.

[81] Soldovieri, F., et al., "Forward-Looking Radar Imaging: A Comparison of Two Data Processing Strategies," *IEEE J. Sel. Top. Appl. Earth Obs. Remote Sens.*, Vol. 10, No. 2, February 2017, pp. 562–571.

[82] Camilo, J. A., L. M. Collins, and J. M. Malof, "A Large Comparison of Feature-Based Approaches for Buried Target Classification in Forward-Looking Ground-Penetrating Radar," *IEEE Trans. Geosci. Remote Sens.*, Vol. 56, No. 1, January 2018, pp. 547–558.

[83] Fu, L., S. Liu, L. Liu, and L. Lei, "Development of an Airborne Ground Penetrating Radar System: Antenna Design, Laboratory Experiment, and Numerical Simulation," *IEEE J. Sel. Top. Appl. Earth Obs. Remote Sens.*, Vol. 7, No. 3, March 2014, pp. 761–766.

[84] Catapano, I., et al., "A Tomographic Approach for Helicopter-Borne Ground Penetrating Radar Imaging," *IEEE Geosci. Remote Sens. Lett.*, Vol. 9, No. 3, May 2012, pp. 378–382.

[85] Gennarelli, G., I. Catapano, and F. Soldovieri, "Reconstruction Capabilities of Down-Looking Airborne GPRs: the Single Frequency Case," *IEEE Trans. Comput. Imaging*, Vol. 3, No. 4, December 2017. pp. 917–927.

[86] Özdemir, C., S. Demirci, E. Yiit, and B. Yilmaz, "A Review on Migration Methods in B-scan Ground Penetrating Radar Imaging," *Math. Prob. in Eng.*, Vol. 2014, pp. 1–16.

[87] Gilmore, C., I. Jeffrey, and J. LoVetri, "Derivation and Comparison of SAR and Frequency-Wavenumber Migration within a Common Inverse Scalar Wave Problem Formulation," *IEEE Trans. Geosci. Remote Sens.*, Vol. 44, No. 6, June 2006, pp. 1454–1461.

[88] Catapano, I., A. Randazzo, E. Slob, and R. Solimene, "GPR Imaging Via Qualitative and Quantitative Approaches," in *Civil Engineering Applications of Ground Penetrating Radar*, A. Benedetto and L. Pajewski (eds.), Cham, Switzerland: Springer International Publishing, pp. 239–280, 2015.

[89] Solimene, R., et al., "SAR Imaging Algorithms and Some Unconventional Applications: A Unified Mathematical Overview," *IEEE Signal Process. Mag.*, Vol. 31, No. 4, July 2014, pp. 90–98.

[90] Bleistein, N., *Mathematics of Multidimensional Seismic Imaging, Migration, and Inversion*, New York: Springer, 2001.

[91] Pike, E. R., and P. C. Sabatier (eds.), *Scattering: Scattering and Inverse Scattering in Pure and Applied Science*, San Diego, California: Academic Press, 2002.

[92] Stolt, R. H., "Migration by Fourier Transform," *Geophysics*, Vol. 43, No. 1, February 1978, pp. 23–48.

[93] Song, J., Q. H. Liu, P. Torrione, and L. Collins, "Two-Dimensional and Three-Dimensional NUFFT Migration Method for landmine Detection Using Ground-Penetrating Radar," *IEEE Trans. Geosci. Remote Sens.*, Vol. 44, No. 6, June 2006, pp. 1462–1469.

[94] Smitha, N., et al., "Kirchhoff and F-K Migration to Focus Ground Penetrating Radar Images," *Int. J. Geo-Eng.*, Vol. 7, No. 1, December 2016.

[95] Soumekh, M., *Synthetic Aperture Radar Signal Processing with MATLAB Algorithms*, Hoboken, NJ: Wiley, 1999.

[96] Robinson, D. A., "Measurement of the Solid Dielectric Permittivity of Clay Minerals and Granular Samples Using a Time Domain Reflectometry Immersion Method," *Vadose Zone J.*, Vol. 3, No. 2, May 2004, pp. 705–713.

[97] Persico, R., "TDR Probes in Frequency Domain in Lossy Cases: Preliminary Results," in *Proc. IEEE MTT-S International Microwave Workshop Series on Advanced Materials and Processes (IMWS-AMP2017)*, Pavia, Italy, 2017.

[98] Di Matteo, A., E. Pettinelli, and E. Slob, "Early-Time GPR Signal Attributes to Estimate Soil Dielectric Permittivity: A Theoretical Study," *IEEE Trans. Geosci. Remote Sens.*, Vol. 51, No. 3, March 2013, pp. 1643–1654.

[99] Minet, J., S. Lambot, E. C. Slob, and M. Vanclooster, "Soil Surface Water Content Estimation by Full-Waveform GPR Signal Inversion in the Presence of Thin Layers," *IEEE Trans. Geosci. Remote Sens.*, Vol. 48, No. 3, March 2010, pp. 1138–1150.

[100] Solimene, R., A. D'Alterio, G. Gennarelli, and F. Soldovieri, "Estimation of Soil Permittivity in Presence of Antenna-Soil Interactions," *IEEE J. Sel. Top. Appl. Earth Obs. Remote Sens.*, Vol. 7, No. 3, March 2014, pp. 805–812.

[101] Solimene, R., et al., "Ground Clutter Removal in GPR Surveys," *IEEE J. Sel. Top. Appl. Earth Obs. Remote Sens.*, Vol. 7, No. 3, March 2014, pp. 792–798.

[102] Alparslan, A., M. I. Aksun, and K. A. Michalski, "Closed-Form Green's Functions in Planar Layered Media for All Ranges and Materials," *IEEE Trans. Microw. Theory Techn.*, Vol. 58, No. 3, March 2010, pp. 602–613.

[103] Chew, W. C., J. L. Xiong, and M. A. Saville, "A Matrix-Friendly Formulation of Layered Medium Green's Function," *IEEE Antennas Wireless Propag. Lett.*, Vol. 5, 2006, pp. 490–494.

[104] Sommerfeld, A., *Partial Differential Equations in Physics.* New York: Academic Press, 1949.

[105] Chew, W. C., *Waves and Fields in Inhomogenous Media*, Hoboken, NJ: Wiley, 1999.

[106] Tai, C., *Dyadic Green's Functions in Electromagnetic Theory*, Scranton, PA: International Textbook Company, 1971.

[107] Wu, B., and L. Tsang, "Fast Computation of Layered Medium Green's Functions of Multilayers and Lossy Media Using Fast All-Modes Method and Numerical Modified Steepest Descent Path Method," *IEEE Trans. Microw. Theory Techn.*, Vol. 56, No. 6, June 2008, pp. 1446–1454.

[108] Fang, D. G., J. J. Yang, and G. Y. Delisle, "Discrete Image Theory for Horizontal Electric Dipoles in a Multilayered Medium," *IEE Proc. H Microw. Antennas Propag.*, Vol. 135, No. 5, October 1988, pp. 297–303.

[109] Michalski, K. A., "On the Efficient Evaluation of Integral Arising in the Sommerfeld Halfspace Problem," *IEE Proc. H Microw. Antennas Propag.*, Vol. 132, No. 5, August 1985, pp. 312–318.

[110] Cui, T. J., and W. C. Chew, "Efficient Evaluation of Sommerfeld Integrals for Tm Wave Scattering by Buried Objects," *J. Electromagn. Waves Appl.*, Vol. 12, No. 5, January 1998, pp. 607–657.

[111] Hochman, A., and Y. Leviatan, "A Numerical Methodology for Efficient Evaluation of 2D Sommerfeld Integrals in the Dielectric Half-Space Problem," *IEEE Trans. Antennas Propag.*, Vol. 58, No. 2, February 2010, pp. 413–431.

[112] Li, D., D. R. Wilton, and D. R. Jackson, "Recent Advances in Evaluating Green's Functions for Multi-Layered Media and Half-Space Problems," in *Proc. 2017 Computing and Electromagnetics International Workshop (CEM)*, Barcelona, Spain, pp. 1–2.

[113] Leone, G., and F. Soldovieri, "Analysis of the Distorted Born Approximation for Subsurface Reconstruction: Truncation and Uncertainties Effects," *IEEE Trans. Geosci. Remote Sens.*, Vol. 41, No. 1, January 2003, pp. 66–74.

[114] Persico, R., R. Bernini, and F. Soldovieri, "The Role of the Measurement Configuration in Inverse Scattering from Buried Objects under the Born Approximation," *IEEE Trans. Antennas Propag.*, Vol. 53, No. 6, June 2005, pp. 1875–1887.

[115] Soldovieri, F., et al., "A Kirchhoff-Based Shape Reconstruction Algorithm for the Multimonostatic Configuration: The Realistic Case of Buried Pipes," *IEEE Trans. Geosci. Remote Sens.*, Vol. 46, No. 10, October 2008, pp. 3031–3038.

[116] Liseno, A., F. Tartaglione, and F. Soldovieri, "Shape Reconstruction of 2-D Buried Objects under a Kirchhoff Approximation," *IEEE Geosci. Remote Sens. Lett.*, Vol. 1, No. 2, April 2004, pp. 118–121.

[117] Bertero, M., and P. Boccacci, *Introduction to Inverse Problems in Imaging*, Bristol, UK : Institute of Physics Pub, 1998.

[118] Lesselier, D., and B. Duchene, "Buried, 2-D Penetrable Objects Illuminated by Line Sources: FFT-Based Iterative Computations of the Anomalous Field," *Prog. Electromagn. Res.*, Vol. 5, 1991, pp. 351–389.

[119] Ambrosanio, M., and V. Pascazio, "A Compressive-Sensing-Based Approach for the Detection and Characterization of Buried Objects," *IEEE J. Sel. Top. Appl. Earth Obs. Remote Sens.*, Vol. 8, No. 7, July 2015, pp. 3386–3395.

[120] Salucci, M., L. Tenuti, G. Oliveri, and A. Massa, "Frequency-Hopping GPR Prospecting of Sparse Scatterers through Bayesian Compressive Sensing," in *Proc. 2017 International Applied Computational Electromagnetics Society Symposium (ACES)*, Florence, Italy, pp. 1–2.

[121] Dong, Q., and C. M. Rappaport, "Microwave Subsurface Imaging Using Direct Finite-Difference Frequency-Domain-Based Inversion," *IEEE Trans. Geosci. Remote Sens.*, Vol. 47, No. 11, November 2009, pp. 3664–3670.

[122] Sun, S., B. J. Kooij, and A. G. Yarovoy, "Linearized 3-D Electromagnetic Contrast Source Inversion and Its Applications to Half-Space Configurations," *IEEE Trans. Geosci. Remote Sens.*, Vol. 55, No. 6, June 2017, pp. 3475–3487.

[123] Özdemir, Ö., and H. Haddar, "Linearized Cauchy Data Inversion Method for Two-Dimensional Buried Target Imaging," *IEEE Trans. Antennas Propag.*, Vol. 61, No. 6, June 2013, pp. 3244–3251.

[124] Zhdanov, M. S., and S. Fang, "Three-Dimensional Quasi-Linear Electromagnetic Inversion," *Radio Sci.*, Vol. 31, No. 4, July 1996, pp. 741–754.

[125] Zhdanov, M. S., and E. Tartaras, "Three-Dimensional Inversion of Multitransmitter Electromagnetic Data Based on the Localized Quasi-Linear Approximation," *Geophys. J. Int.*, Vol. 148, No. 3, January 2002, pp. 506–519.

[126] Zhdanov, M. S., and S. Fang, "Quasi-Linear Approximation in 3-D electromagnetic modeling," *Geophysics*, Vol. 61, No. 3, May 1996, pp. 646–665.

[127] Habashy, T. M., R. W. Groom, and B. R. Spies, "Beyond the Born and Rytov approximations: A Nonlinear Approach to Electromagnetic Scattering," *J. Geophys. Res. Solid Earth*, Vol. 98, No. B2, February 1993, pp. 1759–1775.

[128] Dasgupta, N., N. Geng, T. Dogaru, and L. Carin, "On the Extended-Born Technique for Scattering from Buried Dielectric Targets," *IEEE Trans. Antennas Propag.*, Vol. 47, No. 11, November 1999, pp. 1739–1742.

[129] Torres-Verdín, C., and T. M. Habashy, "Rapid 2.5-Dimensional Forward Modeling and Inversion via a New Nonlinear Scattering Approximation," *Radio Sci.*, Vol. 29, No. 4, July 1994, pp. 1051–1079.

[130] Cui, T. J., W. C. Chew, and W. Hong, "New Approximate Formulations for EM Scattering by Dielectric Objects," *IEEE Trans. Antennas Propag.*, Vol. 52, No. 3, March 2004, pp. 684–692.

[131] Cui, T. J., et al., "Efficient Low-Frequency Inversion of 3-D Buried Objects with Large Contrasts," *IEEE Trans. Geosci. Remote Sens.*, Vol. 44, No. 1, January 2006, pp. 3–9.

[132] Leone, G., R. Persico, and R. Solimene, "A Quadratic Model for Electromagnetic Subsurface Prospecting," *AEU - Int. J. Electron. Commun.*, Vol. 57, No. 1, January 2003, pp. 33–46.

[133] Bozza, G., C. Estatico, M. Pastorino, and A. Randazzo, "Application of an Inexact-Newton Method within the Second-Order Born Approximation to Buried Objects," *IEEE Geosci. Remote Sens. Lett.*, Vol. 4, No. 1, January 2007, pp. 51–55.

[134] Salucci, M., et al., "Electromagnetic Subsurface Prospecting by a Multifocusing Inexact Newton Method within the second-Order Born Approximation," *J. Opt. Soc. Am. A*, Vol. 31, No. 6, June 2014, pp. 1167–1179.

[135] Estatico, C., M. Pastorino, and A. Randazzo, "An Inexact-Newton Method for Short-Range Microwave Imaging within the Second-Order Born Approximation," *IEEE Trans. Geosci. Remote Sens.*, Vol. 43, No. 11, November 2005, pp. 2593–2605.

[136] van den Berg, P. M., and A. Abubakar, "Contrast Source Inversion Method: State of Art," *Prog. Electromagn. Res.*, Vol. 34, 2001, pp. 189–218.

[137] Song, L.-P., Q. H. Liu, F. Li, and Z. Q. Zhang, "Reconstruction of Three-Dimensional Objects in Layered Media: Numerical Experiments," *IEEE Trans. Antennas Propag.*, Vol. 53, No. 4, April 2005, pp. 1556–1561.

[138] Özdemir, Ö., "Cauchy Data Contrast Source Inversion Method," *IEEE Geosci. Remote Sens. Lett.*, Vol. 11, No. 4, April 2014, pp. 858–862.

[139] Li, F., Q. H. Liu, and L.-P. Song, "Three-Dimensional Reconstruction of Objects Buried in layered Media using Born and Distorted Born Iterative Methods," *IEEE Geosci. Remote Sens. Lett.*, Vol. 1, No. 2, 2004, pp. 107–111.

[140] Cui, T. J., W. C. Chew, A. A. Aydiner, and S. Chen, "Inverse Scattering of Two-Dimensional Dielectric Objects Buried in a Lossy Earth Using the Distorted Born Iterative Method," *IEEE Trans. Geosci. Remote Sens.*, Vol. 39, No. 2, February 2001, pp. 339–346.

[141] Cui, T. J., et al., "Three-Dimensional Imaging of Buried Objects in Very Lossy Earth by Inversion of VETEM Data," *IEEE Trans. Geosci. Remote Sens.*, Vol. 41, No. 10, October 2003, pp. 2197–2210.

[142] Chaturvedi, P., and R. G. Plumb, "Electromagnetic Imaging of Underground Targets Using Constrained Optimization," *IEEE Trans. Geosci. Remote Sens.*, Vol. 33, No. 3, May 1995, pp. 551–561.

[143] Pastorino, M., and A. Randazzo, "Buried Object Detection by an Inexact-Newton Method Applied to Nonlinear Inverse Scattering," *Int. J. Microw. Sci. Technol.*, Vol. 2012, 2012, p. 1–7 (637301).

[144] Estatico, C., A. Fedeli, M. Pastorino, and A. Randazzo, "Buried Object Detection by Means of a Lp Banach-Space Inversion Procedure," *Radio Sci.*, Vol. 50, No. 1, January 2015, pp. 41–51.

[145] Estatico, C., A. Fedeli, M. Pastorino, and A. Randazzo, "A Multifrequency Inexact-Newton Method in Lp Banach Spaces for Buried Objects Detection," *IEEE Trans. Antennas Propag.*, Vol. 63, No. 9, Sep. 2015, pp. 4198–4204.

[146] Salucci, M., G. Oliveri, A. Randazzo, M. Pastorino, and A. Massa, "Electromagnetic Subsurface Prospecting by a Fully Nonlinear Multifocusing Inexact Newton Method," *J. Opt. Soc. Am. A*, Vol. 31, No. 12, December 2014, p. 2618.

[147] Firoozabadi, R., E. L. Miller, C. M. Rappaport, and A. W. Morgenthaler, "Subsurface Sensing of Buried Objects under a Randomly Rough Surface Using Scattered Electromagnetic Field Data," *IEEE Trans. Geosci. Remote Sens.*, Vol. 45, No. 1, January 2007, pp. 104–117.

[148] Chiu, C.-C., and Y.-W. Kiang, "Electromagnetic Inverse Scattering of a Conducting Cylinder Buried in a Lossy Half-Space," *IEEE Trans. Antennas Propag.*, Vol. 40, No. 12, December 1992, pp. 1562–1565.

[149] Chiu, C.-C., and Y.-W. Kiang, "Inverse Scattering of a Buried Conducting Cylinder," *Inverse Probl.*, Vol. 7, No. 2, 1991, p. 187.

[150] Wu, Z., "Tomographic Imaging of Isolated Ground Surfaces Using Radio Ground Waves and Conjugate Gradient Methods," *IEE Proc. Radar Sonar Navig.*, Vol. 148, No. 1, February 2001, pp. 27–34.

[151] Salucci, M., G. Oliveri, and A. Massa, "GPR Prospecting Through an Inverse-Scattering Frequency-Hopping Multifocusing Approach," *IEEE Trans. Geosci. Remote Sens.*, Vol. 53, No. 12, December 2015, pp. 6573–6592.

[152] Pastorino, M., *Microwave Imaging*, Hoboken, NJ: Wiley, 2010.

[153] Chien, W., and C.-C. Chiu, "Using NU-SSGA to Reduce the Searching Time in Inverse Problem of a Buried Metallic Object," *IEEE Trans. Antennas Propag.*, Vol. 53, No. 10, October 2005, pp. 3128–3134.

[154] Caorsi, S., et al., "Detection of Buried Inhomogeneous Elliptic Cylinders by a Memetic Algorithm," *IEEE Trans. Antennas Propag.*, Vol. 51, No. 10, October 2003, pp. 2878–2884.

[155] Breard, A., G. Perrusson, and D. Lesselier, "Hybrid Differential Evolution and Retrieval of Buried Spheres in Subsoil," *IEEE Geosci. Remote Sens. Lett.*, Vol. 5, No. 4, October 2008, pp. 788–792.

[156] Chiu, C. C., C. H. Sun, C. L. Li, and C. H. Huang, "Comparative Study of Some Population-Based Optimization Algorithms on Inverse Scattering of a Two-Dimensional Perfectly Conducting Cylinder in Dielectric Slab Medium," *IEEE Trans. Geosci. Remote Sens.*, Vol. 51, No. 4, April 2013, pp. 2302–2315.

[157] Massa, A., M. Pastorino, and A. Randazzo, "Reconstruction of Two-Dimensional Buried Objects by a Differential Evolution Method," *Inverse Probl.*, Vol. 20, No. 6, December 2004, pp. S135–S150.

[158] Chen, X., D. Liang, and K. Huang, "Microwave Imaging 3-D Buried Objects Using Parallel Genetic Algorithm Combined With FDTD Technique," *J. Electromagn. Waves Appl.*, Vol. 20, No. 13, January 2006, pp. 1761–1774.

[159] Caorsi, S., M. Donelli, D. Franceschini, and A. Massa, "A New Methodology Based on an Iterative Multiscaling for Microwave Imaging," *IEEE Trans. Microw. Theory Techn.*, Vol. 51, No. 4, April 2003, pp. 1162–1173.

[160] Salucci, M., L. Poli, N. Anselmi, and A. Massa, "Multifrequency Particle Swarm Optimization for Enhanced Multiresolution GPR Microwave Imaging," *IEEE Trans. Geosci. Remote Sens.*, Vol. 55, No. 3, March 2017, pp. 1305–1317.

[161] Robinson, J., and Y. Rahmat-Samii, "Particle Swarm Optimization in Electromagnetics," *IEEE Trans. Antennas Propag.*, Vol. 52, No. 2, February 2004, pp. 397–407.

[162] Hill, I., *NSGG Geophysics Test Sites*, 2013, www.nsgg.org.uk/test-sites.

[163] Meles, G. A., et al., "A New Vector Waveform Inversion Algorithm for Simultaneous Updating of Conductivity and Permittivity Parameters From Combination Crosshole/Borehole-to-Surface GPR Data," *IEEE Trans. Geosci. Remote Sens.*, Vol. 48, No. 9, Sep. 2010, pp. 3391–3407.

[164] Liu, T., et al., "Estimation of Subsurface Cylindrical Object Properties from GPR Full-waveform Inversion," in *Proc. 2017 9th International Workshop on Advanced Ground Penetrating Radar (IWAGPR)*, Edinburgh, UK, pp. 1–4.

[165] Keskinen, J., et al., "Full-Waveform Inversion of Crosshole GPR data: Implications for Porosity Estimation in Chalk," *J. Appl. Geophys.*, Vol. 140, May 2017, pp. 102–116.

[166] Ernst, J. R., H. Maurer, A. G. Green, and K. . Holliger, "Full-Waveform Inversion of Crosshole Radar Data Based on 2-D Finite-Difference Time-Domain Solutions of Maxwell's Equations," *IEEE Trans. Geosci. Remote Sens.*, Vol. 45, No. 9, Sep. 2007, pp. 2807–2828.

[167] Ernst, J., A. Green, H. Maurer, and K. Holliger, "Application of a New 2D Time-Domain Full-Waveform Inversion Scheme to Crosshole Radar Data," *Geophysics*, Vol. 72, No. 5, August 2007, pp. J53–J64.

[168] Jia, H., T. Takenaka, and T. Tanaka, "Time-Domain Inverse Scattering Method for Cross-Borehole Radar Imaging," *IEEE Trans. Geosci. Remote Sens.*, Vol. 40, No. 7, July 2002, pp. 1640–1647.

[169] Busch, S., J. van der Kruk, J. Bikowski, and H. Vereecken, "Quantitative Conductivity and Permittivity Estimation Using Full-Waveform Inversion of On-Ground GPR Data," *Geophysics*, Vol. 77, No. 6, November 2012, pp. H79–H91.

[170] Klotzsche, A., et al., "Full-waveform Inversion of Cross-Hole Ground-Penetrating Radar Data to Characterize a Gravel Aquifer Close to the Thur River, Switzerland," *Surf. Geophys.*, Vol. 8, December 2010, pp. 631–646.

[171] Yang, H., et al., "Time-Gating-Based Time Reversal Imaging for Impulse Borehole Radar in Layered Media," *IEEE Trans. Geosci. Remote Sens.*, Vol. 54, No. 5, May 2016, pp. 2695–2705.

[172] Moghadasi, S. M., and M. Dehmollaian, "Buried-Object Time-Reversal Imaging Using UWB Near-Ground Scattered Fields," *IEEE Trans. Geosci. Remote Sens.*, Vol. 52, No. 11, November 2014, pp. 7317–7326.

[173] Micolau, G., and M. Saillard, "D.O.R.T. Method as Applied to Electromagnetic Subsurface Sensing," *Radio Sci.*, Vol. 38, No. 3, June 2003, p. 1038.

[174] Micolau, G., M. Saillard, and P. Borderies, "DORT Method as Applied to Ultrawideband Signals for Detection of Buried Objects," *IEEE Trans. Geosci. Remote Sens.*, Vol. 41, No. 8, August 2003, pp. 1813–1820.

[175] Davy, M., et al., "Experimental Study of the Invariants of the Time-Reversal Operator for a Dielectric Cylinder Using Separate Transmit and Receive Arrays," *IEEE Trans. Antennas Propag.*, Vol. 58, No. 4, April 2010, pp. 1349–1356.

[176] Iakovleva, E., S. Gdoura, D. Lesselier, and G. Perrusson, "Multistatic Response Matrix of a 3-D Inclusion in Half Space and MUSIC Imaging," *IEEE Trans. Antennas Propag.*, Vol. 55, No. 9, Sep. 2007, pp. 2598–2609.

[177] Devaney, A. J., "Time Reversal Imaging of Obscured Targets from Multistatic Data," *IEEE Trans. Antennas Propag.*, Vol. 53, No. 5, May 2005, pp. 1600–1610.

[178] Alqadah, H. F., "Multi-Frequency LSM for Buried Object Detection," in *Proc. International Conference on Electromagnetics in Advanced Applications (ICEAA)*, Verona, Italy, 2017, pp. 1323–1326.

[179] Catapano, I., L. Crocco, and T. Isernia, "Improved Sampling Methods for Shape Reconstruction of 3-D Buried Targets," *IEEE Trans. Geosci. Remote Sens.*, Vol. 46, No. 10, October 2008, pp. 3265–3273.

[180] Catapano, I., F. Soldovieri, and L. Crocco, "On the Feasibility of the Linear Sampling Method for 3D GPR Surveys," *Prog. Electromagn. Res.*, Vol. 118, 2011, pp. 185–203.

[181] Gamba, P., and S. Lossani, "Neural Detection of Pipe Signatures in Ground Penetrating Radar Images," *IEEE Trans. Geosci. Remote Sens.*, Vol. 38, No. 2, March 2000, pp. 790–797.

[182] Mydur, R., and K. A. Michalski, "A Neural-Network Approach to the Electromagnetic Imaging of Elliptic Conducting Cylinders," *Microw. Opt. Technol. Lett.*, Vol. 28, No. 5, March 2001, pp. 303–306.

[183] Bermani, E., A. Boni, S. Caorsi, and A. Massa, "An Innovative Real-Time Technique for Buried Object Detection," *IEEE Trans. Geosci. Remote Sens.*, Vol. 41, No. 4, 2003, pp. 927–931.

[184] Sahin, A., and E. L. Miller, "Object Detection Using High Resolution Near-Field Array Processing," *IEEE Trans. Geosci. Remote Sens.*, Vol. 39, No. 1, January 2001, pp. 136–141.

[185] Meschino, S., L. Pajewski, and G. Schettini, "A Direction-Of-Arrival Approach for the Subsurface Localization of a Dielectric Object," *J. Appl. Geophys.*, Vol. 85, October 2012, pp. 68–79.

[186] Meschino, et al., "Detection of Subsurface Metallic Utilities by Means of a SAP Technique: Comparing MUSIC- and SVM-Based Approaches," *J. Appl. Geophys.*, Vol. 97, October 2013, pp. 60–68.

[187] Dou, Q., L. Wei, D. R. Magee, and A. G. Cohn, "Real-Time Hyperbola Recognition and Fitting in GPR Data," *IEEE Trans. Geosci. Remote Sens.*, Vol. 55, No. 1, January 2017, pp. 51–62.

[188] Mertens, L., R. Persico, L. Matera, and S. Lambot, "Automated Detection of Reflection Hyperbolas in Complex GPR Images with No A Priori Knowledge on the Medium," *IEEE Trans. Geosci. Remote Sens.*, Vol. 54, No. 1, January 2016, pp. 580–596.

[189] Pasolli, E., F. Melgani, and M. Donelli, "Automatic Analysis of GPR Images: A Pattern-Recognition Approach," *IEEE Trans. Geosci. Remote Sens.*, Vol. 47, No. 7, July 2009, pp. 2206–2217.

[190] Maas, C., and J. Schmalzl, "Using Pattern Recognition to Automatically Localize Reflection Hyperbolas in Data from Ground Penetrating Radar," *Comput. Geosci.*, Vol. 58, No. Supplement C, August 2013, pp. 116–125.

[191] Omurzakov, A., A. K. Keskin, and A. S. Turk, "Avalanche Transistor Short Pulse Generator Trials for GPR," in *Proc. 8th International Conference on Ultrawideband and Ultrashort Impulse Signals (UWBUSIS)*, Odessa, Ukraine, 2016, pp. 201–204.

[192] Zhou, K., C. L. Huang, and M. Lu, "A Nanosecond Pulse Generator Based on Avalanche Transistor," in *Proc. 16th International Conference on Ground Penetrating Radar (GPR)*, Hong Kong, China, 2016, pp. 1–5.

[193] Öztürk, H., K. Yegin, and H. Nazlı, "High Fidelity Gaussian Monocycle Generation Using Microstrip Lines," *IEEE Microw. Compon. Lett.*, Vol. 24, No. 10, October 2014, pp. 722–724.

[194] Hanawa, M., et al., "Experimental Demonstration of High-Resolution Ultra-Wideband Impulse Radar Based on Electrical-Optical Hybrid Pulse Generation," *IET Microw. Antennas Propag.*, Vol. 4, No. 10, October 2010, pp. 1562–1468.

[195] Yarlequé, M. A., S. Alvarez, H. J. Martínez, and A. C. Canelo, "FMCW GPR Radar for Archaeological Applications: First Analytical and Measurement Results," in *Proc. 32nd General Assembly and Scientific Symposium of the International Union of Radio Science (URSI GASS)*, Montreal, Canada, 2017, pp. 1–3.

[196] Ferrara, V., et al., "Design and Realization of a cheap Ground Penetrating Radar Prototype @ 2.45 GHz," in *Proc. 10th European Conference on Antennas and Propagation (EuCAP)*, Davos, Switzerland, 2016, pp. 1–4.

[197] Hamran, S. E., et al., "RIMFAX: A GPR for the Mars 2020 Rover Mission," in *Proc. 8th Internati/onal Workshop on Advanced Ground Penetrating Radar (IWAGPR)*, Florence, Italy, 2015, pp. 1–4.

[198] Phelan, B. R., et al., "Design of Ultrawideband Stepped-Frequency Radar for Imaging of Obscured Targets," *IEEE Sens. J.*, Vol. 17, No. 14, July 2017, pp. 4435–4446.

[199] Persico, R., D. Dei, F. Parrini, and L. Matera, "Mitigation of Narrowband Interferences by Means of a Reconfigurable Stepped Frequency GPR System," *Radio Sci.*, Vol. 51, No. 8, August 2016, pp. 1322–1331.

[200] Parrini, F., et al., "A Reconfigurable Stepped Frequency GPR (GPR-R)," in *Proc. IEEE International Geoscience and Remote Sensing Symposium*, Vancouver, Canada, 2011, pp. 67–70.

[201] Serhir, M., and D. Lesselier, "Wideband Reflector-Backed Folded Bowtie Antenna for Ground Penetrating Radar," *IEEE Trans. Antennas Propag.*, Vol. PP, No. 99, 2017, p. 1.

[202] Pieraccini, M., N. Rojhani, and L. Miccinesi, "Comparison Between Horn and Bow-Tie Antennas for Ground Penetrating Radar," in *Proc. 9th International Workshop on Advanced Ground Penetrating Radar (IWAGPR)*, Edinburgh, UK, 2017, pp. 1–5.

[203] Congedo, F., G. Monti, and L. Tarricone, "Modified Bowtie Antenna for GPR Applications," in *Proc. XIII International Conference on Ground Penetrating Radar*, Lecce, Italy, 2010, pp. 1–5.

[204] van Coevorden, C. M. J., et al., "GA Design of a Thin-Wire Bow-Tie Antenna for GPR Applications," *IEEE Trans. Geosci. Remote Sens.*, Vol. 44, No. 4, April 2006, pp. 1004–1010.

[205] Lestari, A. A., A. G. Yarovoy, and L. P. Ligthart, "Adaptive Wire Bow-Tie Antenna for GPR Applications," *IEEE Trans. Antennas Propag.*, Vol. 53, No. 5, May 2005, pp. 1745–1754.

[206] Lestari, A. A., et al., "A Modified Bow-Tie Antenna for Improved Pulse Radiation," *IEEE Trans. Antennas Propag.*, Vol. 58, No. 7, July 2010, pp. 2184–2192.

[207] Elsheakh, D. M., and E. A. Abdallah, "Novel Shapes of Vivaldi Antenna for Ground Pentrating Radar (GPR)," in *Proc. 7th European Conference on Antennas and Propagation (EuCAP)*, Gothenburg, Sweden, 2013, pp. 2886–2889.

[208] Alkhalifeh, K., G. Hislop, N. A. Ozdemir, and C. Craeye, "Efficient MoM Simulation of 3-D Antennas in the Vicinity of the Ground," *IEEE Trans. Antennas Propag.*, Vol. 64, No. 12, December 2016, pp. 5335–5344.

[209] Liu, H., J. Zhao, and M. Sato, "A Hybrid Dual-Polarization GPR System for Detection of Linear Objects," *IEEE Antennas Wireless Propag. Lett.*, Vol. 14, 2015, pp. 317–320.

[210] Peñaloza-Aponte, D., J. Alvarez-Montoya, and M. Clemente-Arenas, "GPR Vivaldi Antenna with DGS for Archeological Prospection," in *Proc. IEEE 24th International Conference on Electronics, Electrical Engineering and Computing (INTERCON)*, Cusco, Peru, 2017, pp. 1–4.

[211] Kota, S., and N. Gunavathi, "Design and Analysis of TEM Horn Antenna for GPR Applications," in *Proc. Third International Conference on Sensing, Signal Processing and Security (ICSSS)*, Chennai, India, 2017, pp. 15–22.

[212] Yildiz, S., A. Uysal, and M. Özen, "Shorted TEM Horn Antenna for GPR Applications," in *Proc. 16th Mediterranean Microwave Symposium (MMS)*, Abu Dhabi, United Arab Emirates, 2016, pp. 1–3.

[213] Ahmed, A., et al., "Design of UWB Antenna for Air-Coupled Impulse Ground-Penetrating Radar," *IEEE Geosci. Remote Sens. Lett.*, Vol. 13, No. 1, January 2016, pp. 92–96.

[214] Shao, J., et al., "TEM Horn Antenna Loaded With Absorbing Material for GPR Applications," *IEEE Antennas Wireless Propag. Lett.*, Vol. 13, 2014, pp. 523–527.

[215] Oloumi, D., P. Mousavi, M. I. Pettersson, and D. G. Elliott, "A Modified TEM Horn Antenna Customized for Oil Well Monitoring Applications," *IEEE Trans. Antennas Propag.*, Vol. 61, No. 12, December 2013, pp. 5902–5909.

[216] McFadden, M., and W. R. Scott, "Numerical Modeling of a spiral-Antenna GPR System," in *Proc. IEEE International Geoscience and Remote Sensing Symposium*, Cape Town, South Africa, 2009, p. II-109-II-112.

[217] He, Y., and Y. Akizuki, "FDTD Analysis of a Planar Spiral Antennas for the Ground Penetrating Radar," in *Proc. URSI International Symposium on Electromagnetic Theory*, Berlin, Germany, 2010, pp. 370–372.

[218] Dourthe, C., C. Pichot, J. Y. Dauvignac, and J. Cariou, "Inversion Algorithm and Measurement System for Microwave Tomography of Buried Object," *Radio Sci.*, Vol. 35, No. 5, September 2000, pp. 1097–1108.

[219] Y. J. Kim, L. Jofre, F. D. Flaviis, and M. Q. Feng, "Microwave Reflection Tomographic Array for Damage Detection of Civil Structures," *IEEE Trans. Antennas Propag.*, Vol. 51, No. 11, pp. 3022–3032, November 2003.

[220] Lambot, S., and F. André, "Full-Wave Modeling of Near-Field Radar Data for Planar Layered Media Reconstruction," *IEEE Trans. Geosci. Remote Sens.*, Vol. 52, No. 5, May 2014, pp. 2295–2303.

[221] Maksimovitch, Y., V. Mikhnev, and P. Vainikainen, "Simulation of an Ultra-Wideband Antenna for Step-Frequency Ground Penetrating Radar using Method of Moments," in *Proc. 15th International Crimean Conference Microwave Telecommunication Technology*, Sevastopol, Ukraine, 2005, Vol. 1, p. 385–387.

[222] Hislop, G., et al., "Antenna Calibration for Near-Field Problems with the Method of Moments," in *Proc. 5th European Conference on Antennas and Propagation (EUCAP)*, Rome, Italy, 2011, pp. 2004–2008.

[223] Šušnjara, A., D. Poljak, S. Šesnic, and V. Doric, "Time Domain and Frequency Domain Integral Equation Method for the Analysis of Ground Penetrating Radar (GPR) Antenna," in *Proc. 24th International Conference on Software, Telecommunications and Computer Networks (SoftCOM)*, Split, Croatia, 2016, pp. 1–4.

[224] Venkatarayalu, N. V., C.-C. Chen, F. L. Teixeira, and R. Lee, "Numerical Modeling of Ultrawide-Band Dielectric Horn Antennas Using FDTD," *IEEE Trans. Antennas Propag.*, Vol. 52, No. 5, May 2004, pp. 1318–1323.

[225] Warren, C., L. Pajewski, A. Ventura, and A. Giannopoulos, "An Evaluation of Finite-Difference and Finite-Integration Time-Domain Modelling Tools for Ground Penetrating Radar Antennas," in *Proc. 10th European Conference on Antennas and Propagation (EuCAP)*, Davos, Switzerland, 2016, pp. 1–5.

[226] Sagnard, F., E. Tebchrany, and V. Baltazart, "Evaluation of an UWB Ground-Coupled Radar in the Detection of Discontinuities Using Polarization Diversity: FDTD Modeling and Experiments," in *Proc. 7th International Workshop on Advanced Ground Penetrating Radar*, Nantes, France, 2013, pp. 1–6.

[227] Lampe, B., K. Holliger, and A. Green, "A Finite-Difference Time-Domain Simulation Tool for Ground-Penetrating Radar Antennas," *Geophysics*, Vol. 68, No. 3, May 2003, pp. 971–987.

[228] Klysz, G., X. Ferrieres, J. P. Balayssac, and S. Laurens, "Simulation of Direct Wave Propagation by Numerical FDTD for a GPR Coupled Antenna," *NDT&E Int.*, Vol. 39, No. 4, June 2006, pp. 338–347.

[229] Moghaddam, M., E. J. Yannakakis, W. C. Chew, and C. Randall, "Modeling of the Subsurface Interface Radar," *J. Electromagn. Waves Appl.*, Vol. 5, No. 1, January 1991, pp. 17–39.

[230] Shlager, K. L., G. S. Smith, and J. G. Maloney, "Optimization of Bow-Tie Antennas for Pulse Radiation," *IEEE Trans. Antennas Propag.*, Vol. 42, No. 7, July 1994, pp. 975–982.

[231] Giannakis, I., A. Giannopoulos, and C. Warren, "A Realistic FDTD Numerical Modeling Framework of Ground Penetrating Radar for Landmine Detection," *IEEE J. Sel. Top. Appl. Earth Obs. Remote Sens.,* Vol. 9, No. 1, January, 2016, pp. 37–51.

[232] Warren, C., and A. Giannopoulos, "Creating Finite-Difference Time-Domain Models of Commercial Ground-Penetrating Radar Antennas Using Taguchi's Optimization Method," *Geophysics*, Vol. 76, No. 2, March 2011, pp. G37–G47.

7

Microwave Imaging for Security Purposes

This chapter is devoted to the applications of microwave imaging in the framework of security. In particular, the main focus is on the techniques for through-the-wall imaging (TWI). Moreover, the possible use of imaging methods for personnel surveillance and concealed target detection is also discussed.

7.1 Potentialities and Limitations of Microwave Imaging for Security Applications

Microwaves are acquiring an ever-growing importance in the field of security, where the ability of electromagnetic waves to penetrate dielectric materials opens a wide range of possibilities. Some of the main applications are related to the detection of concealed targets [1–5] and to the inspection and monitoring of inaccessible domains, such as the interior of buildings [6–12]. In the first case, the aim is typically the identification of weapons or other possibly dangerous objects hidden on peoples or inside envelopes/containers. In the latter case, the region to be inspected is obscured by walls, which do not allow the use of more common techniques based on optical cameras. However, with a proper choice of the working frequency band, it is possible to properly illuminate the region of interest and collect the scattering contributions due to the people and other objects located inside the scenario. Moreover, by further processing the measured data, additional information can be extracted, such as breathing Doppler signatures [13–16] or temporal variations of the positions of the objects [17–22]. On one hand, the presence of walls, as well as the impossibility of measuring the scattered field all around the region of interest, can be considered limiting factors. On the other hand, the potential wide range of applications that can benefit from through-the-wall inspections provides a continuous stimulation for research.

In order to reconstruct targets behind walls, beamforming and other qualitative techniques similar to the ones employed in subsurface GPR prospection (see Chapter 6) are usually applied. However, the presence of walls and other obstacles should be properly taken into account in order to obtain correct reconstruction results. This is frequently done in qualitative ways, based on several model approximations, as will be discussed in the next sections. In this framework, inverse scattering approaches may represent valid alternative techniques. In fact, as also discussed in the previous chapters, they are usually based on more rigorous electromagnetic propagation models. In particular, it is possible to straightforwardly take into account the presence of the wall by exploiting the proper Green's function in the scattering equations. However, it is worth noting that the computation of the Green's function strongly depends on the geometrical structure of the wall and on the dielectric properties of the building materials. Therefore, these characteristics should be known or, at least, estimated from the available measurements.

Finally, it is worth mentioning that in this applicative field, more than in others, measurement and computational times represent quite important aspects for the efficacy of microwave imaging techniques, which should provide information about the inspected scenarios in short times. Moreover, portability of imaging systems is another critical point for allowing their use directly in the field. Such constraints clearly pose some additional requirements on both the measurement apparatuses (e.g., in terms of the number of measurements that can be performed and overall size of the devices) and the imaging procedures (e.g., on the computational efficiency), which have to be properly taken into account in the development of microwave imaging systems for security applications.

7.2 Through-the-Wall Imaging

As previously introduced, one of the main security-related applications in which microwave imaging plays an important role is through-the-wall imaging (TWI). In TWI, the aim is to detect, identify, and eventually track targets (which can be human beings or other moving objects) inside buildings [7]. The great interest that has been devoted to TWI in the last few years is related to the potential advantages that this technology can provide in different areas, such as police and military force missions, or firefighting and rescue operations.

A schematic representation of a common TWI configuration is shown in Figure 7.1. For the sake of simplicity, throughout this chapter, two-dimensional (2-D) scenarios are mostly considered. However, the majority of the discussed techniques can be extended to three-dimensional (3-D) settings. A set of S antennas is positioned in front of a wall at a standoff distance h, and is used to illuminate the interior of the building and to collect the electromagnetic field resulting from the interactions with the objects located inside the inspected scene. As can be seen, the imaging configuration is similar to the one usually considered in subsurface prospection, described in Chapter 6. However, there are also significant differences. In fact, the presence of a dielectric wall strongly influences the electromagnetic propagation, requiring proper modifications of the imaging algorithms. Moreover, both monostatic and multistatic configurations can be used in TWI systems. Clearly, multistatic setups allow one to increase the amount of information available for the imaging procedures, although they also increase the complexity and the size of the measurement systems.

TWI techniques usually employ scattered field data in a wide frequency band to build images of the scene under investigation. To this end, pulsed and continuous-wave (e.g., frequency-stepped or modulated) systems can be adopted. Moreover, different types of waveforms can be used to feed the antennas, ranging from standard radar waveforms (such as short sinusoidal pulses and frequency-modulated continuous waves) to noise or pseudo-noise waveforms and UWB signals [23]. In any case, in order to be less affected by the attenuation due to the wall materials, the frequency band of operation is usually contained in the lowest end of microwave frequencies, that is, the range between 0.3 and 4 GHz [7].

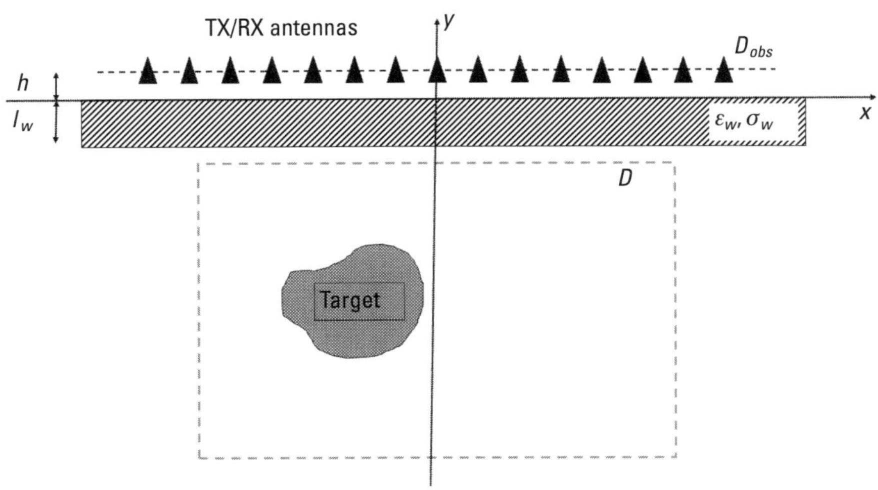

Figure 7.1 Schematic representation of the through-the-wall imaging configuration.

An example of time-domain response of a TWI system (simulated by using a numerical solver based on the FDTD technique [24] in a 2-D setting) is shown in Figure 7.2. Here, the electric field data have been represented in the form of a *B*-scan, as defined in Chapter 6. In this case, it is assumed that a single moving antenna pair (with spacing between the transmitting and receiving elements equal to 4 cm) is used to collect quasi-monostatic measurements on a line with a length of 1.6m located at a stand-off distance of 10 cm from the wall. The wall is assumed to be made of dry concrete (see Section 4.2.1) and it is modeled as a homogeneous dielectric slab with thickness equal to $l_w =$ 20 cm, relative dielectric permittivity $\varepsilon_{w,r} = 4$ and conductivity $\sigma_w = 0.01$ S/m (dispersion is neglected here). A Ricker pulse with central frequency equal to 1.5 GHz is used to feed the transmitting antenna, which has been modeled as an ideal point source. The target is a circular PEC cylinder whose cross section is centered at $\mathbf{r}_{t,c} = (0.2, -0.7)$m (i.e., it is located at a distance of 50 cm from the wall) and has a diameter $d_c = 30$ cm. As expected, in the *B*-scan shown in Figure 7.2(a) it is possible to identify different contributions: The direct wave coming from the transmitting antenna, the reflections from the air/wall and wall/air interfaces, and a hyperbola due to the circular target behind the wall (similar to the one that appears in subsurface configurations, see Section 6.3). Figure 7.2(b) reports the corresponding *z*-components of the scattered field vector by the concealed target only, that is, obtained by subtracting the incident field that can be measured in absence of the target (which takes into account the reflection from the wall). As can be seen, the response is similar to the one of GPR systems for subsurface prospecting. Consequently, as discussed in the following, migration techniques (also referred as synthetic aperture or beamforming approaches in the TWI literature) can be adapted for TWI applications.

7.2.1 Wall Characterization

A crucial point in the development of TWI techniques is the electromagnetic characterization of the walls. Actually, in real scenarios, there is usually a great variability in both the adopted materials and the internal structure of walls. Such characteristics strongly affect the electromagnetic propagation and the available information that can be extracted from the scattered field resulting from the interactions with the targets.

From a geometrical point of view, three main classes of wall structures can be used in the definition of the imaging techniques (Figure 7.3):

1. Homogeneous walls (e.g., for modeling wood, concrete and solid brick walls);
2. Multilayer walls;

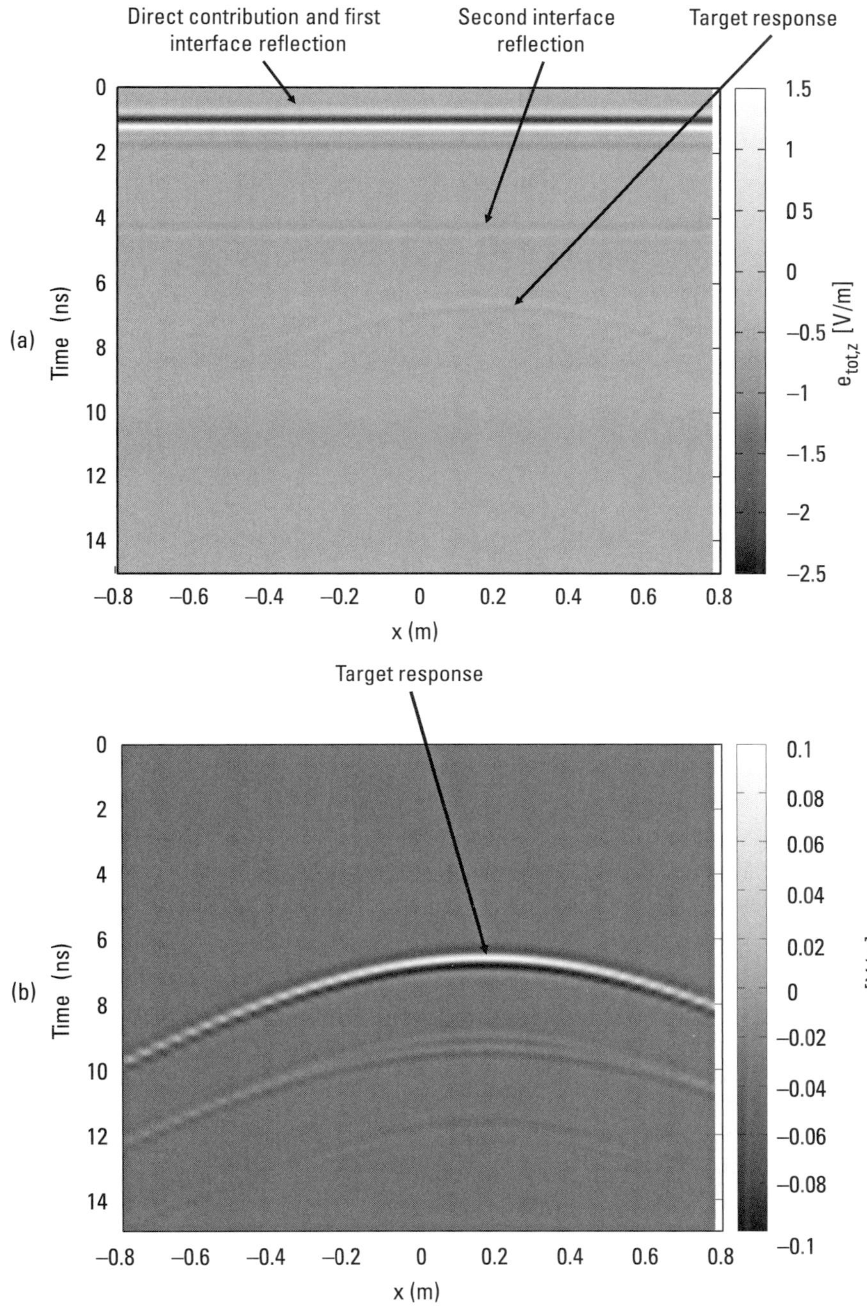

Figure 7.2 Example of a quasi-monostatic *B*-scan in a simplified synthetic TWI configuration. (a) Total and (b) scattered electric field (*z*-components).

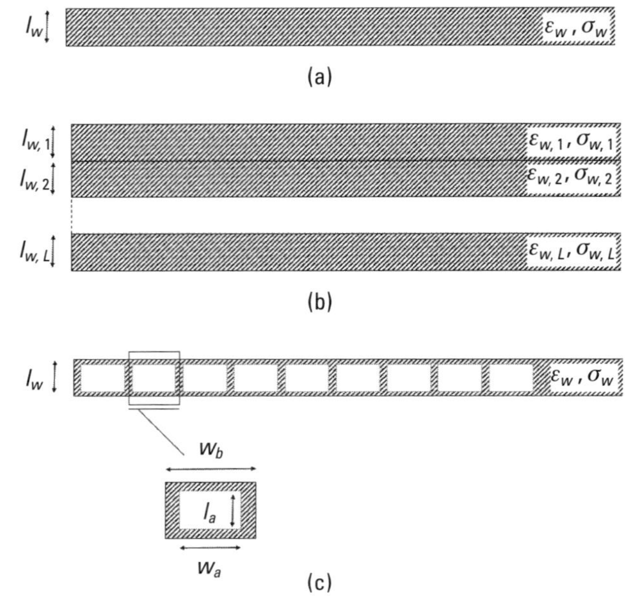

Figure 7.3 Typical wall geometric models. (a) Homogeneous wall, (b) multilayer wall, and (c) inhomogeneous (periodic) wall.

3. Inhomogeneous walls made by elementary building blocks (e.g., cinderblock, walls made of hollow concrete blocks, or reinforced walls).

Materials commonly used in buildings are typically wood, concrete, bricks, and cinderblocks. However, there can be a significant variability in their dielectric properties depending on the construction process and on other environmental parameters (e.g., humidity, temperature, and so forth). Anyway, in the frequency range of interest for TWI applications, some indicative ranges for the dielectric properties of typical wall materials can be identified [7, 25, 26]. For example, for brick walls, the real part of the relative dielectric permittivity is often assumed to be around 4, with values of the electric conductivity between 0.01 and 0.03 S/m. The dielectric properties of concrete, which have been discussed in Section 4.2.1, usually vary in the ranges of 4–10 (for the real part) and 0.1–1.5 (for the imaginary part) in the considered frequency band. In the case of cinderblocks, the values of the equivalent dielectric permittivity are often in the ranges of 3–5 (for the real part) and 0.3–15 (for the imaginary part). Wooden walls (e.g., made of plywood) are usually characterized by low values of the relative dielectric permittivity, for example, around 2 (see also Section 4.2.4). It is worth noting that the dielectric properties of typical building materials are also characterized by a strong dependence on the water content, as previously mentioned with reference to wood and concrete.

Clearly, the actual dielectric properties of the wall strongly affect the electromagnetic propagation and, in particular, the scattered field that can be measured behind the wall. Consequently, the knowledge of the geometrical and dielectric properties of the specific walls is fundamental for developing suitable imaging procedures for through-the-wall sensing. Although some information can be inferred from literature data, the real-world uncertainty is usually too high to obtain data at a sufficient level of accuracy. In fact, the incorrect estimation of dielectric and geometrical wall parameters leads to a significant degradation of the reconstructed images, such as shifted positions and blurring of the detected targets [27, 28]. Consequently, it is often necessary to include in the imaging system a proper technique for extracting the required information from the data with a suitable accuracy. Depending on the considered system architecture, there are two possible approaches:

1. Extraction from time-domain signal [29–31];
2. Extraction from frequency-domain data [32, 33].

In the first case, a pulsed-type radar illumination is used. The information about the average dielectric permittivity and thickness can be extracted by looking at the reflections from the input and output interfaces of the wall. Clearly, it is required that the system exhibits a very high range resolution, in order to correctly separate the reflections from the air/wall interfaces.

An example of simulated response from a homogeneous wall model is shown in Figure 7.4. The same configuration used in Figure 7.2 has been considered. The reported field values correspond to the measurement position in front of the target (i.e., at point (0.2, 0.1)m).

As can be seen in this case, it is clearly possible to identify the contributions of the two interfaces of the wall. Consequently, by estimating the relative times at which they are received, it is possible to approximate the thickness of the wall and some of its dielectric properties. It is worth noting that, in most practical cases, it is not so simple to separate the various contributions, due to their possible overlap and the presence of noise. To address such problems, super-resolution subspace processing techniques, such as the estimation of signal parameters using rotational invariance technique (ESPRIT), the multiple signal classification (MUSIC) method, and support vector machines (SVMs), can be used [30, 31, 34, 35].

Once the time delays are estimated, the dielectric properties of the wall and its thickness can be derived by minimizing a proper residual functional. For example, by adopting the so-called common midpoint processing (which consider a pair of antennas working in bistatic mode), the real part of the relative dielectric permittivity of the wall $\varepsilon'_{r,w}$ and the wall thickness l_w can be estimated as [30, 31]

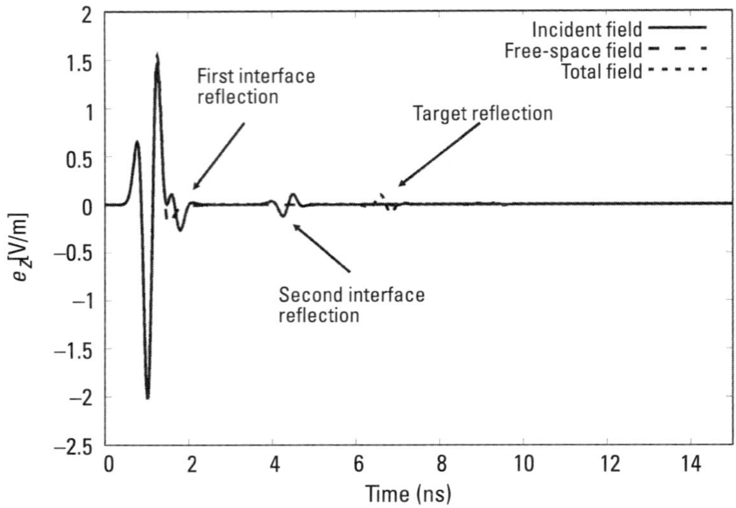

Figure 7.4 Example of a simulated *A*-scan in presence of a homogeneous wall.

$$\left(\varepsilon'_{r,w}, l_w\right) = \arg\min_{\varepsilon'_{r,w}, l_w} \sum_{i=1}^{M} \left| \tau_{meas}^{(i)} - \tau_{calc}^{(i)}\left(\varepsilon'_{r,w}, l_w\right) \right|^2 \tag{7.1}$$

where $\tau_{meas}^{(i)}$ is the time at which the second interface reflection is received, $\tau_{calc}^{(i)}$ is the corresponding estimated time, and M is the number of performed measurements, which are obtained by considering different distances L_i between the transmitting and receiving antennas. It is worth noting that by using such approach it is only possible to retrieve the real part of the relative dielectric permittivity, since the amplitude of the received electromagnetic signal (which is affected by the wall losses) is not taken into account. As an example, in the case of a homogeneous wall, the forward model proposed in [30, 31] can be adopted. In particular, the time at which the second interface reflection is expected can be modeled as (the index i is omitted in the following)

$$\tau_{calc} = \frac{2d_a}{v_0} + \sqrt{\varepsilon'_{w,r}}\,\frac{2d_w}{v_0} \tag{7.2}$$

where d_a and d_w are the lengths of the paths traveled by the wave from the antenna to the wall and inside the wall, which can be obtained by means of simplified geometrical considerations. In particular, it results in [30, 31]

$$d_a = \sqrt{h^2 + d^2},\ d_w = \sqrt{\left(\frac{L}{2} - d\right)^2 + l_w^2} \tag{7.3}$$

where h is the standoff distance and d can found from the following relationship [31]

$$d = \frac{\sqrt{\varepsilon'_{r,w}}\, d_a L / 2}{d_w + d_a\sqrt{\varepsilon'_{r,w}}} \tag{7.4}$$

In frequency-domain estimation techniques, the information contained in the reflection coefficient, whose value depends upon the contributions of all the interfaces, is usually considered. An optimization problem can be defined, in which the mismatch between simulated and measured reflection coefficients is minimized, for example by using the following relationship

$$\left(\varepsilon_{r,w}, l_w\right) = \arg\min_{\varepsilon_{r,w}, l_w} \sum_{\omega \in \Omega} \left|\Gamma_{meas}(\omega) - \Gamma_{calc}\left(\varepsilon_{r,w}, l_w, \omega\right)\right|^2 \tag{7.5}$$

where Ω is the set of considered values of the angular frequency, $\Gamma_{meas}(\omega)$ is the measured value of the reflection coefficient at angular frequency ω, and Γ_{calc} is a direct model that is used to predict the reflection coefficient starting from the dielectric and geometric properties of the wall. Assuming that the wall can be approximated as a homogeneous slab and that the impinging field can be modeled as a plane wave with direction of propagation normal to the wall, the well-known equations for the reflection coefficients by a slab can be used as forward model, that is, [36]

$$\Gamma_{calc}\left(\varepsilon_{r,w}, l_w, \omega\right) = R_{aw} \frac{1 - e^{-2jk_w l_w}}{1 - R_{aw}^2 e^{-2jk_w l_w}} \tag{7.6}$$

where $k_w = \sqrt{\varepsilon_{r,w}}\, k_0$, with $k_0 = \omega\sqrt{\varepsilon_0 \mu_0}$, and

$$R_{aw} = \frac{k_0 - k_w}{k_0 + k_w} \tag{7.7}$$

The above approaches assume that antenna effects can be neglected (e.g., multiple reflections inside the antenna structure or between antenna and wall). In many practical cases however, these effects can be significant and can almost mask the reflections from the wall and from the hidden targets. Consequently, accurate and efficient methods for compensating antenna effects by using proper far-field or near-field calibration procedures have been devised [37, 38].

7.2.2 Beamforming Approaches for TWI

Most of the approaches commonly used in TWI imaging are based on beamforming and synthetic aperture concepts [39–49] and are similar to those used in GPR imaging or medical imaging. However, as discussed below, some modifications are needed in order to deal with the specific TWI configurations.

As mentioned, one of the main issues in the application of such kinds of approaches in TWI applications is related to a proper compensation of the presence of the wall. If the presence of the wall is not taken into account, the obtained image may be defocused and the locations of targets may be shifted [7]. Some different approaches have been proposed in the literature for addressing such problem (for example, see [39, 42, 43, 50, 51]). Let us first consider the case of time-domain migration and beamforming approaches, such as diffraction summation and delay-and-sum methods, for pulsed radar systems. If frequency-stepped or frequency-modulated systems are used, a synthetic pulse can be generated by preprocessing the available frequency-domain data, as for GPR systems. In this case, it is necessary to properly estimate the time delays to be applied to the received fields. If the permittivity and the thickness of the wall are known, it is possible to partially compensate for the defocusing effects caused by the presence of the wall in the beamforming procedure by taking into account the different propagation velocity inside the wall. As an example, let us assume the 2-D configuration sketched in Figure 7.5, in which a homogeneous wall model is adopted and a generic bistatic configuration (e.g., the *m*th transmitting and the *n*th receiving antennas of a multistatic system) is considered.

In this case, the time delay τ_{mn} can be expressed as

$$\tau_{mn}(\mathbf{r}) = \frac{d_m^a(\mathbf{r}) + d_n^a(\mathbf{r})}{v_0} + \frac{d_m^w(\mathbf{r}) + d_n^w(\mathbf{r})}{v_w} \tag{7.8}$$

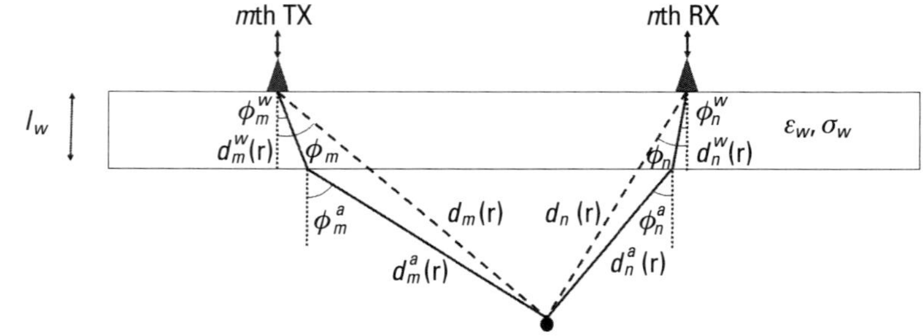

Figure 7.5 Ray paths through a homogeneous dielectric wall.

where $d_m^a(\mathbf{r})$ and $d_n^a(\mathbf{r})$ are the distances travelled in air by the impinging and scattered waves, $d_m^w(\mathbf{r})$ and $d_n^w(\mathbf{r})$ are the parts of the propagation path inside the wall, and v_w is the velocity of propagation inside the wall. Applying basic trigonometric relationships (with reference to Figure 7.5, in which, for sake of simplicity, the antennas are supposed in contact with the wall), it results [43]

$$d_{m/n}^w(\mathbf{r}) = \frac{l_w}{\cos\phi_{m/n}^w(\mathbf{r})} \tag{7.9}$$

$$\begin{aligned} d_{m/n}^a(\mathbf{r}) &= \sqrt{d_{m/n}^2(\mathbf{r}) - d_{m/n}^{w^2}(\mathbf{r})\sin^2\left(\phi_{m/n}^w(\mathbf{r}) - \phi_{m/n}^a(\mathbf{r})\right)} \\ &-d_{m/n}^w(\mathbf{r})\cos\left(\phi_{m/n}^w(\mathbf{r}) - \phi_{m/n}^a(\mathbf{r})\right) \end{aligned} \tag{7.10}$$

According to the Snell's law, the angles $\phi_{m/n}^a$ and $\phi_{m/n}^w$, which depend upon the dielectric properties of the wall, can be obtained by the following relations

$$\phi_{m/n}^{w/a} + \sin^{-1}\left(\frac{d_{m/n}^a(\mathbf{r})}{d_{m/n}(\mathbf{r})}\sin\left(\phi_{m/n}^{a/w}(\mathbf{r}) - \phi_{m/n}^{w/a}(\mathbf{r})\right)\right) = \phi_{m/n} \tag{7.11}$$

$$\sin\phi_{m/n}^{a/w} = \sqrt{\varepsilon_{w,r}}\sin\phi_{m/n}^{w/a} \tag{7.12}$$

The time delay computed by using (7.8) can be subsequently used in beamforming/migration schemes. Similar relationships can also be obtained in the 3-D case [42].

A reconstruction result obtained by using the compensation approach discussed in [42] is reported below. The antennas are located on a planar surface at a distance of 3m from the wall. In particular, 33 × 33 antenna positions separated by 5.7 cm are considered. Stepped-frequency data have been simulated in the range 1–3 GHz. The wall is modeled as a homogeneous slab with $\varepsilon_{r,w} = 4.2$ and a thickness of $l_w = 9$ cm. Two elongated vertical targets of different lengths are located in the inspected scenario at a distance of 1.6m from the wall. Their heights are 52 cm and 1.42m, respectively. The retrieved image is shown in Figure 7.6. As can be seen, both objects are correctly localized and differentiated.

Other strategies can also be followed. For instance, in frequency-domain beamforming/migration approaches, the presence of the wall can be taken into account by using the Green's function for the considered propagation medium [50, 52, 53], which will be further discussed in the next section. However, in these cases, simplifying assumptions on the electromagnetic model are usually

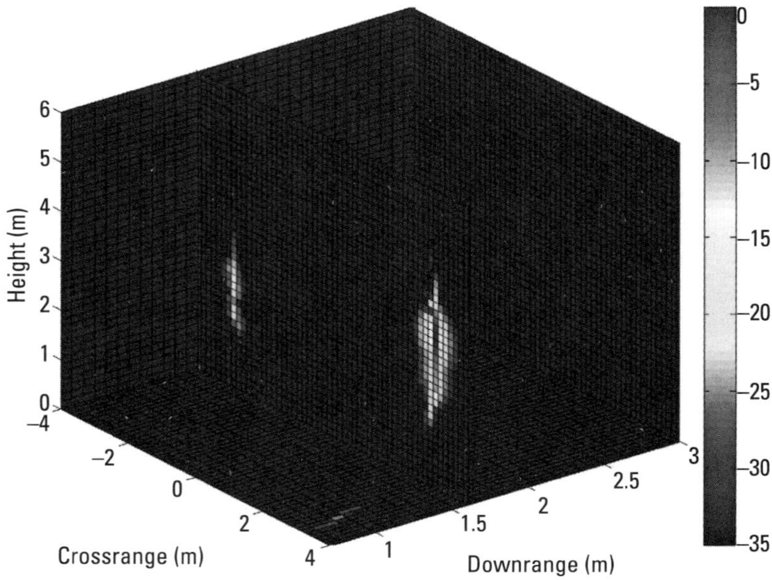

Figure 7.6 Example of reconstruction of two separate targets by using wall compensation to compute the time delays [42]. (© 2008 IEEE.)

made for deriving the imaging schemes. As an example, a possible way for obtaining the reconstructed image is to use the following relationship [50, 53]

$$I(\mathbf{r}_t) = \int_{D_{obs}} \int_{\Omega} E_{scatt,z}(x',\omega)\, g_{tw}^{-2}(\mathbf{r}_t, x'\hat{\mathbf{x}})\, d\omega\, dx', \mathbf{r}_t \in D \quad (7.13)$$

where g_{tw} is the Green's function for the considered wall model (which depends on ω) and a monostatic configuration is assumed. A similar expression can be found in the multistatic case. If the wall is modeled as homogeneous slab (or more in general, as a layered medium) and assuming that the target is in the far-field region of the antennas, the previous relationship can be further simplified by using the asymptotic forms of the Green's function, obtaining [50, 53]

$$I(\mathbf{r}_t) = \int_{D_{obs}} \int_{\Omega} E_{scatt,z}(x',\omega)\, T_w^{-2}(\omega)\, e^{\frac{j\omega}{v_0} 2|\mathbf{r}_t - x'\hat{\mathbf{x}}|}\, d\omega\, dx', \mathbf{r}_t \in D \quad (7.14)$$

where T_w is the transmission coefficient through the multilayer structure modeling the wall [36].

As for subsurface imaging, there is the need for extracting only the contributions of the scattering from the targets behind the wall from the measured data. The most straightforward way to remove the effects of unwanted scattering contributions and avoid the clutter in the reconstructed images is to subtract the incident field due to the target-free scenario. In TWI literature, such approach is often referred as background subtraction. Clearly, such a technique assumes that it is possible to acquire a set of measurements of the actual background scenario. Alternatively, the response of the considered configuration can be simulated (e.g., by solving a forward scattering problem involving a model of the target-free scenario). However, in many practical situations it is usually difficult to proceed in this way due to the unavailability of a full description of the scenario and for the high computational resources needed to simulate large regions.

In order to avoid measurements or simulations of the background scenario, approximated techniques can be used. Under the assumption that the unwanted contributions are mainly due to the wall, they can be removed by using a filtering procedure that basically eliminates the common features that are present in the signals received at different probing locations. One of the simplest filtering approaches is the average subtraction, already discussed in Section 6.6. Assuming a monostatic setup in which the field is collected in a set of points $\mathbf{r}_{meas}^{s}$, $s = 1, \ldots, S$, this can be achieved by approximating the scattered field as

$$e_{scatt,z}\left(\mathbf{r}_{meas}^{s},t\right) \cong e_{tot,z}\left(\mathbf{r}_{meas}^{s},t\right) - \frac{1}{S}\sum_{l=1}^{S} e_{tot,z}\left(\mathbf{r}_{meas}^{l},t\right) \tag{7.15}$$

An example of the application of the above approach is shown in Figure 7.7. The configuration is the same as in Figure 7.2. As can be seen, the wall reflections are eliminated from the *B*-scan and the hyperbola due to the target (a circular PEC cylinder) is preserved. However, comparing the result with the one shown in Figure 7.2(b), which reports the exact scattered field, it is possible to notice that some artifacts are present.

Time-gating procedures and more sophisticated approaches, similar to those developed for subsurface GPR imaging (e.g., those based on subspace and entropy processing), can also be used [7, 54–56].

Moreover, since the number of measurement locations must obey the Nyquist theorem, a large number of transmitting/receiving elements are usually needed. In order to partially overcome such a problem, compressive sensing techniques can be employed when dealing with sparse scenarios [12, 57–60], since they allow (in principle) obtaining good reconstructions even with a reduced number of measurements. In particular, two strategies can be adopted. In the first one, a limited set of scattered field data is acquired and the missing measurements are retrieved by using the compressive sensing theory [61]. Alternatively, compressive sensing can be applied for directly reconstructing the

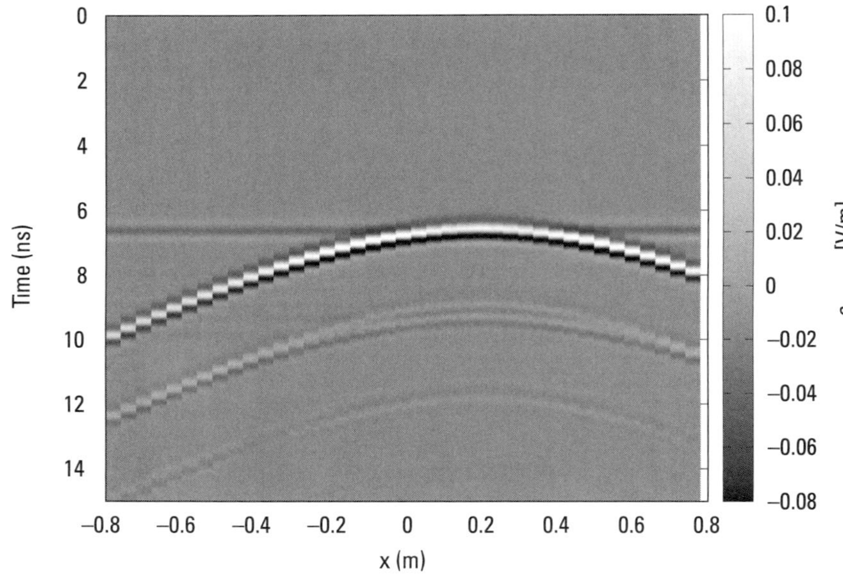

Figure 7.7 Example of estimation of the scattered field by means of average subtraction.

image of the scene under test (providing a proper linear model of the scattering problem is available) starting from the undersampled data [57].

7.2.3 Inverse Scattering Approaches for TWI

Beamforming approaches are widely used, since they are able to produce good results in terms of localization of the targets, and they are usually quite computationally effective for performing reconstructions of large areas. However, as already pointed out in the previous chapters, they may present some limitations. First of all, the propagation is usually described by using simplified scattering models and the presence of the wall is often compensated by using qualitative schemes. Moreover, they do not provide quantitative information about the type of targets that are located inside the inspected scene.

Quantitative inverse scattering-based techniques can be applied, too. Since, in most cases, they work in the frequency domain, frequency-stepped or frequency-modulated measurement systems are usually preferred. However, pulse-based systems can be used by extracting the data at the needed frequencies by means of Fourier transforms (as discussed in Chapter 6). Inverse scattering-based approaches are potentially able to retrieve the dielectric properties of the whole scene under inspection and include physically-based scattering models. In fact, by using the suitable Green's function for the considered configuration, it is possible to rigorously model the propagation and scattering phenomena. If the wall is homogenous, it is possible to employ the Green's function for layered media, which is given by [36]

$$g_{tw,h}\left(x,y,x',y'\right)=$$
$$\frac{j}{4\pi}\int_{-\infty}^{+\infty}\frac{e^{jk_x(x-x')}}{\gamma_0}\begin{cases}e^{-j\gamma_0|y-y'|}+Re^{-j\gamma_0(y+y')}dk_x & y\geq 0\\ Te^{j\gamma_0(y-y'+l_w)}dk_x & y\leq -l_w\end{cases} \tag{7.16}$$

where $\gamma_o=\sqrt{k_0^2-k_x^2}$, k_0 being the wavenumber in vacuum, and R, T are the generalized reflection and transmission coefficients, respectively, given by [36]

$$R=\frac{\rho_{aw}+\rho_{wa}e^{-2j\gamma_w l_w}}{1+\rho_{aw}\rho_{wa}e^{-2j\gamma_w l_w}}=\rho_{aw}\frac{1-e^{-2j\gamma_w l_w}}{1-\rho_{aw}^2e^{-2j\gamma_w l_w}} \tag{7.17}$$

$$T=\frac{\left(1+\rho_{aw}\right)\left(1+\rho_{wa}\right)e^{-j\tilde{a}_w l_w}}{1+\rho_{aw}\rho_{wa}e^{-2j\tilde{a}_w l_w}}=\frac{\left(1-\rho_{aw}^2\right)e^{-j\gamma_w l_w}}{1-\rho_{aw}^2e^{-2j\gamma_w l_w}} \tag{7.18}$$

with $\gamma_w=\sqrt{k_w^2-k_x^2}$ and

$$\rho_{aw}=-\rho_{wa}=\frac{\gamma_0-\gamma_w}{\gamma_0+\gamma_w} \tag{7.19}$$

When dealing with 3-D problems, an analogous expression can be devised for the dyadic Green's function [36]. Similar relationships can also be found in the case of walls modeled as multilayer structures. Some examples of the behavior of the 2-D Green's function for homogeneous walls made of different materials are reported below. In particular, dry concrete ($\varepsilon_{r,w}$ = 4.5, σ_w = 0.007 S/m, see Chapter 4), wet concrete ($\varepsilon_{r,w}$ = 6.2, σ_w = 0.05 S/m, 6.2% moisture content, see Chapter 4) and wood walls ($\varepsilon_{r,w}$ = 3, σ_w = 0.01 S/m) have been considered. The thickness of the wall is equal to l_w = 20 cm and a working frequency of 1 GHz is assumed. The source point is located in $\mathbf{r}_t'$ = (0, 20) cm and the Green's function is evaluated on two horizontal lines located on the opposite sides of the wall, that is, at y = 10 cm and y = –0.45 cm (the reference system shown in Figure 7.1 is assumed). The values of the amplitude and phase of the computed Green's functions are reported in Figures 7.8 and 7.9.

When more complex models of the walls are needed, the Green's function should be evaluated numerically. However, for some configurations, analytic

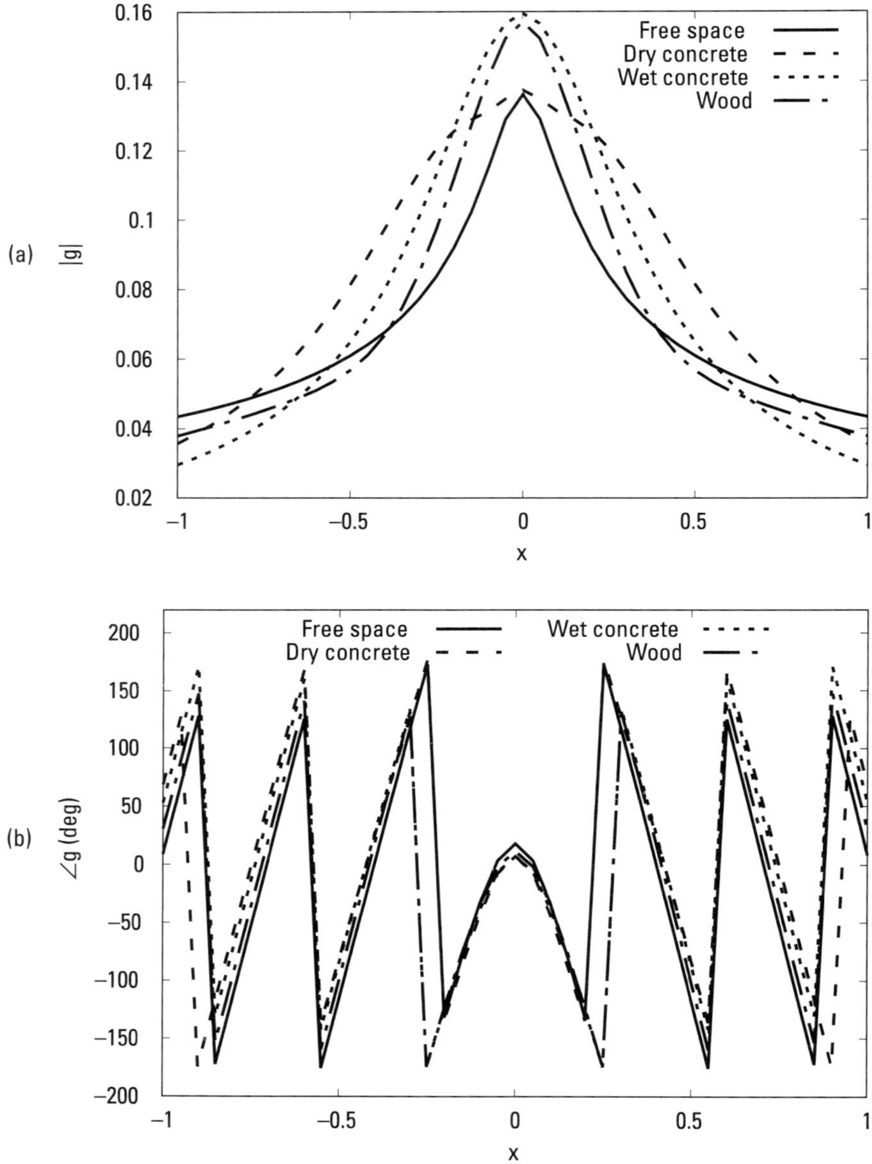

Figure 7.8 Green's function for a homogeneous wall made of different materials. (a) Amplitude and (b) phase for test points located on a horizontal line at $y = 10$ cm and source point in $\mathbf{r}_t' = (0, 20)$ cm.

expressions for the Green's function are still available. For the cinderblock wall represented by the periodic structures shown in Figure 7.3(c), the Green's func-

(a)

(b)

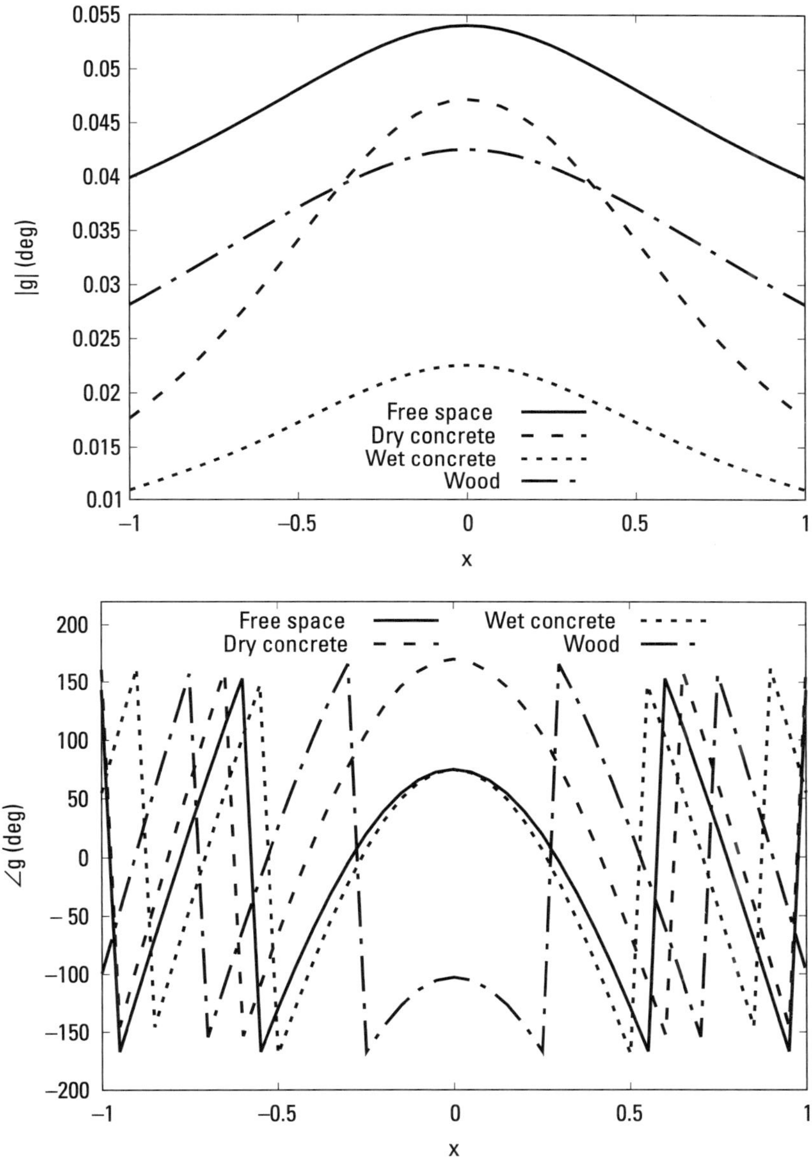

Figure 7.9 Green's function for a homogeneous wall made of different materials. (a) Amplitude and (b) phase for test points located on a horizontal line at y = –45 cm and source point at $\mathbf{r}_t' = (0, 20)$ cm.

tion can be obtained via an expansion in terms of discrete Floquét modes as [62, 63]

$$g_{tw,cb}(x,y,x',y') = \frac{j}{4\pi} \int_{-\frac{\pi}{2}}^{+\frac{\pi}{2}} e^{jk_0(x'\sin\beta + y'\cos\beta)} \left[\sum_{m=-\infty}^{\infty} T_m(\beta) e^{-jk_0(x\sin\beta_m + (y-l_w)\cos\beta_m)} \right] \cos\beta d\beta,\ y \le l_w, y \ge 0 \quad (7.20)$$

where T_m is the transmission coefficient associated to the Floquét mode [64, 62] and β_m is given by

$$\sin\beta_m = \sin\beta + \frac{2\pi m}{k_0 w_b},\ m = 0, \pm 1, \pm 2, \ldots \quad (7.21)$$

7.2.3.1 Techniques Based on Linearized Scattering Models

When the detection and localization of the concealed targets constitutes sufficient information, linearized models can be adopted. In particular, the first-order Born approximation (see Chapter 3) can be used [108], which is valid for weakly scattering objects. When it is a priori known that the targets to be inspected are metallic objects, the Kirchhoff approximation can be used (see [3.27]) [65, 66].

An example of a reconstructed image obtained by a linearized inversion scheme based on the Born approximation is reported below. A 2-D quasi-monostatic setup is considered. The distance between the transmitting and receiving antennas is equal to 4 cm. The standoff distance from the wall has been set equal to h = 10 cm. The probing line is 1.6m long and it has been discretized into 41 illumination/measurement positions. The wall has been modeled as a homogeneous slab with a thickness of l_w = 20 cm with relative dielectric permittivity $\varepsilon_{r,w}$ = 4 and electric conductivity σ_w = 0.01 S/m. The target is a circular cylinder with cross section center $\mathbf{r}_{t,c}$ = (0.2, –0.7)m, radius equal to d_c = 30 cm, and relative dielectric permittivity $\varepsilon_{r,c}$ = 3.5. The electric-field data have been simulated by using the *gprMax* FDTD software [24]. A Ricker pulse [67] with a central frequency of 1.5 GHz has been used to feed the antenna (modeled as an ideal line-current source). Frequency domain data at F = 5 frequencies equally distributed in the range between 1 and 2 GHz have been extracted by the Fourier transform and used in the inversion.

A square investigation domain with side 1m and centered at (0, –0.7)m has been considered. The scattering equations have been discretized by using the method of moments with pulse basis functions and Dirac's delta weighting functions. In particular, N = 30 × 30 square subdomains of side 3.3 cm have been considered in the inversion procedure. The linearized inverse scattering problem has been solved by using a truncated singular value decomposition (TSVD). The truncation threshold has been set equal to 20% of the maximum singular value. The qualitative reconstructed map is provided in Figure 7.10. In particular, the normalized amplitude of the contrast function c, that is, $c_n = |c| / \max|c|$, is shown. As can be seen, the presence of the target and its location are recognizable in the reconstructed image.

Another example concerns the same imaging scenario, in which a PEC cylinder is present. All the other parameters are unchanged. In this case, the Kirchhoff approximation is used and the 2-D map, obtained by using the TSVD, is shown in Figure 7.11. As can be seen in this case as well, the presence of the target is detected. Moreover, as expected, only the part of the boundary visible from the probing line is retrieved.

However, one of the main limitations of inverse scattering approaches is related to the high computational times that are usually required to perform the inversion, especially when dealing with 3-D vector formulations and large investigation domains. In order to avoid the inversion of the full 3-D model, sliced inversion procedures can be employed [10, 32, 68, 69]. In this case, it is assumed that the measurements are collected on a planar surface in front of the wall by using vertically polarized antennas. Subsequently, the measurements obtained at some fixed heights are used to perform several 2-D inversions in order to retrieve the 2-D images related to horizontal slices. Once all the slices have been reconstructed, a 3-D image is constructed by stacking all the 2-D images (eventually performing an interpolation along the vertical direction).

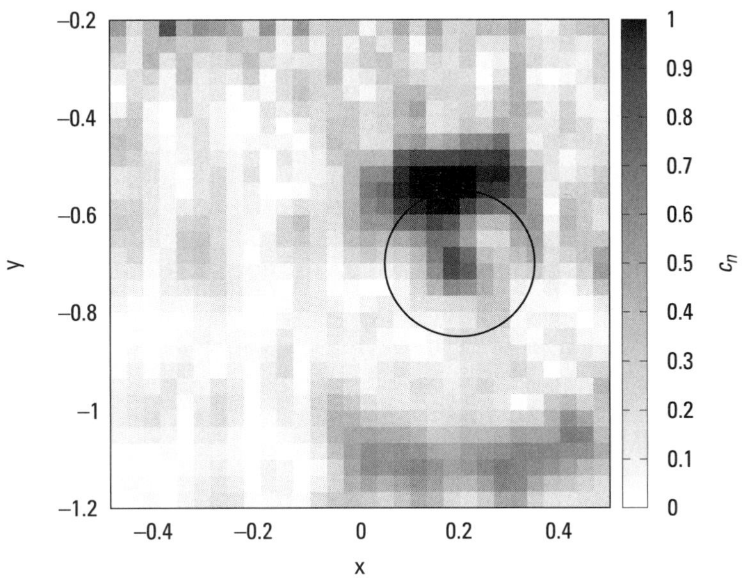

Figure 7.10 Example of a TSVD reconstruction of a circular dielectric cylinder.

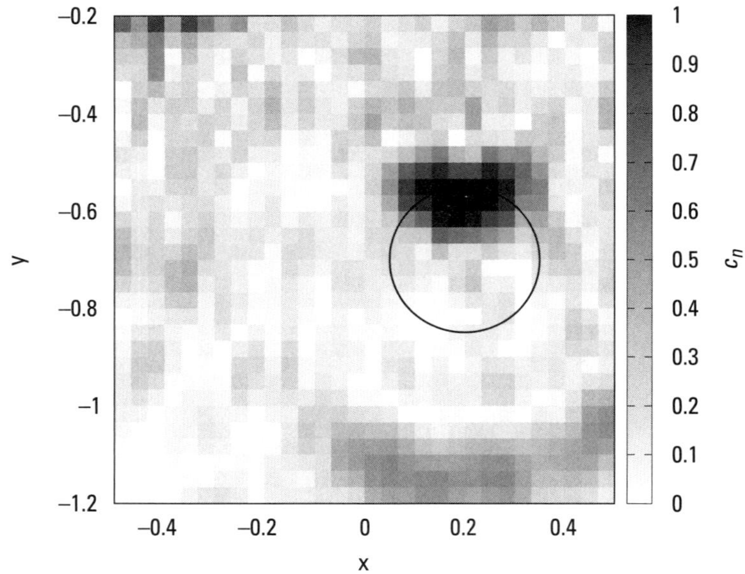

Figure 7.11 Example of TSVD reconstruction of a circular PEC dielectric cylinder.

Diffraction tomography (DT) based on the Born approximation can also be adopted for TWI. The main advantage of DT is that it can be implemented by using the fast Fourier transform, thus allowing one to obtain computationally efficient inversion procedures [70, 71]. In particular, both 2-D [72] and 3-D [11] versions have been developed for TWI applications in the case in which the wall is modeled as a multilayer structure. It is worth remarking that in the 3-D case, it is also possible to exploit different polarizations of the transmitting and receiving antennas in order to increase the available information and to mitigate the effects of the wall [11].

An example of 3-D reconstruction obtained by using the DT algorithm is shown in Figure 7.12 [11]. A homogeneous wall with thickness $l_w = 0.2$m, relative dielectric permittivity $\varepsilon_{w,r} = 6$, and electric conductivity $\sigma_w = 0.01$ S/m is considered. The scattered-field data are collected in 40×40 points uniformly distributed on a planar square surface of side 2m located at a distance of 0.3m from the wall. The target is a realistic human male model as shown in Figure 7.12(a). Synthetic data between 1 and 3 GHz (with step of 36 MHz) have been numerically simulated. The reconstruction obtained by the DT algorithm is shown in Figure 7.12(b).

7.2.3.2 Nonlinear Inversion Schemes

As discussed in Section 7.2.3.1, linearized inverse scattering methods usually allow one to retrieve only limited information about the target. In order to

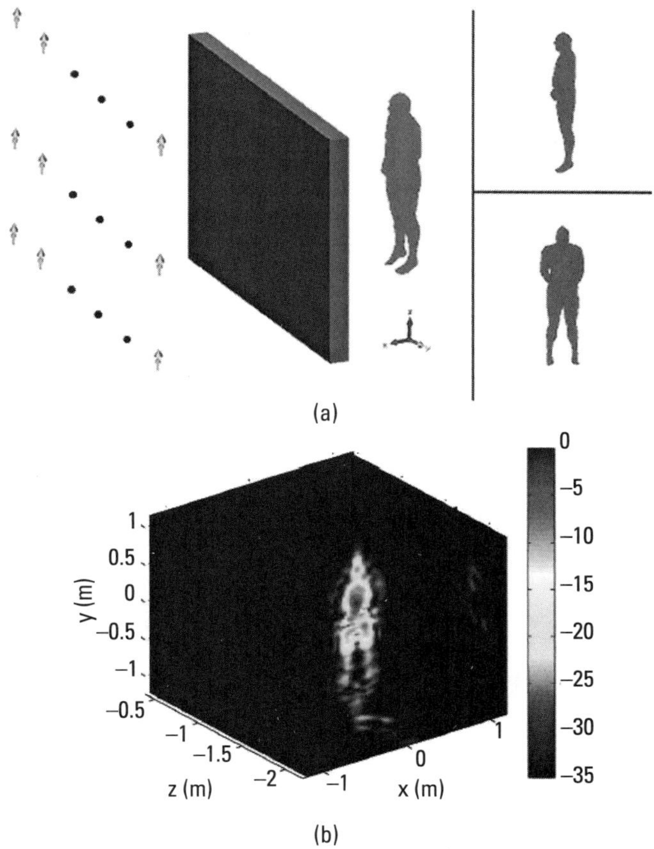

Figure 7.12 Example of a 3-D reconstruction of a realistic human target obtained by using diffraction tomography [11]. (a) Actual configuration and (b) normalized reconstructed values. (© 2013 IEEE.)

obtain quantitative information, it is necessary to fully address the nonlinearity of the scattering problem [73–75]. It is worth noting that in the field of TWI, nonlinear approaches are however still infrequent, mainly due to the high computational resources that are required to process large regions. Nevertheless, some full nonlinear approaches have been proposed in the literature by formulating the TWI inverse problem in terms of the exact scattering equations discussed in Chapter 3 (where the Green's function relevant for the considered TWI configuration must be adopted in place of g_b). For example, in [74, 75] a contrast-source formulation has been used to solve the inverse scattering problem related to the detection of targets hidden behind walls. Subspace-based optimization methods have also been proposed [76], as well as formulations based on electric-field integral equations.

An example of reconstruction obtained by using the nonlinear inversion scheme based on the L^p Banach-space procedure discussed in Chapter 3 is reported in the following. The imaging configuration is the one shown in Figure 7.1, with $l_w = 20$ cm, $\varepsilon_{r,w} = 4$, and $\sigma_w = 0.005$ S/m. The antennas are located on a probing line of length 2m at a distance of 1 cm from the wall. In particular, 41 positions along this line are considered. A subset of $S = 11$ source positions (equally distributed along the probing line) is used to sequentially illuminate the scene, whereas all the other points are used to collect the scattered field. The scattered field data have been numerically computed by means of a solver based on the method of moments [77]. A working frequency of 300 MHz (corresponding to the lower end of the frequency band considered in common TWI systems) has been used in this case. The investigation area is a rectangular domain of sides 2m × 1m, which has been discretized into $N = 1{,}800$ square subdomains of side 3.3 cm. The parameters of the inversion procedure have been set equal to: maximum number of outer iterations $N_{IN} = 10$; maximum number of inner iterations $N_{LW} = 30$.

Two cases have been considered. In the first case, a single circular target of diameter equal to $d_c = 30$ cm, relative dielectric permittivity $\varepsilon_{r,c} = 2$, and with cross section centered at $\mathbf{r}_{t,c} = (-0.2, -0.55)$ m has been assumed. The reconstructed distribution of the relative dielectric permittivity obtained with $p = 1.2$ is shown in Figure 7.13. As can be seen, the shape and positions of the target are retrieved with rather good accuracy. Moreover, the dielectric properties are also correctly estimated. A more complex situation has also been considered. In this second case, two separate targets are present inside the inspected region. The first one is again a circular cylinder with the same dielectric properties and dimensions as before, but centered at $\mathbf{r}_{t,c} = (-0.2, -0.65)$m . The second target is

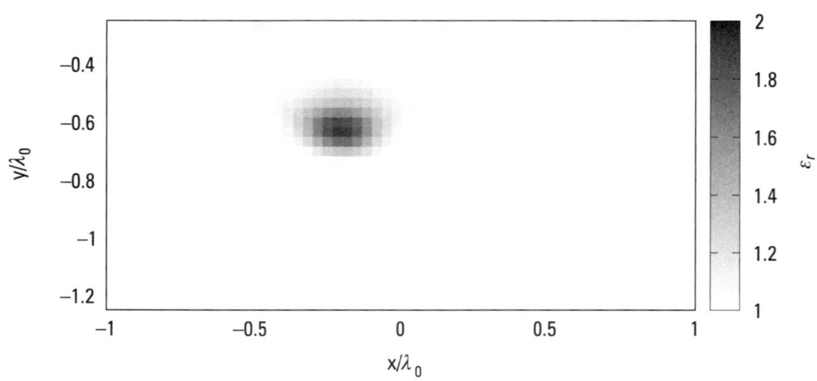

Figure 7.13 Example of 2-D reconstruction by using a nonlinear TWI procedure based on the Banach-space inversion method described in Chapter 3 (with p = 1.2). Target is a single circular dielectric cylinder.

a square cylinder of sides 30 cm × 10 cm, relative dielectric permittivity equal to $\varepsilon_{r,s} = 3$, and located at $\mathbf{r}_{t,s} = (0.4, -0.55)$m. The reconstructed distribution of the relative dielectric permittivity provided by the considered nonlinear inversion procedure with $p = 1.2$ is shown in Figure 7.14. As can be observed, the two targets are correctly located. Moreover, the values of the dielectric properties are correctly differentiated, allowing one to infer that two different kinds of targets are present. However, in this case, the shapes of the objects, especially for what concerns the rectangular one, are not fully retrieved (although some information about their extent can be still obtained).

7.2.3.3 Other Inversion Approaches

Other imaging approaches can also be used in TWI applications. In this framework, different time-reversal techniques, based on TR-MUSIC and DORT, modified in order to use the Green's function for multilayer structures, have been proposed for obtaining qualitative images of the inspected scenario [78–82]. Stochastic inversion procedures can also be adopted to retrieve the external shape of the target. For example, in [83] a method based on the differential evolution algorithm [84, 85] is used for reconstructing the shape of metallic targets, modeled by using *B*-splines. The LSM has also been proposed for retrieving the shape of targets in TWI applications [86, 87]. Similarly to other applications, it has potentially the ability to reconstruct the support of the target, but its performance may be degraded by the availability of only limited-view data. An example of an application of the LSM is reported below. The single-cylinder configuration considered in Section 7.2.3.2 for the Banach-space inversion example is used. A full multistatic setup, with $S = 21$ antennas located on a probing line of length 2m at a distance of 1 cm from the wall, is considered. Each

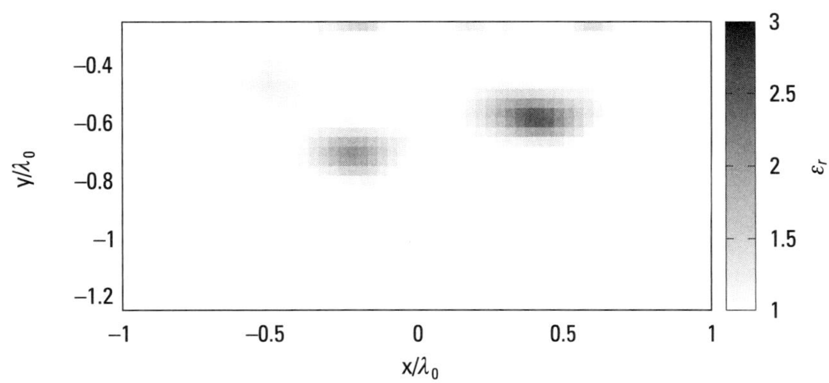

Figure 7.14 Example of 2-D reconstruction by using a nonlinear TWI procedure based on the Banach-space inversion method described in Chapter 3 (with p = 1.2). Two targets, a circular and a rectangular dielectric cylinder, are present.

antenna acts in turn as transmitter, whereas the remaining $M = 20$ ones are used to collect the field produced by the interaction with the target. In this case as well, the scattered field data have been numerically computed by using a solver based on the method of moments [77]. The indicator function provided by the LSM is shown in Figure 7.15. Clearly, in this case, the LSM is able to correctly locate the target and to provide an indication of its extent.

Finally, it should be mentioned that learning-based approaches have also been considered for TWI. For example, in [88, 89] it is assumed that the target is a circular cylinder and the aim is to retrieve the cross section center, the radius, and the dielectric properties from scattered field measurements collected in a multistatic configuration. Such a problem can be cast as the identification of a vector function $\mathbf{F}(\mathbf{x})$ describing the inverse relationship between the measurements (arranged in the array $\mathbf{x}$) and the parameters of interest. To this end, a black-box approach is used, in which the unknown function $\mathbf{F}$ is approximated by using a regression procedure based on SVMs [90, 91]. In particular, in the support vector regression, the ith component f_i of $\mathbf{F}$ is approximated as

$$f_i(\mathbf{x}) = \sum_{n=1}^{N_{SV}} (\alpha_n - \alpha'_n)\psi(\mathbf{x}, \mathbf{x}_n) + b \tag{7.22}$$

where N_{SV} is the number of support vectors $\mathbf{x}_n$ and ψ is a kernel function (e.g., a Gaussian function is often employed) [90, 91]. The parameters α_n, α'_n, and b are found by solving a minimization problem starting from a set of training data, that is, a set of arrays of measured data for which the target configurations are known. After the SVMs are trained, (7.22) can be used to estimate the geometrical and dielectric parameters of the target from new measured data.

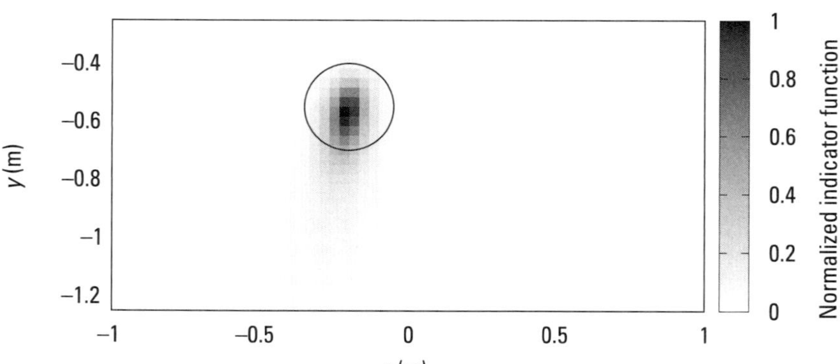

Figure 7.15 Example of a 2-D reconstruction obtained by using the LSM. Target is a single circular dielectric cylinder.

7.2.4 Overview of Practical Implementations of TWI Systems

Similarly to subsurface imaging, TWI apparatuses are typically based on radar systems. In particular, as already discussed, both pulsed- and frequency-modulated apparatuses can be used. Some examples of working modalities and frequency ranges of recently proposed TWI prototypes are reported in Table 7.1. In most cases, the used frequency bands are in the range between 0.3 and 3 GHz. Furthermore, Table 7.2 reports some examples of commercial TWI systems. Concerning the commercial devices, systems providing full imaging capabilities are available. Simplified apparatuses, able to provide only a detection of the presence of targets behind the wall (eventually with indication of their movements derived from Doppler measurements), are also often provided by the manufacturers. Concerning the radiating structures used in TWI systems, due to the necessity of having high gain and large bandwidth, Vivaldi antennas are often adopted [92–96]. Horn antennas can be also used [97, 98], although their bigger size could increase the overall dimensions of the system. Specifically-designed patch antennas have also been applied [95].

7.3 Concealed Weapon Detection

As discussed at the beginning of the Chapter, personnel surveillance and concealed target detection are important areas in which imaging systems play a fundamental role [2, 5, 103]. In this framework, microwaves are attracting a

Table 7.1
Operational Data of Some Recently Proposed TWI Systems and Laboratory Prototypes

Frequency Band	Waveform Type	Reference
2–4 GHz	FMCW	[99]
0.3–3 GHz	Pulsed (UWB)	[96, 100]
0.7–3.1 GHz	SFCW	[54, 101]
0.5–2 GHz,	FMCW	[102]
1–2.1 GHz	SFCW	[97]

Table 7.2
Examples of Commercially Available TWI Systems

Brand	Model	Reference
Camero	Xaver series	UWB pulse system (Frequency range: 3–10 GHz). Life detection (Xaver 100) and imaging (Xaver 400 and 800) capabilities.
L3 Cyterra	Range-R	SFCW radar technology. Target and motion detector.
Cambrate	PRISM 200	Frequency range: 1.7–2.2 GHz. Imaging capabilities.

growing interest, thanks to their ability to penetrate dielectric materials (e.g., clothes or envelopes) while at the same time remaining nonhazardous for the screened people. Some of the main requirements in surveillance applications are related to the need for a quick reconstruction procedure and with a high resolution. Concerning the latter, frequencies ranging from a few GHz to over 100 GHz are typically considered. For the reconstruction procedure, migration and synthetic aperture processing techniques (e.g., based on the F-K migration discussed in Chapter 6) are often used [2, 104, 105]. Specific modifications of the basic algorithms have also been proposed in order to increase the quality of the retrieved images (for example in [3, 106, 107]). In particular, in the considered application the scattered field data are usually collected on a 2-D planar aperture, in order to create 3-D images of the inspected scenario.

Some examples of imaging of different targets obtained by using migration methods are discussed in [107]. The electric field data are collected on a 2-D square planar surface with sides equal to 75 cm (with a separation between two adjacent measurement positions of 1 cm) by using two vertically polarized Vivaldi antennas separated by 5.5 cm. The measurements have been acquired for 161 frequencies in the range 4–20 GHz by using a vector network analyzer. The considered targets are shown in Figure 7.16. In particular, five different objects (a knife, a laser measure, a handgun, a bottle of water, and a set of keys) are attached to a foam board at 60 cm from the measurement plane. As an example,

Figure 7.16 Experimental setup for weapon detection with five different targets [107]. (© 2015 IEEE.)

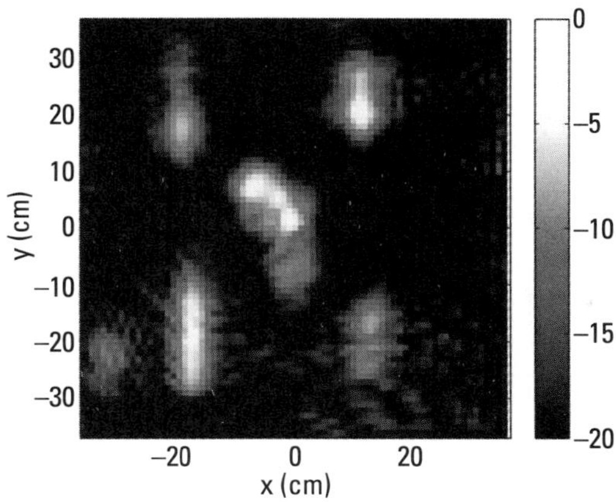

Figure 7.17 Image of the weapon detection setup in Figure 7.16 obtained by using the F-K migration algorithm [107]. (© 2015 IEEE.)

the image obtained by applying the F-K migration algorithm is shown in Figure 7.17. As can be seen, the five objects are correctly located and their shape can be identified, although as expected some blurring is present.

References

[1] Farhat, N. H., and W. R. Guard, "Millimeter Wave Holographic Imaging of Concealed Weapons," *Proc. IEEE,* Vol. 59, No. 9, September 1971, pp. 1383–1384.

[2] Sheen, D. M., D. L. McMakin, and T. E. Hall, "Three-Dimensional Millimeter-Wave Imaging for Concealed Weapon Detection," *IEEE Trans. Microw. Theory Techn.*, Vol. 49, No. 9, September 2001, pp. 1581–1592.

[3] Tan, W., et al., "Three-Dimensional Microwave Imaging for Concealed Weapon Detection Using Range Stacking Technique," *Int. J. Antennas Propag.*, Vol. 2017, 2017, pp. 1–11.

[4] Catapano, I., and L. Crocco, "An Imaging Method for Concealed Targets," *IEEE Trans. Geosci. Remote Sens.*, Vol. 47, No. 5, May 2009, pp. 1301–1309.

[5] Daniels, D. J., *EM Detection of Concealed Targets,* Hoboken, NJ: Wiley, 2010.

[6] Baranoski, E. J., "Through-Wall Imaging: Historical Perspective and Future Directions," *J. Frankl. Inst.*, Vol. 345, No. 6, September 2008, pp. 556–569.

[7] Amin, M. G. (ed.), *Through-the-Wall Radar Imaging,* Boca Raton, FL: CRC Press, 2011.

[8] Peabody, J. E., G. L. Charvat, J. Goodwin, and M. Tobias, "Through-Wall Imaging Radar," *Linc. Lab. J.*, Vol. 19, No. 1, 2012, pp. 62–72.

[9] Amin, M. G., and F. Ahmad, "Wideband Synthetic Aperture Beamforming for Through-the-Wall Imaging," *IEEE Signal Process. Mag.*, Vol. 25, No. 4, July 2008, pp. 110–113.

[10] Soldovieri, F., R. Solimene, and F. Ahmad, "Sparse Tomographic Inverse Scattering Approach for Through-the-Wall Radar Imaging," *IEEE Trans. Instrum. Meas.*, Vol. 61, No. 12, December 2012, pp. 3340–3350.

[11] Zhang, W., and A. Hoorfar, "Three-Dimensional Real-Time Through-the-Wall Radar Imaging with Diffraction Tomographic Algorithm," *IEEE Trans. Geosci. Remote Sens.*, Vol. 51, No. 7, July 2013, pp. 4155–4163.

[12] Wu, Q., Y. D. Zhang, F. Ahmad, and M. G. Amin, "Compressive-Sensing-Based High-Resolution Polarimetric Through-the-Wall Radar Imaging Exploiting Target Characteristics," *IEEE Antennas Wireless Propag. Lett.*, Vol. 14, 2015, pp. 1043–1047.

[13] DellAversano, A., A. Natale, A. Buonanno, and R. Solimene, "Through the Wall Breathing Detection by Means of a Doppler Radar and MUSIC Algorithm," *IEEE Sens. Lett.*, Vol. 1, No. 3, June 2017, pp. 1–4.

[14] Debnath, S. B. C., "TTW Life Sign Detection by Means of CW X-Band Radar, Homeland Security and Rescue Applications," in *Proc. USNC-URSI Radio Science Meeting*, San Diego, CA, 2017, pp. 89–90.

[15] Xu, Y., et al., "A Novel Method for Automatic Detection of Trapped Victims by Ultrawideband Radar," *IEEE Trans. Geosci. Remote Sens.*, Vol. 50, No. 8, August 2012, pp. 3132–3142.

[16] Chen, K.-M., Y. Huang, J. Zhang, and A. Norman, "Microwave Life-Detection Systems for Searching Human Subjects under Earthquake Rubble or Behind Barrier," *IEEE Trans. Biomed. Eng.*, Vol. 47, No. 1, January 2000, pp. 105–114.

[17] Ahmad, F., and M. G. Amin, "Through-the-Wall Human Motion Indication Using Sparsity-Driven Change Detection," *IEEE Trans. Geosci. Remote Sens.*, Vol. 51, No. 2, February 2013, pp. 881–890.

[18] Ram, S. S., and H. Ling, "Through-Wall Tracking of Human Movers Using Joint Doppler and Array Processing," *IEEE Geosci. Remote Sens. Lett.*, Vol. 5, No. 3, July 2008, pp. 537–541.

[19] Hunt, A. R., "Use of a Frequency-Hopping Radar for Imaging and Motion Detection Through Walls," *IEEE Trans. Geosci. Remote Sens.*, Vol. 47, No. 5, May 2009, pp. 1402–1408.

[20] Amin, M. G., and F. Ahmad, "Change Detection Analysis of Humans Moving Behind Walls," *IEEE Trans. Aerosp. Electron. Syst.*, Vol. 49, No. 3, July 2013, pp. 1410–1425.

[21] Soldovieri, F., R. Solimene, and R. Pierri, "A Simple Strategy to Detect Changes in Through the Wall Imaging," *Prog. Electromagn. Res. M*, Vol. 7, 2009, pp. 1–13.

[22] Gennarelli, G., et al., "Multiple Extended Target Tracking for Through-Wall Radars," *IEEE Trans. Geosci. Remote Sens.*, Vol. 53, No. 12, December 2015, pp. 6482–6494.

[23] Zhang, X., X. Xi, M. Li, and D. Wu, "Comparison of Impulse Radar and Spread-Spectrum Radar in Through-Wall Imaging," *IEEE Trans. Microw. Theory Techn.*, Vol. 64, No. 3, March 2016, pp. 699–706.

[24] Warren, C., A. Giannopoulos, and I. Giannakis, "gprMax: Open Source Software to Simulate Electromagnetic Wave Propagation for Ground Penetrating Radar," *Comput. Phys. Commun.*, Vol. 209, December 2016, pp. 163–170.

[25] Thajudeen, C., A. Hoorfar, F. Ahmad, and T. Dogaru, "Measured Complex Permittivity of Walls with Different Hydration Levels and the Effect on Power Estimation of Twri Target Returns," *Prog. Electromagn. Res. B*, Vol. 30, 2011, pp. 177–199.

[26] Stavrou, S., and S. R. Saunders, "Review of Constitutive Parameters of Building Materials," in *Proc. Twelfth International Conference on Antennas and Propagation (ICAP 2003),* Vol. 1, Exeter, UK, 2003, pp. 211–215.

[27] Wang, G., and M. G. Amin, "Imaging Through Unknown Walls Using Different Standoff Distances," *IEEE Trans. Signal Process.*, Vol. 54, No. 10, October 2006, pp. 4015–4025.

[28] Li, L., W. Zhang, and F. Li, "A Novel Autofocusing Approach for Real-Time Through-Wall Imaging Under Unknown Wall Characteristics," *IEEE Trans. Geosci. Remote Sens.*, Vol. 48, No. 1, January 2010, pp. 423–431.

[29] Fereidoony, F., S. A. Mirtaheri, and S. Chamaani, "M-Sequence Sensor and Continuous Basis Estimator for Wall Parameter Estimation Utilizing Through the Wall Sensing," *IEEE Sens. J.*, Vol. 17, No. 13, July 2017, pp. 4083–4091.

[30] Protiva, P., J. Mrkvica, and J. Machac, "Estimation of Wall Parameters from Time-Delay-Only Through-Wall Radar Measurements," *IEEE Trans. Antennas Propag.*, Vol. 59, No. 11, November 2011, pp. 4268–4278.

[31] Thajudeen, C., and A. Hoorfar, "A Hybrid Bistatic–Monostatic Radar Technique for Calibration-Free Estimation of Lossy Wall Parameters," *IEEE Antennas Wireless Propag. Lett.*, Vol. 16, 2017, pp. 1249–1252.

[32] Solimene, R., F. Soldovieri, G. Prisco, and R. Pierri, "Three-Dimensional Through-Wall Imaging Under Ambiguous Wall Parameters," *IEEE Trans. Geosci. Remote Sens.*, Vol. 47, No. 5, May 2009, pp. 1310–1317.

[33] Chang, P. C., R. J. Burkholder, and J. L. Volakis, "Adaptive CLEAN with Target Refocusing for Through-Wall Image Improvement," *IEEE Trans. Antennas Propag.*, Vol. 58, No. 1, January 2010, pp. 155–162.

[34] Sun, M., C. Le Bastard, Y. Wang, and N. Pinel, "Time-Delay Estimation Using ESPRIT with Extended Improved Spatial Smoothing Techniques for Radar Signals," *IEEE Geosci. Remote Sens. Lett.*, Vol. 13, No. 1, January 2016, pp. 73–77.

[35] Zhang, H. M., et al., "An Efficient Method Based on Machine Learning for Estimation of the Wall Parameters in Through-the-Wall Imaging," *Int. J. Remote Sens.*, Vol. 37, No. 13, July 2016, pp. 3061–3073.

[36] Chew, W. C., *Waves and Fields in Inhomogenous Media*, Hoboken, NJ: Wiley, 1999.

[37] Lambot, S., et al., "Modeling of Ground-Penetrating Radar for Accurate Characterization of subsurface Electric Properties," *IEEE Trans. Geosci. Remote Sens.*, Vol. 42, No. 11, November 2004, pp. 2555–2568.

[38] Lambot, S., and F. André, "Full-Wave Modeling of Near-Field Radar Data for Planar Layered Media Reconstruction," *IEEE Trans. Geosci. Remote Sens.*, Vol. 52, No. 5, May 2014, pp. 2295–2303.

[39] Yoon, Y.-S., M. G. Amin, and F. Ahmad, "MVDR Beamforming for Through-the-Wall Radar Imaging," *IEEE Trans. Aerosp. Electron. Syst.*, Vol. 47, No. 1, January 2011, pp. 347–366.

[40] Catapano, I., et al., "On the Reconstruction Capabilities of Beamforming and a Microwave Tomographic Approach," *IEEE Geosci. Remote Sens. Lett.*, Vol. 12, No. 12, December 2015, pp. 2369–2373.

[41] Zhang, W., and A. Hoorfar, "Three-Dimensional Synthetic Aperture Radar Imaging Through Multilayered Walls," *IEEE Trans. Antennas Propag.*, Vol. 62, No. 1, January 2014, pp. 459–462.

[42] Ahmad, F., Y. Zhang, and M. G. Amin, "Three-Dimensional Wideband Beamforming for Imaging Through a Single Wall," *IEEE Geosci. Remote Sens. Lett.*, Vol. 5, No. 2, April 2008, pp. 176–179.

[43] Ahmad, F., M. G. Amin, and S. A. Kassam, "Synthetic Aperture Beamformer for Imaging Through a Dielectric Wall," *IEEE Trans. Aerosp. Electron. Syst.*, Vol. 41, No. 1, January 2005, pp. 271–283.

[44] Solimene, R., et al., "SAR Imaging Algorithms and Some Unconventional Applications: A Unified Mathematical Overview," *IEEE Signal Process. Mag.*, Vol. 31, No. 4, July 2014, pp. 90–98.

[45] Boudamouz, B., P. Millot, and C. Pichot, "Through the Wall Radar Imaging with MIMO Beamforming Processing—Simulation and Experimental Results," *Am. J. Remote Sens.*, Vol. 1, No. 1, February 2013, pp. 7–12.

[46] Charvat, G. L., L. C. Kempel, E. J. Rothwell, C. M. Coleman, and E. L. Mokole, "A Through-Dielectric Radar Imaging System," *IEEE Trans. Antennas Propag.*, Vol. 58, No. 8, August 2010, pp. 2594–2603.

[47] Hantscher, S., A. Reisenzahn, and C. G. Diskus, "Through-Wall Imaging with a 3-D UWB SAR Algorithm," *IEEE Signal Process. Lett.*, Vol. 15, 2008, pp. 269–272.

[48] Dehmollaian, M., M. Thiel, and K. Sarabandi, "Through-the-Wall Imaging Using Differential SAR," *IEEE Trans. Geosci. Remote Sens.*, Vol. 47, No. 5, May 2009, pp. 1289–1296.

[49] Zhang, P., and X. Zhang, "A Modified Phase Shift Migration Algorithm Applied to 2D and 3D Through-the-Wall Imaging," *IEEE J. Sel. Top. Appl. Earth Obs. Remote Sens.*, Vol. 7, No. 5, May 2014, pp. 1700–1708.

[50] Dehmollaian, M., and K. Sarabandi, "Refocusing Through Building Walls Using Synthetic Aperture Radar," *IEEE Trans. Geosci. Remote Sens.*, Vol. 46, No. 6, June 2008, pp. 1589–1599.

[51] Ahmad, F., M. G. Amin, and G. Mandapati, "Autofocusing of Through-the-Wall Radar Imagery Under Unknown Wall Characteristics," *IEEE Trans. Image Process.*, Vol. 16, No. 7, July 2007, pp. 1785–1795.

[52] Jofre, L., et al., "UWB Tomographic Radar Imaging of Penetrable and Impenetrable Objects," *Proc. IEEE*, Vol. 97, No. 2, February 2009, pp. 451–464.

[53] Zhang, W., A. Hoorfar, C. Thajudeen, and F. Ahmad, "Full Polarimetric Beam-Forming Algorithm for Through-the-Wall Radar Imaging," *Radio Sci.*, Vol. 46, No. 5, October 2011, p. RS0E16.

[54] Tivive, F. H. C., A. Bouzerdoum, and M. G. Amin, "A Subspace Projection Approach for Wall Clutter Mitigation in Through-the-Wall Radar Imaging," *IEEE Trans. Geosci. Remote Sens.*, Vol. 53, No. 4, April 2015, pp. 2108–2122.

[55] Yoon, Y.-S., and M. G. Amin, "Spatial Filtering for Wall-Clutter Mitigation in Through-the-Wall Radar Imaging," *IEEE Trans. Geosci. Remote Sens.*, Vol. 47, No. 9, September 2009, pp. 3192–3208.

[56] Solimene, R., and A. Cuccaro, "Front Wall Clutter Rejection Methods in TWI," *IEEE Geosci. Remote Sens. Lett.*, Vol. 11, No. 6, June 2014, pp. 1158–1162.

[57] Huang, Q., L. Qu, B. Wu, and G. Fang, "UWB Through-Wall Imaging Based on Compressive Sensing," *IEEE Trans. Geosci. Remote Sens.*, Vol. 48, No. 3, March 2010, pp. 1408–1415.

[58] Amin, M. G., and F. Ahmad, "Compressive Sensing for Through-the-Wall Radar Imaging," *J. Electron. Imaging*, Vol. 22, No. 3, July 2013, p. 030901.

[59] Li, G., and R. J. Burkholder, "Hybrid Matching Pursuit for Distributed Through-Wall Radar Imaging," *IEEE Trans. Antennas Propag.*, Vol. 63, No. 4, April 2015, pp. 1701–1711.

[60] Zhang, W., and A. Hoorfar, "A Generalized Approach for SAR and MIMO Radar Imaging of Building Interior Targets With Compressive Sensing," *IEEE Antennas Wireless Propag. Lett.*, Vol. 14, 2015, pp. 1052–1055.

[61] Yoon, Y. S., and M. G. Amin, "Through-the-Wall Radar Imaging Using Compressive Sensing Along Temporal Frequency Domain," in *Proc. IEEE International Conference on Acoustics, Speech and Signal Processing*, Dallas, Texas, 2010, pp. 2806–2809.

[62] Chang, P. C., R. J. Burkholder, and J. L. Volakis, "Model-Corrected Microwave Imaging through Periodic Wall Structures," *Int. J. Antennas Propag.*, Vol. 2012, Article ID 948365, pp. 1–7.

[63] Gennarelli, G., and F. Soldovieri, "Radar Imaging through Cinderblock Walls: Achievable Performance by a Model-Corrected Linear Inverse Scattering Approach," *IEEE Trans. Geosci. Remote Sens.*, Vol. 52, No. 10, October 2014, pp. 6738–6749.

[64] Honcharenko, W., and H. L. Bertoni, "Transmission and Reflection Characteristics at Concrete Block Walls in the UHF Bands Proposed for Future PCS," *IEEE Trans. Antennas Propag.*, Vol. 42, No. 2, February 1994, pp. 232–239.

[65] Pierri, R., A. Liseno, R. Solimene, and F. Soldovieri, "Beyond Physical Optics SVD Shape Reconstruction of Metallic Cylinders," *IEEE Trans. Antennas Propag.*, Vol. 54, No. 2, February 2006, pp. 655–665.

[66] Pierri, R., A. Liseno, and F. Soldovieri, "Shape Reconstruction from PO Multifrequency Scattered Fields via the Singular Value Decomposition Approach," *IEEE Trans. Antennas Propag.*, Vol. 49, No. 9, September 2001, pp. 1333–1343.

[67] Daniels, D. J., *Ground Penetrating Radar*, 2nd ed., London: Institution of Electrical Engineers, 2004.

[68] Solimene, R., A. Brancaccio, R. Di Napoli, and R. Pierri, "3D Sliced Tomographic Inverse Scattering Experimental Results," *Prog. Electromagn. Res.*, Vol. 105, 2010, pp. 1–13.

[69] Soldovieri, F., F. Ahmad, and R. Solimene, "Validation of Microwave Tomographic Inverse Scattering Approach via Through-the-Wall Experiments in Semicontrolled Conditions," *IEEE Geosci. Remote Sens. Lett.*, Vol. 8, No. 1, January 2011, pp. 123–127.

[70] Deming, R. W., and A. J. Devaney, "Diffraction Tomography for Multi-Monostatic Ground Penetrating Radar Imaging," *Inverse Probl.*, Vol. 13, No. 1, 1997, p. 29.

[71] Cui, T. J., and W. C. Chew, "Novel Diffraction Tomographic Algorithm for Imaging Two-Dimensional Targets Buried under a Lossy Earth," *IEEE Trans. Geosci. Remote Sens.*, Vol. 38, No. 4, July 2000, pp. 2033–2041.

[72] Zhang, W., and A. Hoorfar, "Two-Dimensional Diffraction Tomographic Algorithm for Through-the-Wall Radar Imaging," *Prog. Electromagn. Res.*, Vol. 31, 2011, pp. 205–218.

[73] Meng, Q. Y., et al., "Two-Dimensional through-wall Imaging Based on Nonlinear Inversion Algorithms," in *Proc. International Applied Computational Electromagnetics Society Symposium (ACES)*, Suzhou, China, 2017, pp. 1–2.

[74] Song, L.-P., Chun Yu, and Qing Huo Liu, "Through-Wall Imaging (TWI) by Radar: 2-D Tomographic Results And Analyses," *IEEE Trans. Geosci. Remote Sens.*, Vol. 43, No. 12, December 2005, pp. 2793–2798.

[75] Abubakar, A., W. Hu, P. M. van den Berg, and T. M. Habashy, "A Finite-Difference Contrast Source Inversion Method," *Inverse Probl.*, Vol. 24, No. 6, December 2008, p. 065004.

[76] Lu, T., K. Agarwal, Y. Zhong, and X. Chen, "Through-Wall Imaging: Application of Subspace-Based Optimization Method," *Prog. Electromagn. Res.*, Vol. 102, 2010, pp. 351–366.

[77] Harrington, R. F., *Field Computation by Moment Methods*, Piscataway, NJ: IEEE, 1993.

[78] Zhang, W., A. Hoorfar, and L. Li, "Through-the-Wall Target Localization with Time Reversal MUSIC Method," *Prog. Electromagn. Res.*, Vol. 106, 2010, pp. 75–89.

[79] Maaref, N., et al., "Electromagnetic Imaging Method Based on Time Reversal Processing Applied to Through-the-Wall Target Localization," *Prog. Electromagn. Res.*, Vol. 1, 2008, pp. 59–67.

[80] Davy, M., et al., "Detection and Imaging of Human Beings Behind a Wall Using the Dort Method," *Prog. Electromagn. Res.*, Vol. 110, 2010, pp. 353–369.

[81] Yoon, Y.-S., and M. G. Amin, "High-resolution Through-the-Wall Radar Imaging Using Beamspace MUSIC," *IEEE Trans. Antennas Propag.*, Vol. 56, No. 6, June 2008, pp. 1763–1774.

[82] Odedo, V. C., et al., "Time Reversal Technique Based on Spatiotemporal Windows for Through the Wall Imaging," *IEEE Trans. Antennas Propag.*, Vol. 65, No. 6, June 2017, pp. 3065–3072.

[83] Dehmollaian, M., "Through-Wall Shape Reconstruction and Wall Parameters Estimation Using Differential Evolution," *IEEE Geosci. Remote Sens. Lett.*, Vol. 8, No. 2, March 2011, pp. 201–205.

[84] Price, K. V., "An Introduction to Differential Evolution," in *New Ideas in Optimization*, Maidenhead, UK: McGraw-Hill, pp. 79–108, 1999.

[85] Qing, A., and C. K. Lee, *Differential Evolution in Electromagnetics*, Springer, 2010.

[86] Catapano, I., and L. Crocco, "A Qualitative Inverse Scattering Method for Through-the-Wall Imaging," *IEEE Geosci. Remote Sens. Lett.*, Vol. 7, No. 4, October 2010, pp. 685–689.

[87] Eskandari, M. R., R. Safian, and M. Dehmollaian, "Three-Dimensional Near-Field Microwave Imaging Using Hybrid Linear Sampling and Level Set Methods in a Medium with Compact Support," *IEEE Trans. Antennas Propag.*, Vol. 62, No. 10, October 2014, pp. 5117–5125.

[88] Wang, F.-F., and Y.-R. Zhang, "The Support Vector Machine for Dielectric Target Detection through a Wall," *Prog. Electromagn. Res.*, Vol. 23, 2011, pp. 119–128.

[89] Wang, F.-F., and Y.-R. Zhang, "A Real-Time Through-Wall Detection Based on Support Vector Machine," *J. Electromagn. Waves Appl.*, Vol. 25, No. 1, January 2011, pp. 75–84.

[90] Vapnik, V. N., *Statistical Learning Theory*, Hoboken, NJ: Wiley, 1998.

[91] Smola, A. J., and B. Schölkopf, "A Tutorial on Support Vector Regression," *Stat. Comput.*, Vol. 14, No. 3, August 2004, pp. 199–222.

[92] Yang, Y., Y. Wang, and A. E. Fathy, "Design of Compact Vivaldi Antenna Arrays for UWB See through Wall Applications," *Prog. Electromagn. Res.*, Vol. 82, 2008, pp. 401–418.

[93] Yang, Y., C. Zhang, and A. E. Fathy, "Development and Implementation of Ultra-Wideband See-Through-Wall Imaging System Based on Sampling Oscilloscope," *IEEE Antennas Wireless Propag. Lett.*, Vol. 7, 2008, pp. 465–468.

[94] Charvat, G. L., et al., "A Through-Dielectric Ultrawideband (UWB) Switched-Antenna-Array Radar Imaging System," *IEEE Trans. Antennas Propag.*, Vol. 60, No. 11, November 2012, pp. 5495–5500.

[95] Fioranelli, F., S. Salous, I. Ndip, and X. Raimundo, "Through-The-Wall Detection with Gated FMCW Signals Using Optimized Patch-Like and Vivaldi Antennas," *IEEE Trans. Antennas Propag.*, Vol. 63, No. 3, March 2015, pp. 1106–1117.

[96] Ressler, M., et al., "The Army Research Laboratory (ARL) synchronous impulse reconstruction (SIRE) forward-looking radar," in *Proc. SPIE 6561, Unmanned Systems Technology IX*, 2007, pp. 1–12.

[97] Akduman, I., L. Crocco, and F. Soldovieri, "Experimental Validation of a Simple System for Through-the-Wall Inverse Scattering," *IEEE Geosci. Remote Sens. Lett.*, Vol. 8, No. 2, March 2011, pp. 258–262.

[98] Karanth, A., et al., "Through-Wall Imaging System Using Horn Antennas," in *Proc. 4th International Conference on Advanced Computing and Communication Systems (ICACCS)*, Coimbatore, India, 2017, pp. 1–6.

[99] Ralston, T. S., G. L. Charvat, and J. E. Peabody, "Real-Time Through-Wall Imaging Using an Ultrawideband Multiple-Input Multiple-Output (MIMO) Phased Array Radar System," in *Proc. IEEE International Symposium on Phased Array Systems and Technology (ARRAY)*, Waltham, MA, pp. 551–558, 2010.

[100] Le, C., T. Dogaru, L. Nguyen, and M. A. Ressler, "Ultrawideband (UWB) Radar Imaging of Building Interior: Measurements and Predictions," *IEEE Trans. Geosci. Remote Sens.*, Vol. 47, No. 5, May 2009, pp. 1409–1420.

[101] Stiefel, M., et al., "Distributed Greedy Signal Recovery for Through-the-Wall Radar Imaging," *IEEE Geosci. Remote Sens. Lett.*, Vol. 13, No. 10, October 2016, pp. 1477–1481.

[102] Fioranelli, F., S. Salous, and X. Raimundo, "Frequency-Modulated Interrupted Continuous Wave as Wall Removal Technique in Through-the-Wall Imaging," *IEEE Trans. Geosci. Remote Sens.*, Vol. 52, No. 10, October 2014, pp. 6272–6283.

[103] Chen, H.-M., et al., "Imaging for Concealed Weapon Detection: A Tutorial Overview of Development in Imaging Sensors and Processing," *IEEE Signal Process. Mag.*, Vol. 22, No. 2, March 2005, pp. 52–61.

[104] Sakamoto, T., T. Sato, P. Aubry, and A. Yarovoy, "Frequency-Domain Kirchhoff Migration for Near-Field Radar Imaging," in *Proc. IEEE Conference on Antenna Measurements Applications (CAMA)*, Chiang Mai, Thailand, 2015, pp. 1–4.

[105] Zhuge, X., and A. G. Yarovoy, "A Sparse Aperture MIMO-SAR-Based UWB Imaging System for Concealed Weapon Detection," *IEEE Trans. Geosci. Remote Sens.*, Vol. 49, No. 1, January 2011, pp. 509–518.

[106] Sakamoto, T., T. Sato, P. Aubry, and A. Yarovoy, "Fast and Accurate UWB Radar Imaging Using Hybrid of Kirchhoff Migration and Stolt's F-K Migration with Inverse Boundary Scattering Transform," in *Proc. IEEE International Conference on Ultra-WideBand (ICUWB)*, Paris, France, 2014, pp. 191–196.

[107] Sakamoto, T., T. Sato, P. J. Aubry, and A. G. Yarovoy, "Ultra-Wideband Radar Imaging Using a Hybrid of Kirchhoff Migration and Stolt F-K Migration with an Inverse Boundary Scattering Transform," *IEEE Trans. Antennas Propag.*, Vol. 63, No. 8, August 2015, pp. 3502–3512.

[108] Soldovieri, F., and R. Solimene, "Through-Wall Imaging via a Linear Inverse Scattering Algorithm," *IEEE Sci. Remote Sens. Lett.*, Vol. 4, No. 4, October 2007, pp. 513–517.

8

New Trends and Future Developments

In the previous chapters, several approaches to microwave imaging have been discussed. As it has been shown, they are usually specifically designed and implemented for the intended application. This consideration holds true for the apparatuses used for illuminating the target under test and receiving the scattered radiation, as well as for the algorithms needed to invert the collected data in order to reconstruct the unknown object or structure. It has been stressed that microwave imaging techniques exhibit unique features that make them of paramount importance in nondestructive evaluations and imaging. First of all, the ability to directly retrieve the dielectric properties of the target under test, which can be correlated to some other physical parameters or to the state of a structure or an object. At the same time, the capabilities of microwaves for penetrating dielectric structures allow for the inspection of the internal inclusions of dielectric targets or the detection of defects in metallic structures covered by dielectric coatings or layers.

However, several limiting factors have not yet completely been overcome by existing or currently under development approaches. These factors are essentially due to the intrinsic nature of the electromagnetic phenomena involved in microwave imaging. First of all, we can mention the significant attenuation associated with the wave propagation at microwave frequencies inside dispersive or lossy materials. The second element is the complex mechanism of the wave scattering by targets whose dimensions are comparable with the wavelengths of the incident radiations. In any case, the information content of the received signal may be quite poor, in terms of dynamic range, signal-to-noise (*SNR*) ratio, and so forth.

Moreover, when strong scatterers (with respect to the host medium) are inspected, the measured values of the scattered field at the receiving positions strongly depend on multiple scattering effects inside the structure. This results in nonlinear relationships relating the dielectric properties of the target to the measurements. In addition, the field is not obtained by the receiving antennas directly, but it is usually derived from measurements of the *S*-parameters. Therefore, suitable models and calibration procedures are required [1]. In addition, as previously mentioned, the equations relating the scatterer properties (position, shape, distributions of physical parameters) to the measurements, that is, the equations that should be inverted in order to solve the electromagnetic inverse problem, are usually strongly ill-posed. Therefore, the solving procedure may often be quite complex, except in the cases in which sharp simplifying assumptions can be adopted. As mentioned in Chapter 2, for a better understanding of these intrinsic limitations, due to the scattering phenomena, the reader is referred to publications in which the inversion process is addressed mainly from a mathematical point of view.

It is clear that all these limiting factors result in the need for further studies toward improved and advanced microwave systems and techniques. As a matter of fact, despite the mentioned limitations, the research activity in this field evolves fast, and it is not simple to indicate the main directions along which new microwave imaging systems and techniques will be developed. However, some considerations can be drawn following the most recent studies reported in the scientific literature.

First of all, it should be mentioned that questions raised by solving the inverse scattering problem involved in several microwave imaging approaches are not completely resolved even from a strictly mathematical point of view. Therefore, much work is still needed to obtain new and effective inversion techniques in order to better face the nonlinearity and ill-posed nature of this problem. For example, some research activities are focused in constructing inverse scattering solutions in nonconventional spaces, such as the L^p Banach spaces (mentioned in Chapter 3), in which exploiting the norms of these spaces seems to lead to more accurate reconstructions with a reduction of the over-smoothing effects in the final images [2]. This can be important for better understanding the final reconstruction results when the inspection is performed by operators or practitioners not necessarily well-acquainted with the mathematical aspects of the inversion process.

Adopting the L^1 norm is another fundamental opportunity when the sparse nature of some reconstruction problems encountered in several applications can be successfully exploited. Various examples, already reported in the scientific literature, have been mentioned in the previous chapters. These research directions should be further followed, together with the use of compressive sensing techniques, which represent powerful tools when applicable.

It is also of great interest to further consider procedures aimed at focusing on specific areas inside the body under inspection, which can be a very effective from a computational point of view. Some interesting results have been already reported, for example with reference to multiscale methods. In such approaches, the inspected domain is iteratively refined by performing subsequent reconstructions at different scales. At any scale step, the regions containing the unknown scatterers are identified in the reconstructed images. Such regions are then used as the new investigation domains of the subsequent inversions, thus allowing one to focus only on the targets of interest [3].

The possibility of using only the amplitude values of the measured scattered field is another challenge that can be further investigated. In fact, the phaseless approach could greatly simplify the imaging apparatuses, but it poses concerns on the additional complexity introduced in the reconstruction procedure. Nevertheless, several proposals have been already reported in the scientific literature (see, for example, [4–6]).

Recently, some novel approaches have been also developed by using the so-called framework of virtual experiments, that is, exploiting the linearity of the electromagnetic scattering phenomena with respect to the primary sources for properly recombining the measured data with a posteriori procedures. This way, it is possible to enforce some particular properties of the scattered fields or of the contrast sources that allow the development of specific and efficient inversion strategies. For instance, the coefficients used for recombining the measured scattered field data can be found by solving the LSM equation with reference to some predefined points (called pivot points) around which some symmetries are enforced. However, it is worth noting that such virtual experiments should be properly designed in order to fully exploit all the available information. Different inversion algorithms based on this concept have been developed (for example, see [7–9]). An example of such techniques is the microwave imaging approach that uses the distorted iterated virtual experiments (DIVE) [10]. As an example, a very accurate experimental reconstruction of the *TwoDielDM* target of the Frésnel dataset [11] obtained by means of this method is reported in Figure 8.1.

Those mentioned are just few examples of techniques that can be adopted to face the general requirements of speeding up the reconstruction problem and improving the accuracy of the results. Actually, except for linearized techniques, which still exhibit limitations about the targets that can be inspected, the reconstruction process suffers from excessively long computational times and the possibility of real or quasi-real time processing of data, which is highly desirable, can be achieved only in a very limited number of practical cases.

Another important research line whose success will depend, of course, on the availability of fast reconstruction procedures is represented by the possibility of 3-D imaging. Although a lot of works have been published concerning qualitative and quantitative 3-D methods (some of them mentioned in the previous chapters), computational aspects remain a challenge. The vector

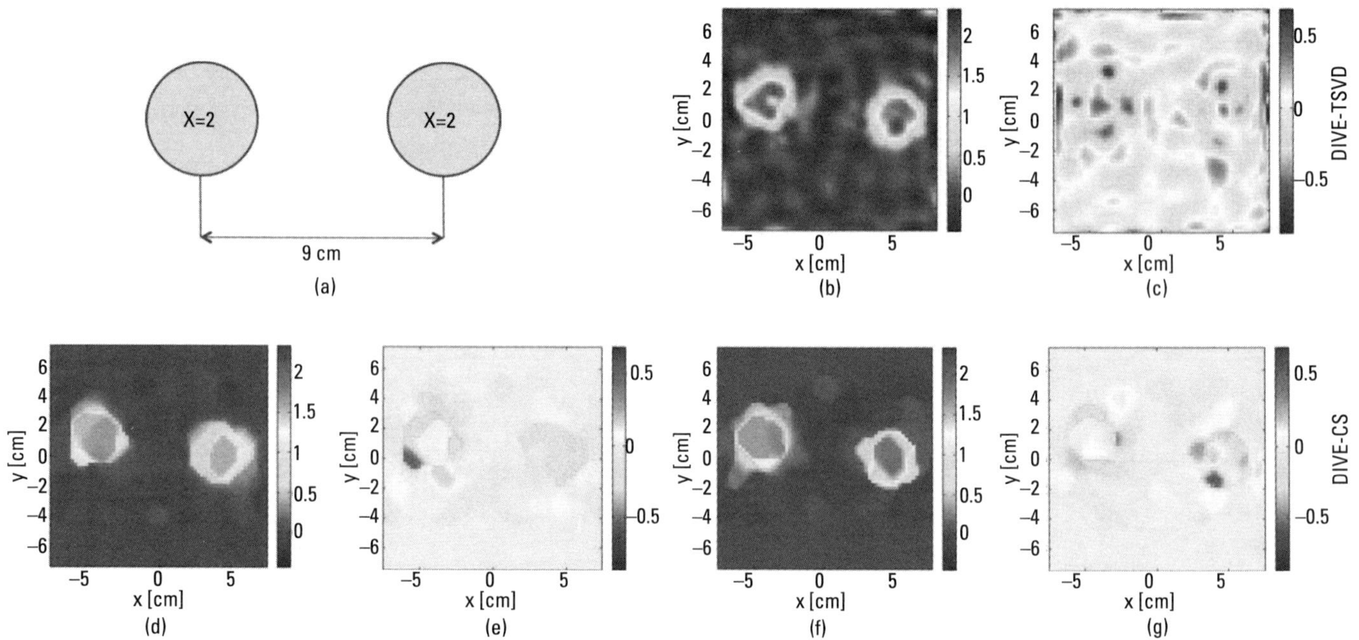

Figure 8.1 Frésnel *TwoDielTM* target at 6 GHz. (a) Reference profile. (b) Real part and (c) imaginary part of the retrieved contrast function with DIVE-TSVD; (d) and (e) are the same as (b) and (c) in the case of DIVE with sparsity promoting regularization; (f) and (g) are the same as (d) and (e) for reduced number of processed data [10].(© 2017 IEEE.)

nature of the electromagnetic interaction poses problems that go beyond the simple dimensionality issue, although, as already mentioned, the possibility of directly extending (in principle) 2-D solutions to 3-D imaging is one of the main opportunities offered by microwave imaging techniques. An additional complexity associated with 3-D imaging is related to measurement apparatuses, which should be able to operate with non-co-polarized field vector components outside the region under test. This fact has stimulated the research of new and specific sensors, antenna elements, and arrays, especially in the field of nondestructive testing and evaluations and in biomedical areas. In most cases, there is limited available space around the target to perform the suitable sampling of the scattered field. Typically, receiving elements have to be small, which is a constraint that can conflict with other electromagnetic requirements, such as efficiency and wideband operations. Therefore, a lot of work is still needed toward the development of miniaturized and effective antennas for the transmitting/receiving subsystems.

These developments could have a significant impact in several new applicative fields, for example, in aerospace applications where microwave and millimeter wave imaging techniques have already been found to be effective, for example, in nondestructive testing of external fuel tank insulating foam in space shuttles [12].

In the biomedical area, as mentioned in Chapter 5, there is a continuously increasing interest in studying new and much more effective methods for breast imaging and cancer detection, which pose several issues concerning the development of both apparatuses and inversion procedures. Brain stroke imaging is key example. In particular, this application is extremely interesting since it represents a case in which a spatial resolution not so high (in the order of few millimeters) could still provide sufficient information, that is, for discriminating between ischemic and hemorrhagic strokes. At present, very effective approaches and imaging systems have been developed. However, the imaging of brain strokes still represents a very challenging problem. The shielding effect due to the human skull on the electromagnetic wave propagation makes the scattering contributions originated by the intracerebral discontinuities (e.g., internal bleeding sites) extremely weak. This effect is even more severe for the necrotic tissues associated with ischemic strokes. Therefore, extremely effective illumination and measurement apparatuses are needed. Many efforts are required toward this direction, as confirmed by the great variety of experimental solutions that research teams have studied and realized.

The biomedical field may also offer new opportunities for exploiting microwave imaging techniques. Some different applications not devoted to breast or brain diagnostics have been reported in Section 5.6. Here we just wish to mention the challenge represented by the possibilities of using microwave imaging techniques to localize in-body sources inside the human body, such as those associated with wireless capsule endoscopy [13, 14].

Poor spatial resolution, low contrast of the dielectric properties in some applications and their consequent low *SNRs* at the measurement antennas, high nonlinearity, and a severely ill-posed nature of the associated inverse scattering problem remain the leading limiting factors for the practical applicability of microwave imaging techniques in several scenarios. Therefore, it is not a paradox that, despite such difficulties, the challenges associated with them have stimulated the research activity in this field worldwide.

Experts in various areas related to the multidisciplinary world represented by microwave imaging are providing important contributions toward making microwave imaging approaches more effective. An example is related to the almost continuous designing and testing of new antennas for microwave imaging purposes. Some of them have been mentioned throughout this book with reference to specific applications. The interested reader can refer to the various cited publications for further details about the specific proposals. Figure 8.2 reports an example of an antenna recently developed for brain stroke imaging.

The following point is quite accepted nowadays: there are no general-purpose solutions for interrogating the target under test, acquiring the scattering data, and performing the reconstruction process. On the contrary, a priori information about the imaging configuration, the ranges of dielectric parameters, the best frequency bands, the estimated values of the *SNR* at the measurement points, and so forth, should be taken into account as much as possible, even in the development of reconstruction procedures and numerical codes.

In the biomedical field, poor contrasts between healthy and pathological tissues pose several questions concerning the effectiveness of using microwave radiations for diagnostic purposes. The possible usage of contrast agents to improve this contrast (which represents a common choice in many of the currently adopted diagnostic modalities) seems to be particularly interesting. The use of nanoparticles represents a significant example of the research in this area [15, 16]. Interestingly enough, as shown in [15], these particles can be

Figure 8.2 Example of an antenna for microwave brain stroke imaging.

functionalized by means of proper molecular groups, and therefore they can be selectively delivered to the cancerous tissues via systemic administration, without requiring knowledge about the tumor location. The design and realization of effective measurement systems for biomedical imaging is another important topic in which we expect further developments. Several apparatuses and prototypes have been already described in Chapter 5. As a relevant example, we wish to mention the very promising diagnostic system recently reported in [17, 18] (see Figure 8.3).

However, it seems that other apparatuses could be developed to allow an early medical evaluation in the ambulance, since a preliminary differential diagnosis between ischemic and hemorrhagic strokes (as previously mentioned) could be extremely important for the patient's treatment and enhanced survival possibility as well.

This is an example of the need for the development of portable microwave devices, which holds true not only in the biomedical field, but also in other areas such as nondestructive testing. The development of portable devices places severe constraints related to the system dimensions, including electronic circuitry and antenna design. Some interesting proposals have been already described in the scientific literature. An example is represented by the microwave camera developed in [19], which is shown in Figure 8.4.

Other examples are related to the development of radar-type systems, which are widely adopted in the field of GPR, TWI, and security applications. In these cases, for easily transporting and placing in the field, different portable implementations of imaging systems have been proposed [20–22]. As an example, Figure 8.5 shows a portable 5.8-GHz hybrid interferometry/FMCW radar

(a) (b)

Figure 8.3 A general view of the brain imaging system called BRIMG1, developed by EMTensor GmbH [18]. (© 2017 IEEE.)

(a) (b)

Figure 8.4 Picture of the fully assembled microwave video camera developed in [19], with a total size of 26 cm × 21 cm × 18 cm and a weight of 4.8 kg. (a) Front side showing the imaging array aperture, and (b) Back side showing the source board and first stage mux board. (© 2017 IEEE.)

Figure 8.5 Portable 5.8-GHz hybrid interferometry/FMCW radar [22]. (© 2017 IEEE.)

[22], which can be used in imaging applications involving SAR and beamforming techniques.

The availability of illumination/measurement apparatuses characterized by low weights and small sizes also opens new possible applications of microwave imaging techniques. An example is related to the use of unmanned aerial vehicles (UAVs) or drones. As also discussed in Chapter 6, airborne or helicopter-borne imaging systems based on ground penetrating radar (GPR) are already used in hydrology and geophysics [23, 24]. However, the need for manned aerial vehicles limits the possible range of applications for such techniques. The use of remotely operated systems would also allow the use of microwave techniques in cases where the personnel cannot directly reach the area to be inspected, such as in some landmine detection applications [25] and in rescue operations over disaster scenarios [26]. Recently, the use of drones has also been proposed for through-the-wall applications [27]. Clearly, apart from the needs related to portability and the energetic efficiency of the measurement apparatuses, there are also significant challenges concerning the development of imaging algorithms. Proper propagation models and inversion procedures are needed in order to correctly take into account that the measurement system is located on the UAV, which is usually in movement with respect to the inspected scene. Moreover, imaging techniques usually also need a very precise definition of the positions of the measurement locations, which may be difficult when the antennas are located on drones. Despite such difficulties, preliminary studies aimed at assessing the feasibility of UAVs and drones for microwave imaging have been recently presented in the literature [28, 29] and the development of imaging procedures able to work with the data that can be acquired by such systems is being pursued by the scientific community [28–30].

References

[1] Beaverstone, A. S., D. S. Shumakov, and N. K. Nikolova, "Frequency-Domain Integral Equations of Scattering for Complex Scalar Responses," *IEEE Trans. Microw. Theory Techn.*, Vol. 65, No. 4, April 2017, pp. 1120–1132.

[2] Estatico, C., M. Pastorino, and A. Randazzo, "A Novel Microwave Imaging Approach Based on Regularization in Lp Banach Spaces," *IEEE Trans. Antennas Propag.*, Vol. 60, No. 7, July 2012, pp. 3373–3381.

[3] Franceschini, D., M. Donelli, R. Azaro, and A. Massa, "Dealing with Multifrequency Scattering Data Through the IMSA," *IEEE Trans. Antennas Propag.*, Vol. 55, No. 8, August 2007, pp. 2412–2417.

[4] Yurduseven, O., et al., "Frequency-Diverse Computational Microwave Phaseless Imaging," *IEEE Antennas Wireless Propag. Lett.*, Vol. 16, 2017, pp. 2808–2811.

[5] Costanzo, S., G. Di Massa, M. Pastorino, and A. Randazzo, "Hybrid Microwave Approach for Phaseless Imaging of Dielectric Targets," *IEEE Geosci. Remote Sens. Lett.*, Vol. 12, No. 4, April 2015, pp. 851–854.

[6] Álvarez, Y., et al., "Inverse Scattering for Monochromatic Phaseless Measurements," *IEEE Trans. Instrum. Meas.*, Vol. 66, No. 1, January 2017, pp. 45–60.

[7] Bevacqua, M. T., L. Crocco, L. D. Donato, and T. Isernia, "Microwave Imaging of Nonweak Targets via Compressive Sensing and Virtual Experiments," *IEEE Antennas Wireless Propag. Lett.*, Vol. 14, 2015, pp. 1035–1038.

[8] Donato, L. D., M. T. Bevacqua, L. Crocco, and T. Isernia, "Inverse Scattering Via Virtual Experiments and Contrast Source Regularization," *IEEE Trans. Antennas Propag.*, Vol. 63, No. 4, April 2015, pp. 1669–1677.

[9] Bevacqua, M. T., L. Crocco, L. Di Donato, and T. Isernia, "An Algebraic Solution Method for Nonlinear Inverse Scattering," *IEEE Trans. Antennas Propag.*, Vol. 63, No. 2, February 2015, pp. 601–610.

[10] Palmeri, R., et al., "Microwave Imaging via Distorted Iterated Virtual Experiments," *IEEE Trans. Antennas Propag.*, Vol. 65, No. 2, February 2017, pp. 829–838.

[11] Belkebir, K., and M. Saillard, "Special Section: Testing Inversion Algorithms Against Experimental Data," *Inverse Probl.*, Vol. 17, No. 6, December 2001, pp. 1565–1571.

[12] Kharkovsky, S., et al., "Millimeter-Wave Detection of Localized Anomalies in the Space Shuttle External Fuel Tank Insulating Foam," *IEEE Trans. Instrum. Meas.*, Vol. 55, No. 4, August 2006, pp. 1250–1257.

[13] Chandra, R., A. J. Johansson, M. Gustafsson, and F. Tufvesson, "A Microwave Imaging-Based Technique to Localize an In-Body RF Source for Biomedical Applications," *IEEE Trans. Biomed. Eng.*, Vol. 62, No. 5, May 2015, pp. 1231–1241.

[14] Chandra, R., H. Zhou, I. Balasingham, and R. M. Narayanan, "On the Opportunities and Challenges in Microwave Medical Sensing and Imaging," *IEEE Trans. Biomed. Eng.*, Vol. 62, No. 7, July 2015, pp. 1667–1682.

[15] Bellizzi, G., O. M. Bucci, and I. Catapano, "Microwave Cancer Imaging Exploiting Magnetic Nanoparticles as Contrast Agent," *IEEE Trans. Biomed. Eng.*, Vol. 58, No. 9, September 2011, pp. 2528–2536.

[16] Bellizzi, G., et al., "Optimization of the Working Conditions for Magnetic Nanoparticle-Enhanced Microwave Diagnostics of Breast Cancer," *IEEE Trans. Biomed. Eng.*, 2017, in print, p. 1.

[17] Semenov, S., B. Seiser, E. Stoegmann, and E. Auff, "Electromagnetic Tomography for Brain Imaging: From Virtual to Human Brain," in *Proc. IEEE Conference on Antenna Measurements Applications (CAMA)*, Antibes Juan-les-Pins, France, 2014.

[18] Tournier, P.-H., et al., "Numerical Modeling and High-Speed Parallel Computing: New Perspectives on Tomographic Microwave Imaging for Brain Stroke Detection and Monitoring," *IEEE Antennas Propag. Mag.*, Vol. 59, No. 5, October 2017, pp. 98–110.

[19] Ghasr, M. T., M. J. Horst, M. R. Dvorsky, and R. Zoughi, "Wideband Microwave Camera for Real-Time 3-D Imaging," *IEEE Trans. Antennas Propag.*, Vol. 65, No. 1, January 2017, pp. 258–268.

[20] Li, C., et al., "A Review on Recent Progress of Portable Short-Range Noncontact Microwave Radar Systems," *IEEE Trans. Microw. Theory Techn.*, Vol. 65, No. 5, May 2017, pp. 1692–1706.

[21] Boutte, D., H. Lee, V. Radzicki, and A. Hunt, "A Portable SIMO Radar for Through Wall Detection and Imaging," in *Proc. IEEE Military Communications Conference (MILCOM2015)*, Tampa, FL, 2015, pp. 204–209.

[22] Peng, Z., et al., "A Portable FMCW Interferometry Radar with Programmable Low-IF Architecture for Localization, ISAR Imaging, and Vital Sign Tracking," *IEEE Trans. Microw. Theory Techn.*, Vol. 65, No. 4, April 2017, pp. 1334–1344.

[23] Lambot, S., et al., "Analysis of Air-Launched Ground-Penetrating Radar Techniques to Measure the Soil Surface Water Content," *Water Resour. Res.*, Vol. 42, No. 11, November 2006, p. W11403.

[24] Catapano, I., et al., "A Tomographic Approach for Helicopter-Borne Ground Penetrating Radar Imaging," *IEEE Geosci. Remote Sens. Lett.*, Vol. 9, No. 3, May 2012, pp. 378–382.

[25] Sipos, D., P. Planinsic, and D. Gleich, "On Drone Ground Penetrating Radar for Landmine Detection," in *Proc. First International Conference on Landmine: Detection, Clearance and Legislations (LDCL)*, Beirut, Lebanon, 2017, pp. 1–4.

[26] Câmara, D., "Cavalry to the Rescue: Drones Fleet to Help Rescuers Operations Over Disasters Scenarios," in *Proc. IEEE Conference on Antenna Measurements Applications (CAMA)*, Antibes Juan-les Pins, France, 2014, pp. 1–4.

[27] Karanam, C. R., and Y. Mostofi, "3D Through-Wall Imaging with Unmanned Aerial Vehicles Using WiFi," in *Proc. 16th ACM/IEEE International Conference on Information Processing in Sensor Networks (IPSN)*, Pittsburg, PA, 2017, pp. 131–142.

[28] Altdorff, D., et al., "UAV-borne electromagnetic induction and ground-penetrating radar measurements: a feasibility test," in *Proc. 74th Annual Meeting of the Deutsche Geophysikalische Gesellschaft*, Karlsruhe, Germany, 2014.

[29] Ludeno, G., et al., "A Micro-UAV-Borne System for Radar Imaging: A Feasibility Study," in *Proc. 9th International Workshop on Advanced Ground Penetrating Radar (IWAGPR)*, Edinburgh, UK, 2017, pp. 1–4.

[30] Catapano, I., G. Ludeno, and F. Soldovieri, "An Approximated Imaging Approach for GPR Surveys with Antennas Far from The Interface," in *Proc. 9th International Workshop on Advanced Ground Penetrating Radar (IWAGPR)*, Edinburgh, UK, 2017, pp. 1–5.

About the Authors

Matteo Pastorino is a full professor of electromagnetic fields at the University of Genoa, Italy. He has been the director of the Department of Biophysical and Electronic Engineering (DIBE) from 2008 to 2011 and the director of the Department of Electrical, Electronic, Telecommunications Engineering and Naval Architecture (DITEN) from 2011 to 2013. He has coauthored more than 450 papers in international journals and conference proceedings. His current research interests include microwave imaging, direct and inverse scattering problems, industrial and medical applications, and analytical and numerical methods in electromagnetism. He is an IEEE Fellow for his contribution on the analysis of the electromagnetic scattering.

Andrea Randazzo received a laurea degree in telecommunication engineering from the University of Genoa, Italy, in 2001 and a Ph.D. degree in information and communication technologies from the same university in 2006. Currently, he is an associate professor of electromagnetic fields at the Department of Electrical, Electronic, Telecommunication Engineering, and Naval Architecture (DITEN) at the University of Genoa, where he is also vice-director for research and technology transfer. His primary research interests are in the field of microwave imaging, inverse scattering techniques, numerical methods for electromagnetic scattering and propagation, and smart antennas. He is a coauthor of more than 200 papers published in journals and conference proceedings.

Index

Artech House Microwave Library

Behavioral Modeling and Linearization of RF Power Amplifiers, John Wood

Chipless RFID Reader Architecture, Nemai Chandra Karmakar, Prasanna Kalansuriya, Randika Koswatta, and Rubayet E-Azim

Control Components Using Si, GaAs, and GaN Technologies, Inder J. Bahl

Design of Linear RF Outphasing Power Amplifiers, Xuejun Zhang, Lawrence E. Larson, and Peter M. Asbeck

Design Methodology for RF CMOS Phase Locked Loops, Carlos Quemada, Guillermo Bistué, and Iñigo Adin

Design of CMOS Operational Amplifiers, Rasoul Dehghani

Design of RF and Microwave Amplifiers and Oscillators, Second Edition, Pieter L. D. Abrie

Digital Filter Design Solutions, Jolyon M. De Freitas

Discrete Oscillator Design Linear, Nonlinear, Transient, and Noise Domains, Randall W. Rhea

Distortion in RF Power Amplifiers, Joel Vuolevi and Timo Rahkonen

Distributed Power Amplifiers for RF and Microwave Communications, Narendra Kumar and Andrei Grebennikov

Electric Circuits: A Primer, J. C. Olivier

Electronics for Microwave Backhaul, Vittorio Camarchia, Roberto Quaglia, and Marco Pirola, editors

EMPLAN: Electromagnetic Analysis of Printed Structures in Planarly Layered Media, Software and User's Manual, Noyan Kinayman and M. I. Aksun

An Engineer's Guide to Automated Testing of High-Speed Interfaces, Second Edition, José Moreira and Hubert Werkmann

Envelope Tracking Power Amplifiers for Wireless Communications, Zhancang Wang

Essentials of RF and Microwave Grounding, Eric Holzman

Frequency Measurement Technology, Ignacio Llamas-Garro, Marcos Tavares de Melo, and Jung-Mu Kim

FAST: Fast Amplifier Synthesis Tool—Software and User's Guide, Dale D. Henkes

Feedforward Linear Power Amplifiers, Nick Pothecary

Filter Synthesis Using Genesys S/Filter, Randall W. Rhea

Foundations of Oscillator Circuit Design, Guillermo Gonzalez

Frequency Synthesizers: Concept to Product, Alexander Chenakin

Fundamentals of Nonlinear Behavioral Modeling for RF and Microwave Design, John Wood and David E. Root, editors

Generalized Filter Design by Computer Optimization, Djuradj Budimir

Handbook of Dielectric and Thermal Properties of Materials at Microwave Frequencies, Vyacheslav V. Komarov

Handbook of RF, Microwave, and Millimeter-Wave Components, Leonid A. Belov, Sergey M. Smolskiy, and Victor N. Kochemasov

High-Efficiency Load Modulation Power Amplifiers for Wireless Communications, Zhancang Wang

High-Linearity RF Amplifier Design, Peter B. Kenington

High-Speed Circuit Board Signal Integrity, Second Edition, Stephen C. Thierauf

Integrated Microwave Front-Ends with Avionics Applications, Leo G. Maloratsky

Intermodulation Distortion in Microwave and Wireless Circuits, José Carlos Pedro and Nuno Borges Carvalho

Introduction to Modeling HBTs, Matthias Rudolph

An Introduction to Packet Microwave Systems and Technologies, Paolo Volpato

Introduction to RF Design Using EM Simulators, Hiroaki Kogure, Yoshie Kogure, and James C. Rautio

Introduction to RF and Microwave Passive Components, Richard Wallace and Krister Andreasson

Klystrons, Traveling Wave Tubes, Magnetrons, Crossed-Field Amplifiers, and Gyrotrons, A. S. Gilmour, Jr.

Lumped Elements for RF and Microwave Circuits, Inder Bahl

Lumped Element Quadrature Hybrids, David Andrews

Microstrip Lines and Slotlines, Third Edition, Ramesh Garg, Inder Bahl, and Maurizio Bozzi

Microwave Circuit Modeling Using Electromagnetic Field Simulation, Daniel G. Swanson, Jr. and Wolfgang J. R. Hoefer

Microwave Component Mechanics, Harri Eskelinen and Pekka Eskelinen

Microwave Differential Circuit Design Using Mixed-Mode S-Parameters, William R. Eisenstadt, Robert Stengel, and Bruce M. Thompson

Microwave Engineers' Handbook, Two Volumes, Theodore Saad, editor

Microwave Filters, Impedance-Matching Networks, and Coupling Structures, George L. Matthaei, Leo Young, and E. M. T. Jones

Microwave Imaging Methods and Applications, Matteo Pastorino and Andrea Randazzo

Microwave Material Applications: Device Miniaturization and Integration, David B. Cruickshank

Microwave Materials and Fabrication Techniques, Second Edition, Thomas S. Laverghetta

Microwave Materials for Wireless Applications, David B. Cruickshank

Microwave Mixer Technology and Applications, Bert Henderson and Edmar Camargo

Microwave Mixers, Second Edition, Stephen A. Maas

Microwave Network Design Using the Scattering Matrix, Janusz A. Dobrowolski

Microwave Power Amplifier Design with MMIC Modules, Howard Hausman

Microwave Radio Transmission Design Guide, Second Edition, Trevor Manning

Microwave and RF Semiconductor Control Device Modeling, Robert H. Caverly

Microwave Transmission Line Circuits, William T. Joines, W. Devereux Palmer, and Jennifer T. Bernhard

Microwaves and Wireless Simplified, Third Edition, Thomas S. Laverghetta

Modern Microwave Circuits, Noyan Kinayman and M. I. Aksun

Modern Microwave Measurements and Techniques, Second Edition, Thomas S. Laverghetta

Modern RF and Microwave Filter Design, Protap Pramanick and Prakash Bhartia

Neural Networks for RF and Microwave Design, Q. J. Zhang and K. C. Gupta

Noise in Linear and Nonlinear Circuits, Stephen A. Maas

Nonlinear Microwave and RF Circuits, Second Edition, Stephen A. Maas

On-Wafer Microwave Measurements and De-Embedding, Errikos Lourandakis

Passive RF Component Technology: Materials, Techniques, and Applications, Guoan Wang and Bo Pan, editors

Practical Analog and Digital Filter Design, Les Thede

Practical Microstrip Design and Applications, Günter Kompa

Practical Microwave Circuits, Stephen Maas

Practical RF Circuit Design for Modern Wireless Systems, Volume I: Passive Circuits and Systems, Les Besser and Rowan Gilmore

Practical RF Circuit Design for Modern Wireless Systems, Volume II: Active Circuits and Systems, Rowan Gilmore and Les Besser

Production Testing of RF and System-on-a-Chip Devices for Wireless Communications, Keith B. Schaub and Joe Kelly

Q Factor Measurements Using MATLAB®, Darko Kajfez

QMATCH: Lumped-Element Impedance Matching, Software and User's Guide, Pieter L. D. Abrie

Radio Frequency Integrated Circuit Design, Second Edition, John W. M. Rogers and Calvin Plett

Reflectionless Filters, Matthew A. Morgan

RF Bulk Acoustic Wave Filters for Communications, Ken-ya Hashimoto

RF Design Guide: Systems, Circuits, and Equations, Peter Vizmuller

RF Linear Accelerators for Medical and Industrial Applications, Samy Hanna

RF Measurements of Die and Packages, Scott A. Wartenberg

The RF and Microwave Circuit Design Handbook, Stephen A. Maas

RF and Microwave Coupled-Line Circuits, Rajesh Mongia, Inder Bahl, and Prakash Bhartia

RF and Microwave Oscillator Design, Michal Odyniec, editor

RF Power Amplifiers for Wireless Communications, Second Edition, Steve C. Cripps

RF Systems, Components, and Circuits Handbook, Ferril A. Losee

Scattering Parameters in RF and Microwave Circuit Analysis and Design, Janusz A. Dobrowolski

The Six-Port Technique with Microwave and Wireless Applications, Fadhel M. Ghannouchi and Abbas Mohammadi

Solid-State Microwave High-Power Amplifiers, Franco Sechi and Marina Bujatti

Spin Transfer Torque Based Devices, Circuits, and Memory, Brajesh Kumar Kaushik and Shivam Verma

Stability Analysis of Nonlinear Microwave Circuits, Almudena Suárez and Raymond Quéré

Substrate Noise Coupling in Analog/RF Circuits, Stephane Bronckers, Geert Van der Plas, Gerd Vandersteen, and Yves Rolain

System-in-Package RF Design and Applications, Michael P. Gaynor

Technologies for RF Systems, Terry Edwards

Terahertz Metrology, Mira Naftaly, editor

TRAVIS 2.0: Transmission Line Visualization Software and User's Guide, Version 2.0, Robert G. Kaires and Barton T. Hickman

Understanding Microwave Heating Cavities, Tse V. Chow Ting Chan and Howard C. Reader

Understanding Quartz Crystals and Oscillators, Ramón M. Cerda

Vertical GaN and SiC Power Devices, Kazuhiro Mochizuki

For further information on these and other Artech House titles, including previously considered out-of-print books now available through our In-Print-Forever® (IPF®) program, contact:

Artech House Publishers
685 Canton Street
Norwood, MA 02062
Phone: 781-769-9750
Fax: 781-769-6334
e-mail: artech@artechhouse.com

Artech House Books
16 Sussex Street
London SW1V 4RW UK
Phone: +44 (0)20 7596 8750
Fax: +44 (0)20 7630 0166
e-mail: artech-uk@artechhouse.com

Find us on the World Wide Web at: www.artechhouse.com

Stability Analysis of Nonlinear Microwave Circuits, Almudena Suárez and Raymond Quéré

Substrate Noise Coupling in Analog/RF Circuits, Stephane Bronckers, Geert Van der Plas, Gerd Vandersteen, and Yves Rolain

System-in-Package RF Design and Applications, Michael P. Gaynor

Technologies for RF Systems, Terry Edwards

Terahertz Metrology, Mira Naftaly, editor

TRAVIS 2.0: Transmission Line Visualization Software and User's Guide, Version 2.0, Robert G. Kaires and Barton T. Hickman

Understanding Microwave Heating Cavities, Tse V. Chow Ting Chan and Howard C. Reader

Understanding Quartz Crystals and Oscillators, Ramón M. Cerda

Vertical GaN and SiC Power Devices, Kazuhiro Mochizuki